NEVE'S
THE CITY AND COUNTRY PURCHASER
AND BUILDER'S DICTIONARY

NEVE'S
The City and Country Purchaser and Builder's Dictionary

A Reprint of the Work subtitled
The Compleat Builders Guide, published 1726

by

RICHARD NEVE

DAVID & CHARLES REPRINTS

7153 4654 7

The City and Country Purchaser, and Builder's Dictionary: or, the Compleat Builders Guide

This is a facsimile reproduction of the second edition.
The work was first published in 1703 under the authorship of T N Philomath, a pseudonym for Richard Neve; another pseudonym was T Snow. This edition, with additions, was published in 1726 and a third edition followed in 1736.
Little is known about Neve himself, but his other works include:

Apopiroscopy: or a Compleat and Faithful History of Experiments and Observations (under the pseudonym of T Snow), 1702

Art's Improvement: or Choice Experiments and Observations in Building, Husbandry, Gardening (also under the pseudonym of T Snow), 1703

Mathematicks Made Plain, 1708

Baroscopologia: or a discourse of the Baroscope, or Quicksilver Weather-Glass, 1708

The Merry Companion, 1716

This facsimile edition published in 1969

Printed in Great Britain by
Latimer Trend & Company Limited Whitstable
for David & Charles (Publishers) Limited
South Devon House Railway Station
Newton Abbot Devon

THE
City and *Country Purchaser*,
AND
Builder's Dictionary:
OR, THE
Compleat Builders Guide.

SHEWING

The *Qualities*, *Quantities*, *Proportions*, and *Rates* or *Value* of All *Materials* relating to *Building*; with the *Best Method* of *Preparing* many of them.

ALSO

The *Customs*, and *Methods* of *Measuring* of all *Artificers Work*, concern'd in *Building* together with the *City* and *Country Prices*, not only of *Workmanship*, but of *Materials* also: The which will be extraordinary Useful in *Making* of *Bargains*, or *Contracts* betwixt the *Work-Master* and *Work Man*; and likewise in *Computing* the *Value* (or *Charge* of *Erecting*) of any *Fabrick*, Great or Small.

LIKEWISE

The *Explanation* of the *Terms* of *ART* used by most *Workmen*.

TOGETHER WITH

Aphorisms, or *Necessary Rules* in *Building*, as to *Situation*, *Contrivance*, *Compactness*, *Uniformity*, *Conveniency*, *Firmness*, and *Form*, &c.

By RICHARD NEVE, *Philomath*.

The *Second* EDITION, with ADDITIONS.

LONDON:

Printed for *D. Browne*, at *Temple-Bar*; *J.* and *B. Sprint* at the *Bell*, *G. Conyers* at the *Ring* in *Little-Britain*; and *Ch. Rivington* at the *Bible* and *Crown* in St. *Paul's* Church-yard. 1726.

ADVERTISEMENT

Concerning this NEW EDITION

OF THE

BUILDER's DICTIONARY.

ARCHITECTURE *coming every Day, more and more, into Practice, the* Booksellers *were of Opinion, That a* NEW EDITION *of the* BUILDER'S DICTIONARY *would, at this Time, be not at all Unseasonable; especially, if by proper* Additions *It were made as fit for* Gentlemen's *Use, as the former* Edititon *was for* Workmen: This, *among other* Improvements *of it, I have endeavour'd to perform.*

THE

THE small Speculative Knowledge of the Builder's ART, *which I think I have gain'd by being concern'd in Translating the famous* Alberti's *Treatise of* ARCHITECTURE, *(which Sir* Henry Wotton *so highly praises, and which was never yet put into* English, *but will now soon see the Light) enabled me to distinguish, in the General, what was* right *from what was* wrong *in this* DICTIONARY, *tho' I don't pretend to be a Master of every individual Matter contain'd therein. And now I mention that Ornament of* England, *Sir* Henry Wotton, *I was surpriz'd and sorry to find that the Author of this* DICTIONARY *in almost all his Quotations (which are very numerous) out of that Great Man's* Elements *of* Architecture, *has made him write such bad* English, *and lay down such absurd* Maxims *of* Building, *as I am certain never appear'd in Print before. To give an instance of but One in above Twenty, in only Two Pages (the* 10*th and* 11*th.) of the former Edition of this* DICTIONARY, *in the Article of* Arches. *Sir* Henry Wotton's *words are*

are these; As *Semi-circular Arches*, being rais'd upon the total Diameter, are of all other the *roundest*, and consequently the *securest*; so those are the *comeliest*, which keeping precisely the same Heighth (*which, as every body knows, is just half its width*) shall yet be distended one *Fourteenth* Part longer than the said Diameter. *Now this Author, or at least the Printer, makes him say the* Distent (*between the* Feet *of the* Arch) *shou'd be a* Fourth, *not a* Fourteenth *part, which if practis'd, I'm sure, wou'd make mad work, and bring, not an old, but a new House, upon one's Head. There wou'd be no end of enumerating the* dangerous *Errors that occur in this Book, by only putting one Word for another, nay but one Letter, as* Carve *instead of* Curve, Right *instead of* Height; *not to insist upon such as are of* a comical Nature, *such as* Plaisters *for* Pilasters; Vesuvius *for* Vitruvius, &c. *He that transcribes a Period from another, shou'd take care to do it, not only in the Author's own words, but in so many words, neither more nor less, neither adding words of his own, as this* Dictionary-maker *did in the*

for-

former Edition, *nor leaving out any Words of the Author, which he was likewise not seldom guilty of. Thus in the Article of* Mosaic-work *he makes Sir* H. Wotton *say,* Mosaic-work consists of small inlaid Pieces of *Stone*, *Glass*, *Shells*, or other Materials, of various Colours; *figured at Pleasure. Whereas Sir* Henry *says,* Pebbles, Cockles *and* Shells *of sundry Colours*; and *of late Days* likewise *Glass &c. He leaves out* of late Days, *which Sir* H. Wotton *prudently inserted, because (as I take it)* Glass *was unknown to the Antients. It is not indeed non-Sense, as this Transcriber has worded it, but yet it is not the Author's Sense. But where is the Man that can make any Sense at all of what he puts into Sir* Harry's *Mouth, in the Article of* Pedestal? The *Height* of the *Pedestal* in each Order ought to be a third part of the whole Column, comprehending the *Base* and *Capital*, and the upper Adjuncts, as *Architrave*, *Frieze*, and *Cornice*. *Thus he points it, and there he ends it. Whereas Sir* Harry's *Words are these,* The *Heighth* of the *Pedestal* in each Order ought to be a third part of the whole Column,

Column, comprehending the *Base* and *Capital*; and the upper Adjuncts, as *Architrave*, *Frieze*, and *Cornice*, a fourth Part of the said Pillar. *These last Words*, a fourth part of the said Pillar, *this Transcriber leaves entirely out.*

WHATEVER Articles I my self have transcrib'd into this New Edition, *out of other Authors upon this Subject, or their Translators, I have taken Care to do it* verbatim, *unless where they were manifestly mistaken, as in* Sebastian le Clerc's *Translator,* building of Equerries (Ecuries *in* French) *I have made it* building of Stables; *so in* pag. 7. Ionic, *I have made* Doric, *&c.*

THE Translator of M. Perrault's *five Orders of* Architecture, *says, The* Zoccolo *or* Socle *(as we call it) is a square Body placed under the* Bases *of* Pedestals; *he should have said placed under the* Mouldings *of the* Bases *of* Pedestals. *Again, in the Word* Metop, *he is indeed in the right*

in

in saying, It comes from the Greek, Metope, *and is compounded of* Meta *and* Ope; *but then, he is in the Wrong to say, that* Ope *signifies* Three. Meta, *does indeed, signifie* Between : *But* Ope *in Greek, signifies the* Mortise-holes *of Rafters and Planks; and so* Metope *originally and strictly signifies the Space between those* Mortise-holes; *and in* Architecture, *it means the* Space *between the* Triglyphs *of the* Doric Frieze. *These, and several others of the like Nature, the* Reader *will find I have rectified in this* New Edition.

AS for whole additional Articles of my own, *I confess there are not* many, *though* some *there are: But then by* adding *frequently* something *to the Author, I may venture to affirm, I have help'd not a little to* illustrate *Matters: Thus in the Article of that most useful Timber-tree, the* Ash, *I thought it not amiss to quote a few Lines out of our incompa-*

rable

rable Poet old SPENCER, *giving us ſuch a ſuccinct Account of the* Nature *of all or moſt Timber-trees, that 'tis worth every* Builder's *while to get them by heart, which he may, with the more Eaſe, do, as they are in* Rhime.

'TIS hardly to be imagin'd how much only a Word or two now and then, properly inſerted, helps the Reader *to diſtinguiſh Things clearer than otherwiſe he could have done. Thus, for Inſtance, in the Word* Baldachin. *This Author ſays it is a* French *Word, and ſignifies a* Canopy ; *ſo indeed it does ; but what* Canopy ? *Not every* Canopy ; *nor indeed any* Canopy *but only that which is carry'd over the* Holy Sacrament *among the* Roman Catholicks *in Foreign Countries, when the Prieſt is going along the Street with the Hoſt to ſome ſick Perſon.*

SO in Page 2. *he ſpeaks of a* large Caſement, *which moſt* Readers *would take for the* Caſement *of a Win-dow ;*

dow; whereas it is the same Word for another Thing, and in that Place signifies a hollow Moulding.

NO printed Book or Ballad ever contain'd such Multitudes of Typographical Errors as the former Edition *of this* DICTIONARY: *Sometimes a* Negative *for an* Affirmative, *and so* vice versâ, *with a world of* false Pointing, *which makes the Sense* utterly Unintelligible. *All which I have taken no small Pains to rectify, besides* mending *the Language, and shewing the* Derivation *and* Etymology *of Numbers of* Architectonical Terms, *which this Author had not* Literature *enough to explain, though otherwise a very* ingenious *Person, and of* good Mechanical Parts. *Thus, for Instance, in the Word* Architrave; *he says, He supposes it to come from the* French, *for that the Word is purely so, adds he. Whereas, almost every Body knows it is not at all* French, *but part* Greek, *part* Latin,

Latin, *and comes from the Greek* Archos, Chief; *and the Latin* Trabs, *a* Beam; *ſo he confounds* Antick *with* Antique, *and takes them to mean the ſame Thing.* Antick *he dwells much upon, and is Right in his Account of it. But* Antique *is quite a different thing, as I have made appear out of the ingenious Monſieur* Richelet's *Dictionary,* &c.

MANY Terms *of* Art *he* miſpells, *and thereby makes it difficult ſo much as to gueſs at the* Etymology *of them. Thus* Soffita *he calls* Sapheta. *Of this and many other Words, which he leaves un-deriv'd, I have taken Care to* deduce *the* original Meaning *by the Help of Books, and many times, where Authors are at a Loſs, I have endeavour'd to help them out. Thus, for Example, A* Scheam *or* Skeen-Arch, *is a Word in every Body's Mouth, and none are ignorant it is a ſort of a* Flat-Arch; *but no Body has yet diſcover'd whence this and other*

Words

Words in this Book, fetch their Origin. *With Submission to the* Learned, *I take* Scheam-Arch *to come from the Italian*, Arco Scemo. Scemo, *in that Language, signifying* Imperfect, Incomplete; *for such, indeed, is a* Scheam-Arch, *being less than a* Semi-Circle.

For the rest, I refer to the Book it self.

The *Prelude* or *Preface*

TO THIS

TREATISE.

Shewing the *Antiquity*, *Excellency*, *Emolument*, and *Necessity* of

ARCHITECTURE.

AMONGST those many ARTS which *Divine Providence* hath been pleased to endow Mankind with the Knowledge of, this of ARCHITECTURE is none of the *least*, and therefore may well challenge a Place amongst the *Primary*, and most *Necessary*, if not the *Pre-eminency* of *Rank*.

FIRST, For its ANTIQUITY; it being almost Coætaneous with the World it self; for it was practis'd in the very Infancy of the World, by by the firſtborn of Mankind, *viz.* *Cain*, as you may find it Recorded, *Gen.* 4. 17. You may further obſerve, That this is the *Second* ART which the *Divine Pen-Man* hath left us upon Record: For we muſt allow *Agriculture* to be the moſt *Ancient*, and moſt *Neceſſary* ART of all others: It muſt certainly be the *Senior* ART; becauſe we find in *Gen.* 2. 15. That *Almighty GOD* did place *Adam* in the Garden of *Eden* in the Time of his Innocency, and was there ordered to perform the Office of a *Georgic*, [or Husband-man.] *Agriculture* muſt alſo as certainly be the moſt *Neceſſary*, becauſe Men can, and do ſubſiſt (in a great Meaſure) without *Cloathing*, *Houſes*, &c. in ſome Climates and Countries; but there is no part of the World where Men can live without *Food*.

BUT, Nevertheleſs, where People are ſo barbarous as to be quite ignorant of ARCHITECTURE, or BUILDING, they are alſo for the moſt part wholly deſtitute of the *Knowledge* of *Agriculture*, or. *Tillage of the Earth*; for *Agriculture* without *Architecture*, would be but of ſmall Uſe; becauſe it would be but to little purpoſe for the Husbandman to *Plow* and *Sow*, and *Plant* Fruit-trees, and the like, if when Harveſt comes, he hath neither *Barns*, nor *Granaries* to preſerve his Corn in; nor *Conſervatories*, or *Store-houſes* to lay up his Fruit in.

SECONDLY, As to the *Nobility*, or *Excellency* of this ART, it may be obſerved from the following *Circumſtances*, That

THE

THE *Holy Ghoſt* has been pleaſed to Honour this ART ſo far, as to grant it the *Privilege* to be Enroll'd (in the *Holy Scripture*) among the Actions of the firſt Monarchs of the World, *viz.* From *Adam* to *Noah*, there is no mention made of the *Military Actions* of the *Antediluvian* Sovereigns: All that is noted concerning them is, That they lived ſo many Years, and taught their Progeny, *To keep Sheep, To Till the Ground, To Plant Vineyards, To dwell in Tents, To Build Cities, To play on Organs and Harps, To Work in Braſs and Iron*, &c.

THE *Divine Architect* of this World hath been pleaſed to Honour this *Excellent* ART ſo far, as to vouchſafe to give neceſſary *Precepts* and *Rules* concerning ſome *Buildings*, of which I will here give ſome *Inſtances*. And,

Firſt, Of the *Diluvian Ark*, mentioned, *Gen.* 6. 14, 15. and 16 Verſes. Where he was pleaſed firſt to give Directions for *chuſing* the *Materials* for it. *Secondly*, He gives Order for the *Dimenſions*. And, *Thirdly*, For its *Form*, both *External* and *Internal*.

THE *Second* Inſtance ſhall be of the *Sanctuary*, conſiſting of the *Ark of the Covenant*, the *Mercy Seat*, the *Tabernacle*, and the *Altar*, as you may find in *Exodus*, Chap. 25, 26, 27, and 30.

THE *Third* Inſtance that I ſhall produce is, Concerning the *Building* of *Solomon's Temple* at *Jeruſalem*, as you may find it upon Record in 1 *Chron.* 28. 29.

THOUGH

THOUGH ſome have been ſo bold as to aſſert, That the *Grounds* of all ARTS are to be found in the *Holy Scriptures*; yet I think none of the *Civil* ARTS can pretend to ſuch *Documents* from *Scripture*, as the *Art of Building* can: For we may obſerve from the Preceeding *Inſtances*, and ſome other *Texts* of *Scripture*, (which I ſhall by and by refer you to) that the *Omniſcient Architect* of the World did direct Men to Build by Deſign, *viz.* By a *Prototype*, *Model*, *Draught*, or *Pattern*: As you may find in *Exodus* 25. 40. *Num.* 8. 4. 1 *Chron.* 28. 11, 12, 18, 19. *Acts* 7. 44. *Heb.* 8. 5.

IT is alſo obſervable, That the *Divine Architect* did not only direct the Form by a *Pattern*, or *Draught*, but alſo by giving the *Dimenſions* of each *Particular*, as you may obſerve from *Gen.* 6. 15 and 16, and *Exod.* 25. 10. alſo in the 26, 27, and 30th Chapters of *Exodus*, 2 *Chron* 3. 3. &c. He alſo gave *Directions* for *chuſing Materials* for each Particular Uſe, as you may find in the afore-cited ſixth Chapter of *Gen.* and *Exod* 25. 26, 27, and 30 Chapters; not only ſo, but was pleaſed to direct to the *Quantity* by *Weight*, as you may ſee in the afore mentioned Chapters of *Exodus*, and in the 28th of the Firſt Book of *Chronicles*, &c.

AND let me further add, Tho' *Architecture* be contemned, and ſlighted by ſome, becauſe it depends upon *Mechanicks*, or *Handicrafts-men's* Practice; yet it ought not to be deſpiſed, as they would fain make the World believe it is:

For

For I must, and will, tell such Men the Plain Truth, That they must certainly be Infidels, and do not deserve the Title of a *Jew*, and much less a *Christian*: For if such Persons were *Jews*, I think they must needs be very Ignorant of the *Pentateuch*, or Writings of *Moses*, or else they would have been convinced of this their Error from *Exod.* 31. *ver.* 3 and 6. where the *Holy Ghost* stiles the *Mechanick* Knowledge of *Bezaleel*, and *Aholiab*, &c. the *Spirit of God*, *Wisdom*, &c. And if they were *Christians*, they were very mean Ones to be *ignorant* of the *Books* of *Moses*, and the *Gospels* of St. *Matthew*, *Mark*, and *Luke*; for we may learn out of St. *Matthew*, chap. 13. ver. 55. and *Mark*, chap. 6. ver. 3. and *Luke*, chap. 2. ver. 51. that our *Blessed* REDEEMER did not think the *Knowledge* and *Practice* of *Architecture* and *Mechanicks* of *Handicrafts*, so Abject as some would represent it to be; for from these *Texts* we may learn, That He was pleased to exercise this ART of *Architecture*, and to be a *Mechanick*, *viz.* A *Carpenter*, for some part of the Time, that He was here Conversant with Mortals; which I must tell you, is no *small* Honour to the *Mechanicks* and *Architecture*.

I Profess, I can discern no more Reason, Why the *Sordidness* of some one *Workman* or *Mechanick*, should be the Cause of Reproach to *Handicraft Arts*, than *that* most Excellent Invention of Raising Water at *London-bridge* (for the Service of many Families in the City,) should be *disesteemed* and *neglected*, because there

there may ſometimes happen to be a blind, (or otherwiſe *defective*) Horſe employ'd in the Operation.

AND tho' *Mechanick Employments* be by ſome reckon'd ſo very *Ignoble* and *Scandalous*, yet it is very well known that many Gentlemen of good Rank and Quality in this Nation, are often converſant in *Handicraft Employments*: And other Nations exceed us in the Number of ſuch Gentlemen, of which I ſhall give you an Inſtance.

IN *France* the Making of Glaſs at the Glaſs Houſes is performed by Gentlemen of no mean Extraction, *viz.* Moſt of them of the great Glaſs houſe deſcending from Prince *Anthony Broſſard*, Natural Son of *Charles* of *France*.

NONE but Gentlemen are employed in the *Art* of Making Glaſs in *France*, and theſe Gentlemen bear Honourable Coats of Arms, and both they, and their Servants, and Deputies are exempt from *Paying of Taxes*, *Quartering of Soldiers*, &c.

THIS Noble *Art* makes *Marble*, and other *Stones* become the Delights of Men, of which are made our Glorious Palaces, and the Ornaments of our moſt ſplendid Churches, and the moſt durable Monuments which the Ambition of Men could ever invent, whereby to render themſelves and their Grandure known to future Ages.

THIS *Art* hath always been had in great Eſteem; for it is become Familiar in the Courts of Kings and Princes, *&c.* The late King of of *France* Eſtabliſhed an Academy for promoting this Noble *Art*.

ANOTHER Thing which proves the *Excellency* of this *Art*, is, That it is always *Poſſeſſed* and *Practiſed* moſt, by the moſt Civilized and Learned Nations.

ALMOST every Nation (that are *Civiliz'd*, eſpecially) have ſhewed Ample Teſtimonies of the Reſpect they had for this ART. For the *Jews* boaſt much of *Solomon*'s Temple; the *Aſſyrians* of *Great Babylon*; the *Egyptians* of their Pyramids, &c. The *Ionians* of their *Epheſian*, *Diana*'s Temple; and the other *Greeks* of the *Temples of their Gods*: *Rome* boaſts of its *Temples*, *Amphitheaters*, *Triumphant Arches*, and One thouſand *Stately Palaces*, &c. *France* glories in its *Louvre*, *Nôtre-Dame*, *Verſailles* Palace, of St *Germains en lay*, *Fontainbleau*, &c. *Spain*, of its *Eſcureal*, &c. *Holland*, of its *Stately Churches*, *Stadt-houſes*, &c. And *England* of its *Hampton-Court*, *Windſor-Caſtle*, *Weſtminſter-Abby*, *Royal Exchange*, St. *Paul*'s Cathedral, *Saliſbury* Church, &c It would be endleſs to give Inſtances of Things of this Kind; therefore I ſhall conclude with what I have already ſaid, concerning the Excellency of this *Art*, and proceed to ſay ſomewhat,

Secondly, Of the *Emolument* and *Neceſſity* of this ART of BUILDING.

AS to the *Profit* of this ART, I think I need not to ſay much; for all know, (that know any thing) that *Merchants* and other *Traders*, cannot ſubſiſt without their *Ware-houſes*, nor

Tradeſmen

Tradesmen without their *Shops*, nor the *Husbandman* without his *Barns*, *Granaries*, &c. For without these *Buildings* to preserve their Goods in, none of these different Ranks of Men would make much Profit of their Commodities.

Besides, there can be no Pretence to any kind of *Profit* without *Buildings* ; for there are no Nations in the World where the Air is so serene, as not to want *Buildings* to protect the Manufactures of it from the Injuries of corroding *Time* and *Air* : Add to this, That it is no small *Profit* that accrues to a State, or Nation, by those many Trades that depend purely upon *Building*, viz. *Carpenters*, *Joyners*, *Masons*, *Bricklayers*, *Sawyers*, *Glazers*, *Plumbers*, *Painters*, *Carvers*, *Smiths*, *Brick* and *Tile-makers*, *Stone cutters*, &c. For as I have made it appear before, other Trades cannot subsist without these, *viz.* Because they cannot subsist without *Buildings* ; In like manner where there is no *Architecture* in a Nation, there can, by Consequence, be no Princely Government ; for where the People are so Barbarous as to live in Caves, and in Hutts made of Boughs, *&c.* there are no *Mechanicks* amongst them, and by Consequence no *Improvement* of *Manufactures* ; and from thence it will follow, there will be little or no Revenues fit to maintain a Regal Power ; and for the most part, where such a Power is wanting, People are so *Savage* and *Barbarous*, that they live more like Brutes than Rational Men ; living by *Rupine*, *Murder*, &c. So that no Man can properly call any Thing his own, for they live (as it were) in daily Expectation of being

being robb'd, and deprived of that Little which they have, by those which are stronger than themselves. Which also implies a *Necessity* of *Building*, to which may be added that Complaint which *Aristotle* saith was made by some, *viz. That Man was worse dealt with by Nature than other Creatures, whereas they have some of them Hair, some Shells, some Wool, some Feathers, and some Scales to defend them from the Injuries of the Weather, Man alone is born Naked, and without any Covering:* But to this it may be answered, That *Divine Providence* hath endowed every Man with two such Superlative Instruments, *viz.* The *Hands*, (those admirable Instruments of Action) in Conjunction with that Faculty of the Soul, call'd *Reason*, which fully *supply* in Man whatever may or can be supposed to be *wanting* in human Creatures, as 'tis unwisely objected by some, That they are not sent into the World uncovered, and exposed to all Extreams of Weather, as *Heat*, *Cold*, *Drought*, and *Moisture*, &c. By this Instrument of Instruments, the *Hand*, which is adapted (by the *Divine Architect* of the World,) to supply the most *necessary* and *useful* Services of Man's Life, *viz. Building*, *Husbandry*, *Military Actions*, *Chirurgery*, *Writing*, *Engraving*, *Playing* on *Musical Instruments*, and all the *Necessary*, as well as the *Recreative Arts* and *Employments* of human Life. For indeed, if Man had wanted this Member, (as Learned Mr. *Ray* observeth) then we must have lived like Brutes, without *House* or *Shelter*, but what the Woods and Rocks afforded, with-

out

out *Cloaths* or *Covering*, without *Corn*, or *Wine*, or *Oil*, or any other *Drink* but *Water*; without *Warmth* and *Comfort*, or other *Uses* of *Fire*, as *Baked*, *Boiled*, and *Roasted Meats*; but we must have scrambled with the wild Beasts for *Acorns*, *Crabs* and *Nuts*, and such other Fruits as the Earth produces of her own Accord. We had then lain open, and been exposed to Injuries, unable to resist, or defend our selves against, almost, the feeblest Creature. Altho' Men were endowed with this incomparable Instrument the *Hand*; yet History informs us, That in the Infancy of the World, Men lived almost like wild Beasts, in Caves, and fed on Fruits and Roots of the Earth; but after they perceived the necessary Use of Fire against the Extremity of Cold, some began to edifie Cottages with Boughs, &c. and some digged Caves in the Mountains; and finding the Conveniency, and Necessity of it, at last with Practice they attain'd, by Degrees, to a greater Perfection in *Building*. So that now there are but few Nations but practice it in some Degree, having found the Usefulness, and Necessity of it, to protect them from the Injuries of the Weather, (and in some Countries from the Assaults of Rapacious Beasts.) Tho' in some Countries, (where the People are in Subjection to a Governour, and in a great Measure Civiliz'd) their *Buildings*, for want of Art, are very *imperfect* and *defective*, in Comparison of ours, in the learned Part of *Europe*, of which I will give an Instance or two.

At

At *Puna*, an Iſland in the *Weſt Indies*, their Houſes ſtand on Poſts, Ten or Twelve Foot high, with Ladders on the out-ſide to aſcend them. Alſo in the Iſle of *Mindanao*, one of the *Philippine* Iſlands in the *Eaſt Indies*, their Houſes ſtand on Poſts, fourteen, ſixteen, eighteen or twenty Foot high; and they have but one Floor, but many Rooms in it; under the Houſes there is a clear Paſſage like a Piazza, but a filthy one commonly; for ſome make this Place the Draught of their Houſes, but, building by the Rivers moſtly, the Floods cleanſe thoſe Places. At the *Nicobar* Iſlands, their Houſes are built after the Manner of thoſe at *Mindanao*, only the Roofs are Arched, but at *Mindanao* they are ridged; but in the *Nicobar* Iſles, they have but one Room in a Houſe. Theſe Sorts of Buildings are all the Mode amongſt the *Malayans* in the *Eaſt-Indies*.

I Shall add one Inſtance more of the *Neceſſity* of *BUILDING*, and that from the Obſervation of the ignorant *Indians* in *New-England*, &c. Who, ſeeing that there is a Neceſſity of laying up ſome Corn for a Winter-Store, and for Seed for the next Seaſon, make them Barns for that Purpoſe, tho' very poor ones, being only great Holes digged in the Ground, and ceiled with the Barks of Trees.

I Cannot here but commend our compleat Method of *Building* now uſed in *England*, by much to be preferr'd before the ancient Practice here, or that is now uſed in many Nations. The *Principal Qualities* of this Modern

Modern Method, are these, *viz. Compactness, Uniformity* and *Conveniency.*

THIS ART, like most others, hath in this *Age* been much improved; Men being now more *mathematically* inclined than formerly; having likewise better Opportunity to attain it, from the many *Treatises* that have been made Publick of these ARTS from Time to Time, in this last Age. For as an ingenious Man well observes, *There is scarce any Part of the* Mathematicks, *but is some way subservient to* Architecture, Geometry, *and* Arithmetick, *for the due Measure of the several Parts of a* Building, *the* Plans, Models, Computation *of* Materials, Time *and* Charges; *for ordering right its* Arches *and* Vaults, *that they may be both* firm *and* beautiful: Mechanicks *for its* Strength *and* Firmness, transposing *and* raising Materials: *And* Opticks *for* Symmetry *and* Beauty. He further adds, *I would not have any one assume the Character of an* Architect, *without a competent Skill in* All *of these.*

VITRUVIUS requires these, and many more to make a compleat *Architect.*

IN the ensuing TREATISE, I have endeavour'd to assist all such as have a *Desire,* (or have an *Occasion*) to understand the *Grounds* and *Rules* of *Architecture.*

I WAS first induced to undertake this Task of Composing a *Treatise* of this *Nature,* purely for my own Use; for by daily Experience, I found, That *such* a *Tract* was very much wanting, and that if I did compose such a *Piece,* it

might

might prove exceeding uſeful to others, as well as to my ſelf; an Ingenious Bookſeller being informed of my Deſign, was for my carrying it on.

I HAVE been in a great Meaſure excited to it, of late, out of Pity to ſome poor Workmen; for I have been informed of ſeveral, That for Want of *Skill* and *Foreſight*, have undertaken Buildings by *Gueſs*, by which they have been almoſt ruined, or at leaſt kept very low in the World; (tho' they have been very induſtrious in their Callings) and that purely by the Means of *unadviſed Contracts*.

AND then again, on the other Hand, it hath been an Obſervation made by others as well as my ſelf, That, ſome honeſt well-meaning Gentlemen, (and others) who have had Occaſion to Build, *&c.* have been ſtrangely over-reach'd by ſome Fraudulent, Crafty Workmen. But I hope this ſmall *Treatiſe* may be a Means to promote diſtributive Juſtice (in ſuch Things as relate to *Building*) and like a juſt Arbitrator whereby both Maſter and Workmen ſhall have what is *Juſtice* and *Equity* between 'em. For

THE enſuing *Treatiſe* contains not only the *Price* that moſt *Materials* are uſually *valued at*, and *ſold for* in different Parts of the Kingdom, and alſo the uſual *Rates* of all ſorts of *Workmen's Work*, both in the City, and in different Parts of the Country. But likewiſe,

2. IT contains Informations, as to the *Qualities*, *Quantities*, *Proportions*, and *Methods* of

of *preparing* and *making* many of the *Materials* relating to *Building*.

3. ALSO the *Cuſtoms*, and *Methods* of *Meaſuring* all Sorts of Artificers Works.

4. AN *Explanation* of the *Terms* of *Art* made uſe of amongſt *Workmen*.

5. IN the following TRACT, is alſo comprehended, *Conſiderations* as to the *Choice* of a *Surveyour*, *Workmen*, *Model*, or *Draught*, &c.

6. *APHORISMS*, or neceſſary *Rules* in *Building*; as to *Situation*, *Contrivance*, *Receipt*, *Firmneſs*, or *Strength*, and *Form*, or *Figure*, and *Beauty*.

7. THE Method of *Surveying of Buildings*; as to *entering Dimenſions in Pocket-books*, and *making Bills of Meaſurement*, &c.

8. OF *Valuing Buildings* when they are *Erected*.

9. A METHOD of *Cenſuring*, or paſſing one's *Judgment* on a *Building* (that is already compleated or finiſhed) as to its *Commodiouſneſs*, *Firmneſs*, and *Delightfulneſs*, which are the *Principal Qualities* in a good *Fabrick*.

IN the *Compoſing* of this WORK, (beſides my own, and ſome Friends *Obſervations*; which, together,

together, conſiſted of ſeveral Sheets of Paper, which were never yet made publick) I have made uſe of the beſt Authors extant, to the Number of about Fifty, great and ſmall; which I hope may be allowed to be no ſmall Task in *Reading*, *Comparing*, *Chuſing*, *Correcting*, *Reviſing*, *Diſpoſing*, and *Tranſcribing*, in Reſpect of *Matter*, *Form*, and *Order* For, to ſpeak the Truth, I muſt tell you, I found I had no ſmall Difficulties to wreſtle with; deſigning to collect the *Heads* of all ſuch *Things* as were moſt Material, from ſo many *Authors*, and from my own *Notes*, which would have Compoſed a ſmall *Treatiſe* of it ſelf; (for, I muſt tell you, they are not a few, nor have they been no ſmall Time a *Collecting*, nor from none but *experienced* Men, and my every Days *Obſervations* almoſt, my *Buſineſs* being frequently amongſt Workmen of divers Profeſſions, and different Places; ſo that the *Reader* will here have a great *Number* of *Obſervations*, which are wholly *New*. So that I was oftentimes more ſolicitous, and concerned, to conſider *what*, than *what not* to *write*: Yet I have diligently endeavoured to *inſert* the moſt *important Things*, that nothing *material* might be wanting to preſent you with a Satisfactory Account of the ART of BUILDING, in all its *Parts*, ſo as to make good our *Title*. Some, perhaps, may think *it* too *Small*, and the *Diſcourſe* too *Brief* (for a Subject of this *Nature*) indeed, I think ſo too; but

but then I muſt tell you, that it was my Deſign to be as *brief* as the Nature of the Subject would admit of; conſulting, *Friendly* READER, your Advantage therein, by rendering the Book both more *Portable*, and leſs *Chargeable*; ſo that every one that deſires to look into the *Precepts*, &c. of this ART, may here find *Satisfaction* without great *Expence*, either of *Time* or *Money*.

THE *Method* of this TREATISE is wholly *New*, but the whole ART is here ranged under certain *Heads*, and brought to a certain *Method*, and limited to *Practical Rules*, and that ſo *Perſpicuous*, as to be underſtood by the *meaneſt Capacity*.

ONE Great *Reaſon* of my making Choice of this *Method*, of *Compoſing* it under *Alphabetical Heads* was this, *viz.* In a *Subject* of this *Nature* there would *unavoidably* have been a *Neceſſity* to have made Uſe of Abundance of *Terms* which are uſed by *Architects*, and *Workmen*, which would not have been underſtood by *Gentlemen*, and *young* (and *ignorant*) *Handicrafts-man*, (for whom this *Treatiſe* is chiefly deſign'd,) unleſs we had explain'd theſe *Terms* as they fell in, by Conſequence in the *Diſcourſe*; but if I had done ſo, I ſhould too often have been neceſſitated to make large *Excurſions* or *Digreſſions* from the *Matter* in Hand, which would have ſo *disjointed* the *Diſcourſe*, that it would not have been ſo eaſily underſtood by young *Tyro*'s in this ART, eſpecially in a *Compendious Diſcourſe*, as this was deſign'd to

be;

be ; and put the Caſe, any one had wanted at any Time to know the *Meaning* of ſuch *Terms*, when they had heard them uſed, it would have been no ſmall Trouble to have found them in a continued Diſcourſe.

ALL the *Treatiſes* that I ever yet ſaw on this *Subject*, (which are not a few) were either continued *Diſcourſes*, or branched into *Chapters*, (or the like) containing the *Parts*, *Members*, and *Materials* relating to BUILDING ; or elſe *compriſed* under the *Titles* of *Mechanick Trades*, as *Carpenters* Work, *Bricklayers*, &c. or elſe in a *Dialogue*, which I like worſt of all ; becauſe, in asking of Queſtions, commonly there is a *Neceſſity* of uſing more Words than any other Way of Writing : I dare maintain, That none of the aforeſaid *Methods* are ſo fit for a *Subject* of this *Nature* that relates to ſo many *Orders* of *Men* ; *Parts* of *Buildings*, *Members* of *Parts*, and vaſt *Variety* of *Materials*, &c. as this which I have here choſen, (and I have heard ſome others wiſh for it, not only in this, but other *Arts* alſo.) For here, either *Gentleman*, or *Workman* may immediately find (by the Letters at the Top of each Page) any Thing that he hath occaſion to be inform'd about, without the Trouble of reading over whole *Chapters*, or the like.

COURTEOUS READER, I will aſſure you, That you have here *Epitomized* the *Subſtance* or *Marrow*, of all, or moſt of the known Authors that have *Treated* of this ART ; beſides a great many *New* and *Neceſſary Obſervations*,

tions, &c. which I hope will prove *Useful* to the *Publick*, tho' composed after a *new Method*, *viz.* an *Alphabetical Order* ; and upon that Account it may perhaps be objected against as a very *broken* and *imperfect* Subject. But I must here inform such, That many Times each particular *Word* (or *Head* is a *Compleat Discourse*, by it self, and where it is not, you are referr'd to another Place, which will make it *compleat*, by only turning over a few Leaves.

HAVING thus *briefly* hinted at the *Reason* of *Composing it in this* Method, I shall now proceed to give some necessary Directions for the better Apprehension of the *Scope* and *Use* of the BOOK.

First, That it is intended for *Beginners*, and such as have not had occasion to study this ART, and not for *Accomplished* ARCHITECTS, and therefore *adapted* to the *meanest Capacities*.

Secondly, In several Places, the *Derivation* of the Word is hinted at, (which *assists* the *Memory*, and *informs* the *Understanding*) as well as the *Meaning* and *Sense* it is commonly used in ; and several other *Things* or *Circumstances*, according to the *Nature* of the Word ; of which I will here give some *Instances*. As suppose,

First, The Word *Arch*, there you have its *Derivation*, An *Explanation* of five different Kinds, The *Method* of *making* them, Five *Theorems*

Theorems concerning them, The *Method* of *Measuring* them, and *lastly*, The *Price*.

Secondly, The next Instance shall be of the Word *Bricks*, where you have an Account of *what* and *wherefore* made, and then a *Description* of eighteen Sorts, with their *Dimensions*, *Price*, *Weight*, and *Form*; also the *Method* and *Price* of *Making, and Burning*; also *Directions* for *Buying*, *Choosing*, and *Laying* of *Bricks*, &c.

Thirdly, Of the *Material*, call'd *Glass*, after a General *Definition* of it, you have an Account of thirteen Sorts, and the *Price* of most of them, *&c.*

Fourthly, Of *Lead*, where you have fourteen *Particulars*, too many to repeat here.

Fifthly, Of *Mortar*, you have thirteen *Heads*.

Sixthly, Of *Nails*, you have twenty five *Particulars*.

Seventhly, Of *Painting*, where are eighteen distinct *Heads*. All other *Trades* having their proper *Heads* also.

Eighthly, Of *Stair-Cases*, there is an Account of about twenty five Sorts.

Ninthly, Of *Tiles*, There are more than forty *Particulars*.

I SHALL

I SHALL forbear to *Enumerate* any more, but refer you to the *Book* it ſelf, where you ſhall alſo find a *Deſcription* of all the *Members* of the Five *Orders* of *Columns*, with their *Dimenſions* and *Proportions.*

Laſtly, I DO declare, That if any *Thing* which I have ſet down in the enſuing *Treatiſe*, be objected againſt as a *Miſtake*, or that it is not ſo *Plain* and *Expreſs* as it might have been, upon the leaſt *Information* thereof, I ſhall be very ready to *revoke* it. And therefore, If in any *Thing* I have been beſide the *Matter* in *Hand*, or made a falſe *Step* or *Blunder* ; I do earneſtly intreat the Courteous, Good-Natured READER, either to *inform* me of it, (and upon *Eviction* I ſhall freely yield) or elſe, That he would freely *remit* the Fault ; ſince all know

Humanum eſt errare:

THE

CITY and COUNTRY

Purchaſer and *Builder's*

DICTIONARY.

Abacus.

IT is a ſquare Table, Liſt, or Plinth, in the upper part of the Chapiters of Columns, eſpecially thoſe of the *Corinthian* Order, which ſerves inſtead of a Drip or Corona to the Capital. It ſupports the nether Face of the Architrave, and whole Trabeation. In the Corinthian and Compound Orders, the Coronets of it are call'd the *Horns*; the intermediate Sweep and Curvature the *Arch*, which has commonly a Roſe carv'd in the middle.

Abacus comes from the Greek Word *Abax*. and has ſeveral Meanings; ſometimes a ſquare Trencher, and ſometimes a Buffet or Side Board, call'd *Credenza* by the Italians. But in Architecture, (as is ſaid before) it ſignifies a quadrangular Piece, which ſerves as a *Crowning* to the *Capitals* of *Columns*; and in thoſe of the *Corinthian* Order repreſents a kind of ſquare Tile, covering a Basket, ſuppoſed to be encompaſs'd with Leaves.

The Sieur *Mauclerc*, in the Ionick Order, deſigns an O--G with a Fillet over it for an *Abacus*; and this Fillet is half the Latitude of the O--G, the which

which he calls the Fillet of the *Abacus*.

And in the Corinthian Order, he describes the *Abacus* to be one *7th* Part of of the whole Capital, which he divides into three Parts, and the uppermost of these is a Boultin, and $\frac{1}{3}$ of the next third below, is the Fillet of the *Abacus*, and the rest below being 1 and $\frac{2}{3}$, is the Plinth of the *Abacus*

Andrea Palladio in the *Tuscan* Order, calls the Plinth above the Boultin, (or *Echinus*) *Abacus*; which from its Form, saith he, is commonly call'd Dado, or Dye, the which is $\frac{1}{3}$ of the whole Heighth of the Capital.

In the Dorick Order, he also calls the Plinth above the Boultin of the Capital, the *Abacus*; above which he places a Cimatium, for the uppermost Member of the Capital.

In the Ionick Order, he defines it to be the same with the Sieur *Mauclerc*.

In the Corinthian and Composite Order, he described it to be the same almost with the Sieur *Mauclerc*, only his is a large Casement, or hollow Moulding, instead of the Plinth. But *Vincent Scamozzi* gives the Title of *Abacus* to a Casement or Hollow, which is the Capital of the Pedestal of the Tuscan Order. V. *Capital*.

Abreuvoirs.

A Term in Masonry, by which is understood the Intervals, or Spaces between the Stones in laying them, commonly call'd the Joints, wherein the Morter is placed: A French Word, and properly signifies a soaking with Water.

Acanthus.

The Herb whose Leaves are represented in the Capital of the Corinthian Column. See at the End of the Word *Orders*, a short Account of the first Occasion of this most beautiful Ornament.

Acroteria.

Are sharp and spiry Battlements, or Pinacles, that stand in Ranges, with Rails and Ballisters upon flat Buildings. Also Images set on the Tops of Houses, are so call'd by some.

Acroteres.

Are Pedestals upon the Corners and Middle of a Pediment to support Statues; they may properly be call'd Pinacles. The Word in Greek signifies the Extremity of any thing; the Tip of a Finger, a Rock, or the like

Aditon V. *Chauncel.*

Alabaster.

1. *What.*] 'Tis a kind of soft, clear, white Marble; if it be so soft as to be cut, it is call'd *Gypsum*.

2. *Where found, or digg'd*] Some is brought to us out of the *Indies*, and from *Egypt*, *Syria*, &c. There is also some found in *Lincolnshire*, and in *Staffordshire*.

3. It's

3. *It's Use.*] It's chiefly used for Monuments in Churches, *&c*, where there are any Figures in Relief, or in Bass-Relief, *&c.* carved. It's also sometimes used for a Coat of Arms when a Gentleman will have his Coat of Arms cut in Relief, to set in Brick or Stone-work in the Front of his House

Alcove.

By the *Spaniards* call'd *Alcobar*; 'tis a Recess within a Chamber for the setting of a Bed out of the Way; where for State many Times the Bed is advanced upon 2 or 3 Ascents, with a Rail at the Feet. These *Alcoves* are frequent in many Noble-men's Houses in *Spain*, and other Parts beyond Sea. The Word is Arabick; *Elkauf*, a Place to sleep in.

Alder.

1. *What.*] 'Tis an Aquatick Tree, so very common, that it needs no Description.

2. *It's Use.*] Those which were large, were formerly made use of in building of *Boats*: So now are very large *Alders* sought after for such *Buildings* as lie continually under Water, where it will become as hard as a Stone; but being kept in an unconstant Temper, it decays in a little Time.

Vitruvius tells us, That the *Morasses* about *Ravenna* in *Italy* were *pil'd* with this Timber to superstruct upon, and he highly commends it. It was also used under that famous Bridge at *Venice*, the *Rialto*, which passes over the grand Canal, bearing a vast Weight.

3. *Poles*] of this Wood are extraordinary useful for *Pumps*, *Water-pipes*, (*Troughs* and *Sluices* also, if large.) These Poles I have known used (in the Country) for *Water-pipes*, to convey Water thro' Bays, and Dams, and also to carry Water from any Spring, to supply a House with this necessary Element; large Poles of this sort of Wood I have known used for *Ground-guts* to convey the Water out of Stews; the Poles were about 8 or 10 Inches Diameter, and the Cavity in them about 4n, or 4 ½; for boreing, and fitting of which Size they have about 3 *s*. 6 *d* *per* Rod for Workmanship.

4. But for *Water-pipes*] the *Poles* need not be above 4 or 5 Inches Diameter; the Cavity is commonly about 1 ¼, or 1 ½ Inch Diameter.

5. *Of the Method of boring* Alder-Poles.] The Order in which they proceed, in this Operation, is this: Being furnish'd with Poles of a fit Size, not too small (nor too large, if for *Water-pipes*.) They procure Horses, or Trussels, of a fit Altitude, to lay the *Poles*, and rest the Auger on whilst they are boring; they also set up a Lath, to turn the least Ends of the Poles to adapt them to the Cavities of the greater Ends of the others; their Lath

Lath being up, and your Poles cut to the Lengths they will conveniently hold, *viz* 8, 10, or 12 Foot; they proceed to turn the ſmall Ends of the Poles, about 5 or 6 Inches in Length, to the Size they intend to bore the greater Ends, about the ſame Depth, *viz* 5 or 6 Inches; (this you muſt Note is to make the Joint to ſhut each pair of Poles together, the Concave Part is the Female, and the other Part the Male part of the Joint.) In turning of the Male-part, they turn a Channel in it, or ſmall Grove, at a certain Diſtance from the End and in the Female-part they bore a ſmall Hole to fit over this Channel; for what Purpoſe you ſhall hear when they ſet the Poles together: They having thus far proceeded, they then bore thorough their Poles, and to prevent boring out at the Sides, they ſtick up great Nails at each end to guide them right through; but they commonly bore it at both Ends; and therefore if a Pole be crooked one way, they can bore him through, and not ſpoil him: The Poles being bored, they proceed to form them into Pipes in the Ground; for which Purpoſe they have a Trench digg'd, and prepared with Clay, to ram them in the Female part, being prepared with an Iron-Ring round it, to prevent its ſplitting; they drive in the Male-part, till the Grove in it is juſt under the Hole, which is bored in the Upper-ſide of the Female-part; and then having ſome melted Pitch ready, they pour it into the Hole in the Female-part, which flows round in the Grove which was turned in the Male-part; by which Means it is made very ſtanch and cloſe: And thus they proceed till they have placed all the Poles in their Order.

6. *Of the Charge of theſe Pipes.*] For Workmanſhip only, they uſually ask about 2 *s.* 6 *d.* or 3 *s per* Rod, *viz.* only to bore and fit them; but the Charge of all Work, and Materials, *viz.* Boring, Digging the Trench, Laying and Ramming in the Clay, *&c.* And the Charge of Poles, Clay, Pitch, and Iron Rings will be 4 *s.* 6 *d.* 5 *s.* 5 *s.* 6 *d* or 6 *s. per* Rod; according as the Materials can be procur'd.

Amphitheater, or *Amphitheatre.*

Is an Edifice, or Building of an Oval, or Circular Form, with Rows of Seats, one above another, where Spectators might ſit to behold Stage plays, and other publick Spectacles; as Sword-playing, fighting of wild Beaſts, *&c* The *Amphitheatre* of *Veſpaſian*, call'd the *Coliſæum*; that at *Verona* in *Italy*, and that at *Niſmes* in *Languedoc*, are the moſt celebrated that we have now remaining of Antiquity: The Word comes from the Greek *Amphi* a-round, and *Theatron* Theatre.

Ana-

Anabathrum.

A Place that is aſcended to by Steps.

Anchors.

In Architecture, is a certain ſort of Carving, ſomewhat reſembling an Anchor, or Arrow-head; 'tis commonly part of the Enrichments of the Boultins of Capitals of the Tuſcan, Dorick, and Ionick Orders; and alſo of the Boultins of Bed-mouldings, of the Dorick, Ionick, and Corinthian Corniſhes. Theſe Anchors, and Eggs being alternately carved throughout the whole Buildings.

Ancones

The Conſoles (a ſort of Brackets and Shouldering Pieces) are call'd *Ancones* by *Vitruvius*. See *Conſole.*

Annelet, or *Annulet.*

The ſame as *Cincture* from the Lat. *Annulus*, a Ring; in Architecture, 'tis uſed to ſignifie a narrow flat Moulding, (of which v. *Capital*) which is common to divers places of the Columns, as in the Baſes; and Capitals, *&c.* 'Tis the ſame Member as the Sieur *Mauclerc*, from *Vitruvius*, calls a *Fillet*, and *Palladio*, a *Liſtella* or *Cincture*; and *Brown from Scamozzi* a *Supercilium*, *Liſt*, *Tinea*, *Eye-brow*, *Square* and *Rabit*.

Antæ.

Pillars adjoyning to the Wall. See *Paraſtatæ.*

Antechamber.

1. *VVhat.*] From the Lat. *Ante-camera* an Outer, or Fore-chamber; a Room in Noblemen's Houſes, where Strangers ſtay till ſuch Time as the Party to be ſpoke with, is at Leiſure.

2. *Of it's Proportion in Length,* &c.] A well-proportion'd *Ante Chamber*, ought to have in Length, the Diagonal Line of the Square of the Breadth, and not to excel the Breadth and ½ at moſt.

3. *Of their Heighth*] They are made either arched or flat; if they are flat, ⅔ Parts of the Breadth ſhall be the Heighth from the Floor to the Joiſts.

But if you will have it higher, divide the Breadth into 7 Parts, and take 5 for the Heighth. Or divide the Breadth into 4 Parts and three of thoſe ſhall be the Heighth.

In great Buildings, the *Antechamber*, *Hall*, and other Rooms of the firſt Story may be Arched, which will make them handſome, and leſs ſubject to Fire; and in ſuch Buildings, the Heighth may be $\frac{5}{6}$ of the Breadth, which will be the Heighth it ought to be from the Floor to the bottom of the Key of the Arch.

But if this Altitude be thought too dwarfiſh, the Height may be $\frac{7}{8}$ of the Breadth.

Or

Or 1½ of the Breadth, which will make it yet more ſtately.

4. *Of their Situation*] *Ante-chambers*, and others alſo ought to be ſo poſited, that they may be on each ſide of the *Entry*, and of the *Hall*: And likewiſe it ought to be obſerved, that thoſe on the Right-hand, may anſwer, and be equal (or nearly ſo) to thoſe on the Left: to the end that the Buildings may on all Sides bear equally the Burden of the Roof.

Antick.

A Term in *Sculpture*, and *Painting*, being a confuſed Compoſure of Figures of different Natures, and Sexes, *&c.* As of Men, Beaſts, Birds, Flowers, Fiſhes, *&c.* and alſo ſuch like Fancies as are not *in Rerum Natura*. Of which I will give ſome Inſtances, and firſt of Humane Creatures, *viz.* How ſtrangely deform'd and confuſed ſome of the Heathens, repreſented their Gods, either in *Painting*, or *Sculpture*.

And 1. of *Saturn*; he is deſcribed by ſome with 3 Heads, *viz.* A Lion's, a Dog's, and a Wolf's; others pourtrayed him with two Wings on a Humane Head.

2. Of *Jupiter*; Him the *Lacedæmonians* pictur'd with 4 Faces. The *Argives* had his Repreſentation in Sculpture with 3 Eyes, *viz.* One in his Fore-head.

3. Of *Apollo*; Him the *Lacedæmonians* depicted with 4 Hands, and as many Ears. (The *Perſians* deſcrib'd *Phœbus* or *Apollo*) with the Head of a Lion.

The *Egyptians* had his Statue in the Likeneſs of a Man, with the Head of a young Ram, with ſmall Horns on his Shoulders.

4. Of *Mercury*, the Ancients deſcrib'd him like a young Man, with Wings behind his Shoulders and his Ears.

The *Egyptians* fram'd his Image with the Head of a Dog on his Shoulders.

5. Of *Janus*; by ſome he was depicted with two Faces, by others with four. *Numa*, King of the *Romans*, cauſed his Statue to be hewed out with 365 Fingers.

The *Phœnicians* form'd his Image like a Serpent, with her Tail in her Mouth.

6. Of *Neptune*; ſome repreſent him in his upper Part like a Man, and the lower like a Fiſh; in his Right-hand he holds a Trident, or three-pointed Spear.

7 Of *Pan*; the Ancients depicted him from the middle upwards, like a Man with a ruddy Complection, being very Hairy (his Skin and Breaſt covered with the Skin of a ſpotted Doe, or Leopard; holding in one Hand a Shepherd's Hook, in the other a Whiſtle,) and from the middle downwards, the perfect Shape of a Goat.

8. Of *Fauns*, *Sylvans*, *Fairies*, and *Satyrs*, as to their corporal Shape, they were deſcribed like *Pan*, only they had ſhort Horns on their Heads with ſmall Ears and ſhort Tails.

It would be a Task too tedious to enumerate all the Antick Forms, and Fancies by which

which the Heathens did repre ſent their ſeveral Gods, and their Poets. and Painters, and Carvers did deſcribe Them,and the Powers, Paſſions, Vertues, Vices, Nymphs, Muſes, *&c* They not only had ſtrange and monſtrous Fictions of Humane Creatures; (in *Poetry* and *Sculpture*, and *Painting* alſo:) but of Brutes, As, 1. Of the Syrens, or Mermaids, half a Woman, and half a Fiſh; Griffins, half Beaſts, and half Birds; *Pegaſus* was alſo another of the ſame Fictitious Kind; Harpyes alſo which were part Women, and part Birds; Centaurs, half Men, and half Horſes; Sagitaurs, half Men, and half Beaſts; Dragons, alſo part Serpent, and part Birds 2. They had alſo ſome Repreſentions of twiform'd Creatures, as the Amphisbœna, a Serpent with a Head at each end; the Spread Eagle with two Heads on the ſame Neck. And likewiſe they ſometimes have the Repreſentation of divers ſorts of Fruits, and Flowers, growing on the ſame Plant, *&c* With many ſuch like Fictions which we have good Reaſon to believe, there are really no ſuch ſtanding Species of Animals, and Vegetables in Nature; tho' the Belief of ſuch Exiſtences hath been propagated by Orators, upon account of their fitneſs, to be made uſe of, in the way of Similitude

This Work which we call *Antick*, the *Italians* call *Groteſca*, and the *French Groteſque.*) which ſignifies Comical, Pleaſant, apt to make one laugh; alſo ridiculous. And their word *Groteſques*, ſignifies fooliſh, idle Fancies.

See *Antique* lower.

Anticum.

From the Lat. a *Porch* before a Door, the Fore-door, a *Hatch.*

Antipagments.

The Ornaments, or Garniſhing in carved Work, ſet on the Architrave, (Jambs, Poſts, or Puncheons of Doors;) whether of Wood, or Stone; a *Latin Antique* Word (*Antipagmenta.*)

Antique.

This Word is uſed to ſignifie a Building or Statue made at the Time when the Arts were in their greateſt Purity and Perfection among the Ancient *Greeks* and *Romams* We likewiſe ſay, the *Antique Manner*, to ſignifie any thing done according to the ſtrict Rules, and the good Taſte of the Ancients. The ingenious Monſ. *Richelet*, ſays, This Word *Antique* is uſed in general, to expreſs ſuch Works of Painting and Sculpture as were made between the time of *Alexander the Great*, and that of the Emperor Phocas, under whoſe Reign the Noble Arts were extinguiſh'd, about the Year 600.

Apertions, or *Apertures.*

1 *What*] From the Lat ſignifying *opening* But in Architecture 'tis uſed to ſignifie, *Doors*, *Windows* *Stair Caſes*, *Chimnies*, or other *Conducts*: In ſhort,

all *Inlets*, or *Outlets*, of *Men*, *Light Smoak*, &c. To which belong two general Cautions, *viz* 1. Of their Number, and 2d. Their Position.

2 *Of their Number and Dimensions.*] Let them be as few in Number, and as moderate in Dimension, as may possibly consist with other due Respects; for in a word, all *Openings* are *Weaknings*.

3. *Of their Position*] Be sure to let them not approach too near the Angles of the Walls; for it were indeed a most essential Solecism to weaken that Part which must strengthen all the rest.

Aqueduct.

From the Lat. *Aqueductus*, a Conveyance made, for the carrying of Water from one place to another.

Arches.

1 *Whence derived*] It comes from the Latin, *Arcus*, a *Bow*.

2. *What*] In Architecture, 'tis used to signifie an inward Support to the Superstructure; and it is either *Circular*, *Eliptical* or *Straight* Of Circular Arches, there are three Kinds; *Semicircular*, *Scheme*, or *Skeen*, and Arches *of the* 3*d and* 4*th Point*. Of these and of *Elliptical*, and *straight Arches*, I shall treat in their Order.

3. *Semicircular*] These Arches are an exact Semicircle, and have their Centre in the middle of the Diameter, or right Line that may be drawn betwixt the Feet of the Arch. Of this Form the Arches of Bridges, Church-windows, and great Gates, are sometimes made in our modern Buildings.

4. *Scheam*, or *Skeen.*] These consist of less than a Semicircle, and consequently are flatter Arches. Some of these contain an Arch of about 90 Degree, others about 70, and others (which are yet flatter) about 60 Degrees; these last are very flat Now, 'tis very easie to distinguish between Semicircular, and Scheme Aaches; for the *Chord*, (or right Line) drawn between the Feet of a *Semicircular-arch*, is just double to its *heighth*, (measur'd from the middle of the *Chord*, to the Key piece, or top of the Arch;) whereas the *Chord* of a *Scheme-arch of* 96 *Degrees* will be above four time its height, and the *Chord* of a *Scheme-arch* of 60 Degrees, will be above six times its heighth The term *Skeam-arch* is derived from the Italian Word *Arco Scemo Scemo* signifying *imperfect*, *incomplete*. For such indeed is a *Skeam-arch*; being less than a Semi circle. See more of *Arches*, at the latter end of this Article; being a Paragraph taken out of the Famous *Alberti*'s *Architettura*, a Book that never yet appear'd in English, and out of which (because it is so highly prais'd by Sir *Henry Wotton*) I shall now and then make bold to borrow an Article, as I find 'em in a Manuscript Translation of Mr. *Ozell*'s.

In all *Openings*, says he, over which we make *Arches*, we shou'd ſhou'd contrive to have the Arch never leſs than a half Circle, with an Addition of the ſeventh part of half its Diameter: The moſt experienced Workmen having found that Arch to be by much the beſt adapted for enduring in a manner to perpetuity; all other Arches being thought leſs ſtrong for ſupporting the Weight, and more liable to ruin. It is more-over imagin'd, That the half Circle is the only Arch which has no Occaſion either for Chain, or any other Fortification; and all others, if you don't either chain them, or place ſome Weight againſt them for a Counterpoiſe, are found by their own Weight to burſt out and fall to Ruin. I will not omit here, what I have taken notice of among the Antients, a Contrivance certainly very Excellent and Praiſe-worthy: Their beſt Architects placed theſe Apertures, and the Arches of the Roofs of the Temples, in ſuch a manner, that even tho' you took away every Column from under them, yet they wou'd ſtill ſtand firm, and not fall down, the Arches on which the Roof was placed being drawn quite down to the Foundation with wonderful Art, known but to few: So that the Work upheld it ſelf by being only ſet upon Arches; for thoſe Arches having the ſolid Earth for their Chain, no wonder they ſtood firm without any other Support.

5. *Of the 3d and 4th Point.*) So our *Engliſh* Authors call 'em, but the *Tuſcan* Authors call 'em *di terzo, & di quarto acuto*, becauſe they always concur in an acute Angle at the Top. They conſiſt of two Arches of a Circle, (meeting in an Angle at the Top) drawn from the Diviſion of the Chord, into three, or four, or more parts, at pleaſure. The particular Method of drawing which, and all other Arches, and Mouldings, is not the Buſineſs of this Treatiſes. I have obſerv'd many of theſe Acute Arches Arches, in old Stone Buildings, both Houſes and Churches. But I ſay, (ſays that great Architect, Sir *Henry Wotton*) that theſe kind of Arches, (both for the natural Imbecility of their acute Angle, as likewiſe for their Uncomelineſs) ought to be exil'd from all judicious Eyes, and left to their firſt Inventors, the *Goths* and *Lombards*, amongſt other Reliques of that barbarous Age.

6. *Elliptical.*] Theſe kind of Arches conſiſts of a Semi-Ellipſis, and were formerly much us'd inſtead of Mantle-trees in Chimneys. They are commonly deſcrib'd on three Centers; but they may be drawn otherwiſe. Theſe conſiſt of three parts, *viz.* two Hanſes, and a Scheme. Now Workmen call each end of theſe *Arches* the Hanſe, which Hanſes are always the *Arches* of ſmaller Circles than the Scheme, which is the middle part of theſe Arches, and conſiſts of a part of a larger Circle; which is drawn betwixt the two Hanſes to conjoin them all together, to make, as it were one Helical Line,

Line, and by Consequence an *Elliptical Arch*. These Arches have commonly a *Key stone*, and *Chaptrels* (the Key-stone, is that which is the very summity, or top of the Arch, which is equally distant from both ends, and the breadth of this Key-stone at the top, ought to be equal to the height of the Arch, (which is commonly about fourteen Inches, when made of Brick) and *Sommer* (or point with its two edges) to the Centre of the Scheam; the Key-stone should break without the Arch, somuchas the *Chaptrels* project, over the *Jambs*. The *Chaptrils* I understand to be the same which most Architects call Imposts; and tis that on which the Feet of the Arches stand, whose height, or thickness ought to be equal to the breadth of the lower part of the Key-stone, *N. B.* That each other Course in these Arches consists of two *Stretchers*, which are seven Inches long apiece, (when the Arch is fourteen Inches deep) and the other Courses betwixt these of three *Headers*, and two *Closers*; the length of the *Headers* must be 3 ½ Inches and the *Closers* 1 ¾ Inches; thus one Course of the Arch will be divided into two Stretchers and the other alternately into three Headers, and two Closers, throughout the whole Arch.

7. *Strait*] These Arches have a straight, upper, and under edge, as the former had curved ones; and those two edges are parallel; and the ends, and the joints, all point toward one certain Centre: They are generally used over Windows and Doors, and 'tis a certain Rule amongst Workmen, That according to the breadth of the Peers betwixt the Windows, so ought the *Skew back*, or *Sommering* of the Arch to be; for if the Peers be of a good breadth, as three or four Bricks in length, then the straight Arch may be described from the *Oxi* (as it is vulgarly call'd) part of the Word *Oxigonium*, signifying an Equilateral Triangle; but if the Peers are small, as sometimes they are but the lenghth of two Bricks, and sometimes but one Brick and a half, then the breadth of the Window, or more, may be the Perpendicular (to the middle of the under side of the Arch) at whose end below, shall be the Centre for the *Skew back*, or *Sommering* to point to. These *straight Arches* are commonly about 1 ½ Brick, which when rubb'd, makes about twelve Inches high, tho' sometimes they are but eleven Inches, or thereabouts, which answers to four Course of Bricks; but they may be made more or less in height, according as Occasion requires *N.B.* That by the word *Skew-back*, is meant the levelling end of the Arch, and by *Sommering*, is to be understood the level Joints betwixt the Courses of Bricks in the Arch. These Arches commonly consist of a Stretcher, and a Header in height, the *Stretchers* being a whole Brick's length, and the *Headers* a Brick's breadth

Now the whole Business of building Arches, (saith Sir

H. W.) may be reduced into these five following *Theorems*.

8. *Theorem* th 1ſt.] All ſolid Materials, free from Impediment, deſcend Perpendicularly downwards, becauſe Ponderoſity is a natural Inclination to the Center of the Earth, and Nature performeth her Motions by the ſhorteſt Lines.

9. *Theorem* the 2d] Bricks moulded in their ordinary Rectangular Form; if they be laid one by another in a level row, between any Supporters ſuſtaining their two ends then all the pieces between will neceſſarily ſink even by their own natural Gravity; and muſt more if they ſuffer any Preſſure by a ſuperincumbent Weight; becauſe their ſides being parallel, they have room to deſcend Perpendicularly, without Impeachment, according to the former Theorem: Therefore to make them ſtand, either the Poſture, or their Figure, or both, muſt be changed.

10. *Theorem* the 3d] If Bricks moulded, or Stones ſquared, *Cuneatim*, (*i. e.* Wedge-wiſe, broader above, than they are below) ſhall be laid in a row-level, with their ends ſupported, as in the precedent Theorem, pionting all to one Centre; then none of the pieces between, can ſink, 'till the Supporters give way; becauſe they want room in their Figuration to deſcend Perpendicularly. But this is yet a weak piece of Structure, becauſe the Supporters are ſubject to much Impulſion, eſpecially if the Line be long; for which Reaſon this Form, (*viz. ſtraight Arches*) is ſeldom uſed, but over Windows or narrow Doors. Therefore to fortifie the Work, as in this third Theorem, we have ſuppoſed the Figure of all the Materials different from thoſe in the ſecond. So likewiſe we muſt now change the Poſture, as will appear in the following Theorem.

11. *Theorem* the 4th] If the Materials figured Wedge-wiſe, (as in the preceeding *Theorem*) ſhould be diſpoſed in the Form of ſome Arch, or Portion of a Circle, pointing all to the ſame Center, in this caſe, neither the pieces of the ſaid Arch, can ſink downwards for want of room to deſcend, (as in the 1ſt. Theorem) Perpendicularly: Nor the *Supporters*, or *Butments* of this Arch, can ſuffer ſo much Violence, as in the precedent flat Poſture, for the roundneſs will always make the incumbent Weight, rather to reſt upon the Supporters, than to ſhove them; whence may be drawn an evident Corollary, that the ſafeſt of all Arches is the Semicircular, and of all Vaults the Hemiſphere; tho' not abſolutely exempted from ſome natural Weakneſs, (which is the ſole Prerogative of Perpendicular Lines, and right Angles) as *Bernardino Baldi, Abbot* of *Guaſtalla* hath obſerved in his Commentary upon *Ariſtotle's Mechanicks*; where let me note by the way, that when any thing is Mathematically demonſtrated weak, it is much more Mechanically weak: Errors ever occuring more eaſily in the Management of groſs Materials,

Materials, than in Lineal Deſigns.

12. *Theorem* the 5th] As Semicircular *Arches*, or Hemiſpherical *Vaults*, being raiſed upon the total Diameter, be of all other the roundeſt, and conſequently the ſecureſt by the precedent Theorem: So thoſe are the comelieſt, which keeping preciſely the ſame heighth, ſhall yet be diſtended one 14th. part longer than the ſaid Diameter, which Addition of Diſtent will confer much to their Beauty, and detract but little from their Strength.

13. *Of Meaſuring Arches*] In meaſuring of them, whether they are Straight, or Circular; they muſt be meaſured in the middle, *i. e.* If a ſtraight Arch be twelve Inches in height, or depth, the length muſt be meaſured in the middle of the twelve Inches, which length will be no longer than if it were meaſured at the under ſide, next the head of the Window, by ſo much as one ſide of the ſpringing of the Arch is skew'd back from the upright of the Jambs, Peers, or Coins of the Windows.

Alſo in *Circular Arches*, 'tis to be obſerved, that the upper part of the Arch is longer, (being girt about) than the under part, becauſe it is the Segment of a greater Circle, cut off by the ſame right Line that the leſſer is, and therefore it muſt be girt in the middle.

14. *Price.*] For the Workmanſhip of *ſtraight Arches*, well rubb'd, and handſomely ſet; (of Brick) in *London*, about 8 *d.* or 9 *d. per* Foot; but in ſome parts of *Suſſex* and *Kent*, they will not do it under 12 *d. per* Foot, running Meaſure. But in *London*, if the Workmen find Materials, then 'tis about 10 *d.* or 12 *d per* Foot.

Skeen, or *Scheam Arches*, and *Elliptical* ones; of *rubb'd Brick*, are common about the ſame Price with ſtrait ones. But *Sheam Arches* of unrub'd Bricks are commonly included with the plain Work, unleſs the plain Work be done at a reaſonable Price: But you muſt here note, that the Maſter of the Building. (or Owner) is at the charge of the Centers to turn the Arches on; and not the Workman, unleſs he be allow'd for it in the Price of the Work.

Architectonick.

Belonging to the chief Overſeer of Buildings, alſo to an

Architect.

A Maſter-workman in a Building; 'tis alſo ſometimes taken for the *Surveyor* of a Building, *viz.* He that deſigns the Model, or draws the Plot, or Draught of the whole Fabrick; whoſe Buſineſs it is to conſider of the whole Manner, and Method of the Building, and alſo the Charge, and Expence: In the Management of which, he muſt have reſpect to its due *Situation Contrivance*, *Receipt*, *Strength*, *Beauty*, *Form*, and *Materials*. All which are to be duly ly deliberated of by the *Superintendent*, (or *Surveyor*) of a *Building*; it being wholly committed to his Circumſpection, and therefore it will be his Prudence to manage the whole Affair adviſedly, and with great Caution, that all may be ſo order'd, and diſpoſed (in all Circumſtances) that it may anſwer the Deſign, and be conſentaneous to Reaſon. But tho' the whole Fabrick be the Care of the *Superintendent*, yet Sir *H. Wotton*, would have a ſecond *Superintendent*, (or *Officinator*, as *Vitruvius* calls him) whoſe Care it ſhould be to chooſe, (or examine) and ſort all the *Materials* for every part of the Structure.

Architecture.

A Mathematical Science, which teacheth the Art of Building being a Skill obtain'd by the Precepts of Geometry, by which it gives the Rules for deſigning, and raiſing all ſorts of Structures, according to Geometry and Proportion. Containing under it all thoſe Arts that conduce any thing to the Framing Houſes, Temples, *&c* The *Scheme* or *Projection* of a Building, is uſally laid down in three ſeveral Draughts or Deſigns. The firſt is a Plan, which ſhews the Extent, Diviſion, and Diſtribution of the Ground into Apartments, and other Conveniences. The ſecond ſhews the Stories, their Heights, and the outward Appearances of the whole Building; and this we call the Deſign or Elevation. The third, call'd the Section, ſhews the Inſide: And from theſe three Deſigns, the Undertaker forms a Computation of the Expences of the Building, and the Time requir'd to go thro' with it. So much for what is call'd *Civil Architecture*. *Military Architecture*, uſually call'd *Fortification*, has for its Object, the Security of Cities and States: but this part is foreign to my purpoſe.

Archives.

A Place where ancient Records, Charters, and Evidences of a Nation are kept: Alſo the Records themſelves. *Archivum* in *Latin*.

Architrave.

1. *Whence derived*] The word comes from the Greek *Archos*, chief; and the Latin *Trabs* a Beam. 'Tis alſo ſometimes call'd *Epiſtyle* from the Greek *Epi* upon, and *Stylos* a Column.

2. *What.*] 'Tis uſed in Architecture, to ſignifie the Moulding, or Ornament next above the Capital of a Column: It being always the next groſs Member below a Freeze. The word is alſo ſometimes uſed to ſignifie the chief, or principal Beam of a Building. Now I cannot conceive what they mean by the principal Beam in a Building, becauſe I do not ſuppoſe it can properly be applied to all Buildings but only to ſome peculiar Kinds, as what we call *Porticoes*, *Piazza's*

za's, or *Cloysters*, (by which we understand a long kind of *Galleries*, or Walking-places, whose Roof is born, or supported by Columns, or Pillars, at least at one side.) The which have not Arches rising from them to bear the superincumbent part of the Fabrick, but have a Beam resting, or lying ing on the Tops of the Columns, by which the superiour Part of the Edifice is supported; upon which Account I suppose it to be called the chief or principal Beam. In Truth, according as *Perrault* defines it, it is the first Member of the *Entablement*, being that which bears upon the Column, and is made sometimes of a single *Summer*, as appears in most of the ancient Buildings and sometimes of several *Haunses*, as we usually see it in the Works of the Moderns.

In Chimneys, the Architrave is the Mantle. Over the Jambs of Doors, and Lintels of Windows, 'tis called *Hyperthyron*.

There are also *Architrave Doors*, and *Windows*; those are call'd *Architrave Doors*, which have an *Architrave* on the Jambs, or Puncheons, and over the Door, upon the Cap-piece, if straight; or on the Arch, if the Top be curved. The Form of these *Architraves* about Doors, are not always the same; for sometimes they are according to one of the 5 Orders of Architecture. But 'tis sometimes done according to the Workman's Fancy; for I have seen some have put for an *Architrave* round a Door, 1st. next the Door a small Bead, then a broad Plinth, or Fatio, above that an O.-G, and List. There are Stone and Brick *Architraves*, as well as Timber-ones. *Architrave Windows* of Timber, are commonly an O.-G, rais'd out of the solid Timber, and a List above, but sometimes they are stuck, and laid on. Brick *Architraves* are usually cut in the Length of a Brick, but sometimes they are cut in the length of a Brick and ½, then each other Course alternately consists of the Breadth of 2 Bricks; the upper One on which the *O.-G* is cut, and part of the upper *Fatio*, they call *Header*, or *Heading Architrave* and the Breadth, or Head of Bricks on which the lower *Fatio*, and part of the upper one is cut, they call a *Jak* Architrave of Stone. v. Door, N. 4.

3. *Kinds*] *Architects* distinguish them into 5 *Kinds*, *viz.* Tuscan, Dorick, Ionick, Corinthian, and Composite, according to the five Orders of Columns.

4. *Parts, or Members*] Are more numerous than the Kinds, because some of the Orders have two different sorts of *Architraves* and what yet more increases the Number, is, that some Authors differ from others in their Forms of the same *Orders*. Of all which I shall give a particular Account, in the following Order.

5. *Tus-*

5 *Tuſcan*] according to *Vitruvius* ought to be ½ a Model or Module in Altitude; this general Member, he hath deſcribed in two Forms; the 1ſt conſiſting of 3 Parts, or Members, *viz* 2 *Fatio*'s and a *Cimatium*, is thus divided: The whole Height is divided into Six Parts, which is ſubdivided in this Manner, *viz.* the upmoſt 6th Part is the *Cimatium*, which being ſubdivided into 3, the upper part ſhall be the *Fillet*, and the 2 lower ones the O-*G*. The 5 grand Diviſions which remain, muſt be divided into 9, 5 of which ſhall be for the ſuperior *Fatio*, and the other 4 for the inferiour one. His 2d. Form conſiſts of but 3 Members, or Parts, *viz* a large *Plinth*, or *Planchier*, a *Caſement*, and a large *Fillet*, and is thus ſubdivided, the whole Heighth is divided into 6, the upper part is for the Fillet, (which projects in ſquare beyond the *Plinth*) the 5th part is for the *Caſement*, (which riſes from the Plain of the *Plinth* and terminates in a Quadrant, as the lower Corner of the *Fillet*.) The other 4 Parts remaining, are for the *Plinth*, or *Planchier*, or *Fatio*.

Palladio hath alſo 2 diſtinct Shapes for the *Tuſcan Architrave*, the 1ſt, which we ſhall mention, conſiſts of 2 *Faſia*'s, (or *Fatio*'s) and a *Liſt*; the lower *Fatio* is 12 ½ m. high, the upper *Fatio* is 17 ½ m. which terminates with a Quadrantal Caſement, riſing from its Plain, and terminating with the lower Corner of the *Liſt*; the *Liſt* is 5 m. high; ſo the whole heighth of the *Architrave* is 35 m. His 2d *Architrave* is only a plain Fatio of 35 m. high. *Scamozzi*, according to his Delineations, makes the *Tuſcan Architrave* 31 ½ m. high, the which he divides into 4 Parts, or Members, *viz.* 2 *Fatio*'s, a *Liſt*, and a *Plinth*; his 1ſt. *Fatio* he makes 10 m. his 2d. 16½ m. his *Liſt* 1 ½ m. and his Plinth 3 ½ m. all which make 31 ⅓ m. tho', according to this verbal Account of it, he ſaith it muſt be 32 ½ m. except it ſhould be a Typographical Erratum.

Vignola deſcribes it with the ſame Parts, Heighth, and Form with *Vitruvius*'s 2d.

6. *Dorick*] This *Architrave*, according to *Vitruvius*'s, is half a Module in Altitude, the which he delineates in two Forms; the 1ſt, (which I ſhall mention) he divides into 7 Parts, the uppermoſt of which is the *Tenia*, the other 6 remaining Parts, he makes a Faſcia under the *Tenia*; he placeth *Drops*, whoſe Heighth are $\frac{1}{7}$ of the *Architrave*; ¼ of this $\frac{1}{7}$ is the *Fillet*, to which the *Drops* hang; the Drops are 6 in Number, placed under, (and of the ſame Breadth with) the *Trigliphs*. His 2d Figure of his *Architrave*, conſiſts of the ſame Members with the 1ſt, and the whole Heighth is equal to the 1ſt, but he divides the Altitude but into 6 Parts; the upper one of which is his *Tenia*, and the other 5 the *Faſcia*, the uppermoſt of which is the Altitude of his *Drops*, which have a *Liſt*, which is ¼ of their Heighth, as before.

Palladio

Palladio composes this *Architrave* of the same Heighth, with *Vitruvius*, but of a different Fashion; for he makes it to consist of 3 Parts, or Members; *viz.* 2 Fascia's, and a *Tena*, or *Tenia*; he divides the whole Heighth into 6 Parts, one of which being 5 m. he assigns for the *Guttæ*, *Bells*, or *Drops*; the Listella of the *Drops*, is $\frac{1}{5}$ of the whole Height, 1 $\frac{1}{3}$ m. and the Drops 2 $\frac{2}{3}$ m. The *Tenia* above the *Drops*, (or of the *Architrave* rather) he makes 4 $\frac{1}{2}$ m. and the *Prima* (or upper) *Fascia*, he makes 14 $\frac{1}{2}$ m. and the *Secunda*, (or lower) *Fascia*, he allows 11 m. for, in all 30 m. which is the whole Heighth.

Scamozzi, (according to his Portraicture of this *Architrave*) makes it 35 m. in Altitude, and he makes this grand Member, to comprehend 3 pretty Members, *viz.* 2 *Fatio*'s, and a *List*; whose Dimensions are as follows (beginning at the Top, and so descending) the *List* to be 5 m, the upper *Fatio*, 18 m. and the lower one 12 m. in all 35 m. The *Drops*, or *Bells*, he thus divides, the List above them he designs to be 1 $\frac{1}{4}$ m. and the Bells, or Drops themselves 4 $\frac{1}{4}$ m. so that your whole Heighth is 6 m.

Vignola, delineates this *Architrave*, 30 m. in Heighth, the same with *Vitruvius*, and *Palladio*; both which he also imitates in the lesser Member; for he hath 2 distinct Forms, one like *Vitruvius*, containing 2 Members, or Parts, one a *List*, the other a *Fatio*: his other Form is like *Palladio*'s, comprehending 3 petty Members, *viz* a *Tenia*, and 2 *Fatio*'s.

7. *Ionick*.] According to *Vitruvius*'s Order, this grand Member ought to be $\frac{1}{2}$ a Module high; he hath described 2 Forms of *Architraves*, in the *Ionick* Order, *viz* one for the Ionick Column, without a Pedestal, and the other with a Pedestal: and 1st I will describe that without a Pedestal; the which he composes of 4 minuter Parts, *viz.* 3 *Fascia*'s, and a *Cimatium*; which is thus divided; the whole Altitude is divided into 8 Parts, the uppermost of which is allotted to the *Cimatium*, which is subdivided into 3 Parts, the uppermost of which is for the *List*, and the 2 remaining, for the O--G. The other 6 remaining Parts, they divide into 12, 5 of which he makes the upper *Fascia*, 4 the middle one, and 3 the lower. The other for the *Ionick* Column, with a Pedestal, he thus proportions, *viz* He reckons the whole Heighth of the *Architrave*, *Freese*, and *Cornish*, to be 2 Modules, the which he divides into 10 Parts, 3 of which are for the *Architrave*, (which is 36 m.) the which he distinguishes into 6 minuter Parts, or Members; the which he thus names, (beginning at the Top, and so descending) *viz.* A *Fillet*, a *Cima*, a *Thorus*, and 3 *Fascia*'s; all which smaller Members he thus finds, *viz.* He 1st divides the whole Heighth into 6 equal Parts, the uppermost of which Parts he subdivides into 4 Parts, the highest of these 4 is for the *Fillet*,

Fillet, the 2 next of the 4 are alloted to the *Cima*, and the 4th remaining, is for the *Thorus*. The 5 grand Divisions remaining, he subdivides into 12, which are thus distributed, *viz* 5 for the upper, 4 for the middle, and 3 for the lower *Fascia* *Palladio* assigneth 34 m for the Altitude of this *Architrave*, according to his Scheme of this Member, it is composed of 7 Parts, *viz*. A *List*, a *Cima*, 3 *Fascia*'s, and 2 *Astragals* ; the which he thus proportions, *viz* To the *List*, (which is above the *Cima*, for I will descend with the De scription) he allots 2 $\frac{3}{10}$ m. the *Cima* 4 $\frac{3}{5}$ m. To the upper *Fascia*, he allows 10 $\frac{1}{8}$ m. to the *Astragal* at his Foot $\frac{1}{5}$ m. the middle *Fascia*, is to contain 7 $\frac{52}{60}$ m. and the *Astragal* at his foot 1 $\frac{1}{3}$ m. to the lower *Fascia*, he assigns 6 $\frac{9}{10}$ m all which being added into one Sum. amounts to 34 $\frac{1}{2}$ m. *Scamozzi* makes the *Ionick Architrave*, 35 m. high and of the same Shape with *Vitruvius*'s second, *viz* To consist of 6 Parts, *viz* A *List*, *Cima*, *Astragal* (or *Thorus*) and 3 *Fascia*'s, which he thus proportions; he allots to the *List*, 2 $\frac{1}{2}$ m. to the *Cima*, 4 m. to the *Thorus*, 2 m. to the upper *Fascia*, 11 $\frac{1}{2}$ m to the middle one 8 $\frac{1}{2}$ m. and to the lower one 6 $\frac{1}{2}$ m.

Vignola, allows 37 $\frac{1}{2}$ m. to the *Ionick Architrave*, in Heighth, and as to the Form, his is much the same with *Vitruvius*'s 1st. of this Order.

8 *Corinthian*] According to *Vitruvius*, ought to be $\frac{1}{2}$ 2 Module high ; but you must note this is for the *Corinthian* Column, without a *Pedestal* ; this Member he divides into 7 Parts, the uppermost of these is the *Cimatium*, the 6 remaining Parts he divides into 12, 5 of which he allots to the upper *Fascia*, $\frac{1}{8}$ part of this *Fascia* is to be allow'd for a *Bead* at his Foot, 4 of the 12 Parts he allows to the middle *Fascia*, and $\frac{1}{8}$ of this *Fascia*, he makes the Bead of at the Foot, and the 3 Parts remaining, he makes the lower *Fascia*. The Architrave for the *Corinthian* Order with a Pedestal, *Vitruvius* alloweth a larger Altitude, than that without ; it consists of the same Members, both for Number and Form, with the former *Architrave* but they differ in Dimensions. The Division, and Subdivision of which, take as followeth : The whole Heighth of the *Architrave*, ought to be $\frac{1}{4}$ of the Heighth of the Column; (nearly to $\frac{2}{8}$ of the Body of the Column below) which is — to 40 $\frac{1}{2}$ m. This Altitude he divides into 7 equal Parts, and at the uppermost of these 7, he maketh a *Cimatium*, and the 6 remaining, he divides into 12 equal Divisions, 5 of which are allotted to the upper *Fascia*, 4 to the middle one, and 3 to the lower one : The upper, and middle *Fascia*, he subdivides into 8 Parts each, one of these 8ths he allows for a Bead at the Foot of each of these *Fascia*'s.

Palladio makes this *Architrave* to contain 8 Parts, *viz*. 1 *List* 1 *Cime*, 3 *Beads*, and 3 *Fascia*'s ; the Heighth of all which

which, he orders to be 36 m. which he thus ſubdivides *viz.* To the (upper Members, or) *Liſt*, he allows $2\frac{3}{4}$ m. the next in order, is a *Cima* and the next in order is of 2 m. high, at the Foot of the which is a Bead; then follows the upper *Faſcia* with his Bead at his Foot, both which contains about $13\frac{1}{7}$ m. then comes the middle *Faſcia*, and his Bead, which contain $8\frac{1}{8}$ m and laſt of all the lower *Faſcia*, of $6\frac{1}{4}$ m. high.

Scamozzi reckons the whole Heighth of this *Architrave* to be 40 m the which he ſubdivides into 9 ſmall Members, *viz.* (beginning at the Top deſcending) a *Liſt* of 2 m. a *Caſement*, $3\frac{1}{4}$ m and O. G. of $2\frac{1}{4}$ m. a Bead of $1\frac{1}{2}$ m a *Faſcia* of 12 m and his *Bead* of 2 m the middle *Faſcia* $8\frac{1}{2}$ m. and his *Bead* $1\frac{3}{5}$ m and *lower* Faſcia of $6\frac{1}{2}$ m. in all 40. m. as before ſaid.

Vignola makes the *Corinthian Architrave* to be 45 m. high, the which he ſubdivides into eight ſmaller Members, as *Palladio* doth, *viz.* a *Liſt*, a *Cima*, 3 *Beads*, and 3 *Faſcia*'s.

9. *Compoſite*, *Compound*, or *Roman.*] *Vitruvius* makes the *Architrave* in this *Column*, and the *Frieſe* and *Corniſh*, all of an equal Heighth, *viz.* Each of which is equal in Heighth to the Diameter of the *Column* above, juſt under the *Capital*; which is $\frac{10}{12}$ of a Module, — 50 m. This *Architrave*, *Vitruvius* divides into 6 Parts, one of which is for the *Cimatium*, and its *Boultin* under it; this upper 6th Part he divides into 4, and one of theſe 4 he allows for the *Fillet* above the *Cima*, the 2 next for the *Cima* it ſelf, and the 4th remaining he allots for the ſmall *Boultin* under the *Cima*, the other five grand Diviſions, he ſubdivides into 12 minuter Parts, 5 of which Parts he aſſigns for the upper *Faſcia*, 4 for the middle one, and 3 for the lower; the upper and middle *Faſcia*'s he ſubdivides into 8 Parts each, and one of theſe 8ths he allows for a *Bead*, at the Foot of each of theſe *Faſcia*'s.

Palladio makes this *Architrave* 45 m. high, the which he diſtributes amongſt 7 particular minuter Members, which I will thus reckon up in Order (beginning at the Top, and ſo deſcending) And 1ſt, to the *Liſt*, he allows $2\frac{1}{8}$ m. to the *Caſement*, $4\frac{1}{8}$ m. to the O. G. $9\frac{1}{4}$, to the *Bead* $1\frac{1}{4}$ to the upper *Faſcia*, 15 m. to an O. G. at his Foot $2\frac{1}{4}$, and to the lower *Faſcia* 11 m.

Scamozzi makes this 40 m. high, the which he divides amongſt theſe 8 following Members, or Parts, *viz.* (deſcending) 1ſt a *Liſt* of 3 m. 2dly, an O. G. of $4\frac{1}{2}$ m 3dly, an *Aſtragal* of 2 m. 4thly, the upper *Faſcia* of a $11\frac{3}{4}$ m. 5thly, a *Bead* at his Foot of $2\frac{1}{4}$ m. 6thly, the middle *Faſcia* of $8\frac{1}{2}$ m. 7thly his at his Foot $1\frac{1}{2}$ m. 8thly, and laſtly, the lower *Faſcia* of $6\frac{1}{2}$ m.

Vignola makes this *Architrave*, 45 m. in Altitude; the which he divides into 7 Members, a *Liſt*, a *Caſement*, a *Boultin*, a *Fillet*, a *Faſcia*, a *Bead*, and a *Faſcia*.

10 *Mea-*

10. *Measuring.*] As to measuring of *Architraves*, in Buildings, (whether of *Brick* or *Stone*) they are commonly done by the Foot Lineal and therefore the Length being taken in Feet, the Content is alſo had at the ſame Time.

11. *Price.*] As to their Value, they are different according to their Breadth, or Wideneſs: *Architraves* of Stone, about Doors and Windows, Mr. *Wing* ſaith, they are commonly reckon'd 1 *d. per* Inch broad, at 1 Foot: E. G. if it be 9 n. broad, it's worth 9 *d. per* Foot, 10 n. 10 *d.* &c.

Aræoſtyle.

This Word *Vitruvius* uſes to ſignify the greateſt Interval or Diſtance which can be made between the Columns; conſiſting of eight Modules, or four Diameters. It comes from the Greek *Araios*, thin-ſet, or *Rare* and *Stylos*, a Column.

Aſh.

Of Sawing.] In ſome Places they have 3 *s. per* Hundred, in others 3 *s.* 6 *d.* and other ſome 4 *s.* The Price varies in this, as it doth in other Buſineſs, *viz.* according to the Cuſtom of the Place; but it is certainly worth 6 *d.* per Hundred (at leaſt) more than 'tis to ſaw Oak.

Now I am upon the Subject of one of the moſt uſeful Timber-trees (the *Aſh*) it may not be amiſs to quote a few Lines out of our incomparable Poet, old *Spencer*, giving us ſuch a ſuccinct Account of the Nature of all (or moſt) Timber-trees that tis worth every Builder's While to get them by Heart, which he may the more eaſily do, as they are in Rhime.

Much can they praiſe the Trees ſo ſtraight and high,
The ſailing Pine, *the* Cedar *proud and tall,*
The Vine-prop Elm, *the* Poplar *never dry,*
The Builder Oak, *ſole King of Foreſts all,*
The Aſpin *good for Staves, the* Cypreſs *Funeral.*
The Laurel, *meed of mighty Conquerors,*
And Poets ſage; the Firr, *that weepeth ſtill,*
The Willow *worn of forlorn Paramours;*
The Eugh *obedient to the Bender's Will;*
The Birch *for Shafts, the* Sallow *for the Mill;*
The Myrrhe *ſweet, bleeding in the bitter Wound;*
The warlike Beech, *the* Aſh *for Nothing ill,*
The fruitful Olive, *and the* Platane *round,*
The Carver Holme, *the* Maple *ſeldom inward ſound.*

Aſhlar

Ashlar.

1. *What.*] I understand by Workmen, that by this Word, they mean common, or free Stones, as they come out of the Quarry of different Lengths and Thicknesses Mr. *Leybourn* saith, that 9 n. is the common Thickness.

2. *Price.*] Mr. *Wing* saith, in *Rutland*, they commonly value them at 3 *d. per* Foot at the Quarry.

About us (in *Sussex* and *Kent*) they sell them, being a common ordinary sort of Stone by the Load, about 18 or 20 Foot makes a Load, which costs, if they come rough from the Quarry about 3 *d. per* Foot; being laid down at the Place, where they are to be used; but if they are ready scapted, then they are valued at about 4*d. per* Foot. But if they are bought rough at the Quarry then they are valued about 2 *d per* Foot; but if scapted, then about 3 *d. per* Foot But in some other Places in *Kent* and *Sussex*, I have known them sold rough at the Quarry for about 1 ½ *d per* Foot, and for 2 ½ *d. per* Foot scapted; but if they were laid down at the Place for Use rough, then they were usually valued at about 2 ½ *d. per* Foot, and if ready scapted, at about 3 ½ *per* Foot.

But as to the real value of *Stones*, or *Ashlar*, in all Places, it's impossible to give a certain Rule to know it: Because the Price differs; 1st, According to the different Customs of the Places. And 2dly, The Circumstances of the Quarry. And 3dly, Goodness of the *Ashlar*. To all which 3 Heads I shall briefly say something. 1st, Of the Customs of Places, by which I mean as to Carriage; I have known *Stones* carried above a Mile for 1 *s* 8 *d. per* Load, at one Place; and again, at another Place the usual Price to carry a Load but about ½ a Mile was 2 *s.* which is 4 *d. per* Load more than at the 1st Place tho' they were carried but ½ so far.

2dly. As to the Circumstances of the Quarry, which I shall consider under 2 Heads. And 1st, Whether the Stones are drawn in inclosed Land, or on the Lord's Waste, (*viz.* In the High-ways, or on Commons, *&c.*) For if they are drawn within Land (as they commonly call it) he that is the Proprietor of the Land, will be paid well for damaging his Ground, both by drawing, and carrying the Stones out of his Land. Whereas, if they are drawn on the Lord's Waste, the Lord hath only (commonly) a small Acknowledgment (by the Load, or so forth) for Trespassing upon his Waste.

3dly. As to the Goodness of Stones, that may be consider'd of under 2 Heads, *viz.* Durability, and Magnitude. And 1st, Of Durability; this wholly depends on Experience; for none can certainly tell when a new Quarry is first opened, how the *Stones* may prove; for some Stones, when they are taken out of the Quarry, are very soft and friable, and

Leng

being but a few Years exposed to the Weather, moulder into Sand; tho' some of these soft *Stones* are indurated by being exposed to the open Air; but as for hard *Stones*, they are generally durable, being of a more solid and firm Consistence. 2dly. As to their Magnitude, I need not say much, for all know that large *Stones* must needs be better, and make firmer Works than small which are only fit for filling Work in thick Walls, or to use in such Places where the Country affords no better. But 'tis too often, through the Stone-drawer's Carelessness, or Ignorance, that *Stones* are broken up so small in the Quarry; and therefore to promote (in some Measure) so useful an Art; I shall, when I come to speak of *Stones*, lay down some Precepts to be observ'd in drawing of *Stones*, as I receiv'd it from an ancient experienced *Stone Drawer*, who always drew the best *Stones* on our Side the Country.

I might have added a third Head to the Circumstances of the Quarry; which because I did not think of then, I shall here annex it, tho' it be not so exactly in its proper Place; Which third Head is this, *viz.* There is a great Difference as to Qnarries, in respect to the Position of the *Stones* in the Ground, which may be again considered under two Heads, *viz.* As to their Depth in the Ground, and their Manner of lying there And first, as to their Depth: When they lie a considerable Depth, it requires a great deal of Labour to remove the Earth over the *Stones*, or uncope it, as Workmen call it. 2dly, If they lie almost even with the Surface of the Ground, then it will require but little work to uncover them. And 2dly, As to their manner of lying in the Ground that is also different, and that in two Respects; for if the Quarry consist of a Rock, it will require more Labour to raise the *Stones*, and break them fit for Use, than if the Stones lie separate, and dis-united All which forecited Circumstances, being duly consider'd, will make the Price of *Stones* very uncertain; for I have known *Stones* drawn for 9 *d per* Load, and I have known 3 *s per* Load given.

Before I conclude this of *Ashlar*, I shall add something to the third General Head, of the Goodness of Stones and that shall be the third Division of that Head, *viz.* The Form of *Stones* ought to be consider'd, as to their being raised Square, and not with obtuse, and acute Angles, which requires more Work in Scapting and wasts more of the Stones, V. P. Stones of drawing.

Ashlering.

Quartering (to tack to) in Garrets about 2 ½. or 3 Foot high, perpendicular to the Floor, up to the under side of the Rafters; 'tis from 4 to 6 *s. per* Square Workmanship.

Astragal

Astragal.

A little round Moulding, which encompasses the Top of the Fust or Shaft of a Column. It comes from the Greek *Astragalos*, the Bone of the Heel. The Shaft always terminates a-top with an *Astragal*, and at Bottom with a *Fillet*, which in this Place is call'd *Ozla*.

Attick.

In building, a little Order, placed upon another much greater; for instead of Pillars, this Order has Nothing but Pilasters, with a Cornice Architraved for an Entablement; as that for Instance, in the Castle of *Versailles* above the *Ionick*, on the Side of the Garden.

Attick, or *Athenian Base.*

The same as Palladio's Ionick Base, which see.

B.

Back, *Baguette.*

A Kind of *Astragal* or *Hip molding* is a Term in Carpentry, by which they signifie the outward Angle or the Hips, or Corners of a Roof; which in square Frames, where the Roof is $\frac{3}{4}$ Pitch, contains an Angle of 116 Degrees, 12 m.

It's also a Term used by Iron-mongers to signifie a certain sort of Nails, V. Nails N. 2.

Bake-House.

1. *What.*] It's a Room of Office, in all Noble Buildings, where the Oven is placed &c.

2. *Its Position*] According to the Rules of Sir *Henry Wotton*, it ought to be placed on the South-side of any Building.

Balcony.

1. *What.*] Is a kind of open Gallery, (without the Walls of a House or Building) for People to stand in, and behold any Action, as Pageants, and the like, in Cities, or to take the Air, &c. This Jutty, or projective Building, commonly is in the midst of the Front of a House, if there be but one Balcony to it; and for the most part level with the 1st. Floor above Stairs. And they are sometimes of Wood, and sometimes of Iron; the wooden consist of Rails and Ballisters, and so do the Iron ones sometimes, but not always, for they are sometimes made of cast Iron of various Figures in semi-Relief, and others of wrought Iron, in crail'd Work, or Flourishes, of different Shapes, according to the Fancy of the Master, or Workman.

2. *Price*] Wooden *Balconies* are commonly done by the Yard, *viz.* From 3 to 5 *s. per* Yard, Workmanship, according as what Work the Carpenter bestows on it.

Iron Balconies are commonly done by the lb. (*viz.* wrought ones)

ones) from 4 *d.* to 8 *d per* lb. according to the Curiosity of the Work.

Here I can't but take Notice what Sir *Henry Wotton* says of all Inlets or Outlets (such as Windows, Balconies, &c.) that they should not approach too near the Angles of the Walls; for that it were indeed a most essential Solecism, to weaken that Part which must strengthen all the rest: A Precept (says he) well recorded, but ill practised by the Italians themselves, particularly at *Venice*, where he had observ'd divers *Pergoli*, or *Mencina* (as *Vitruvius* seems to call 'em, which are certain ballised Outstandings, to satisfy Curiosity of Sight) very dangerously set forth, upon the very Point it self of the Mural Angle.

Baldachin.

It's a perfect *French* Word, and they pronounce it *Baldaquin*, which properly signifies a Canopy carry'd over the Holy Sacrament, among the Roman Catholicks. 'Tis used by Architects to signifie a piece of Architecture, built in fashion of a Canopy, or Crown, supported by several Pillars to serve for a covering to an Altar; some also use it to signifie a Shell over the Front Door of a House.

Balkes.

Great pieces of Timber coming from beyond the Seas by Floats.

Ballon.

French, a Term in Architecture, signifying the round Globe of the Top of a Peer, or Pillar.

Ballustrade.

A Term in Architecture, used to signifie a Row of little turn'd Pillars, call'd *Balusters*, made of Marble, Iron, Wood, or Stone, so high, as for a Man to rest his Elbows, fixed upon a *Terrass*, or upon the Top of a Building, or to make any Separation.

Ballulster, (or *Bannister* by Corruption.)

1. *What.*] It's a small Column, or Pilaster, of different Sizes, *viz.* from 1 $\frac{3}{4}$ Inches, to 4 In Square, or Diameter: Their Dimensions and Forms are various, according to the Fancy of the Workmen. They are adorn'd with Mouldings. The Word comes from the Latin *Balaustrum*, which it self is borrow'd from the Greek *Balaustin*, the Flower of the wild Pomegranate, which it very much resembles.

2 *Their Use.*] They with Rails are placed on Stairs in the Fronts of Galleries, in Churches, *&c.* Round Altar-pieces in Churches, on Terrass Walks, and in Balconies, and Plat-forms, *&c.*

3. *Their Price.*] With Rails, *&c* of Wood on Balconies, Platforms, Stairs, *&c* according

ing to the Work, *viz.* About 3 *d per* Yard, running Measure.

4. *Of Turning them only.*] 1 *d per* Inch Workmanship, is the usual Allowance.

5. *Of Painting them*] They, with what belongs to them, are usually painted by the Yard; the Custom of Measuring which is this, *viz.* Both sides of the Ballisters are measured as if it were flat Measure, including the Vacuity betwixt the Ballisters; which being cast up in Feet and Parts, it's reduced into Yards as other plain Painting is. Mr. *Leybourn* saith, That he hath seen the Experiment tryed, by girting the Ballisters, to find the Difference betwixt that Way, and measuring them, and the Vacuity on both Sides, as if it were flat; and he found that the Difference would not countervalue the trouble of Girting. But this stands to Reason, it should be nearly the same, because it's the Custom to set them but their Square or Diameter asunder, and then the Flanks make good the Vacuities.

Band.

In Architecture, is any flat Member that is broad, and not very deep; the Word *Face*, from the Latin *Fascia*, is sometimes used to signify the same Thing.

Bandelets.

It is derived from the *French*, *Bandelette* a little *Fillet*, or Band; 'tis used by Architects, to signifie the 3 Parts that compose an *Architrave*.

Bannister.

V. *Ballister.*

Barbican.

A Term in *Architecture*, derived from the *French*, and made use of to signifie an Outwork in a Building.

Bargecourse.

It is a Term used by Workmen, by which they signify a part of the *Tyling*, which projects over without the principal Rafters in all Buildings where there is either a Gable or a *Kirkin-Head*.

Barn.

Is a Word that needs no Explanation, because 'tis a Building that every one knows, they being so common: But I shall add 2 Things concerning them, and the

1st. Shall be what Mr. *Worlidge* advises concerning placing them, which is this, That it is very Inconvenient to build *Barns*, or *Stables*, or such like Places, too near to a House, because Cattle, Poultry, &c. require to be kept near

near to Barns, *&c.* which would then annoy a Houſe.

2. *Of the Price of Framing*, &c.] I have known the Carcaſs of a *Barn* for 3 *s.* 6 *d. per* Square, Carpenters Work only, and I have known 8 *s. per* Square given for Carpenters Work, the Felling. Hewing, and Sawing his Timber and Boards, and finding Nails.

I have been inform'd by ſome Workmen, that the Charge of a Square of *Building* of the Timber-work of a Timber-Barn, may be thus computed, *viz.* 4 *s. per* Square for Sawing the Boards, (conſidering their lapping one over another, and the Staving the Logs 2 *s. per* Square for Sawing the Timber Members, 3 *s.* 6 *d. per* Square for *framing* the Carcaſs, and from 4 *s* to 7 *s. per* Square for the value of the Timber, reckoning the Price of the Timber from 12 *s.* to 21 *s. per* Tun; and one Tun to make 3 *s.* Square of Frame in *Barn work.* He reckon'd rough Timber, *viz.* Unhew'd, or Squar'd, and that a Tun of rough Timber (which is equal to a Load of hew'd: From theſe Computations, we may reckon the whole Value of a Square of ſuch Timber-work to be worth from 3 *s.* 6 *d.* to 16 *s.* 6 *d. per* Square.

Bars.

Of Iron for upright ones for Windows, their uſual Price is 3 *d* ½, or 4 *d. per* ℔. in *London.*

Bar-poſts.

Are a neceſſary ſort of *Poſts*, much uſed in the Country, 2 of which, and 5 Rails or Bars, ſerve inſtead of a Gate, for an Inlet to Fields, and other Incloſures; each of theſe Poſts conſiſt of 5 Mortices, and thoſe *Poſts* are commonly about 6 Foot, or 6 ½ Foot long, 4 Foot of which ſtands above Ground. Theſe *Poſts* are in ſome Places made by the Piece, *viz.* 1 *d.* or 1 *d.* ½ *per* Poſt Hewing, and ½ *d. per* Hole for Morticing them.

Baſe.

1. From the Greek, *Baſis*, a Reſt or Support. Any Body, which bears up another; but is particularly applied to the Bottoms of Columns and Pedeſtals.

2. The *Baſe* of a Column is alſo ſometimes call'd *Spire*, from the Latin *Spiræ*, which ſignifies the Fold of a Serpent laid at reſt, or a Coil of a Cable, *&c.* Theſe making a Figure not unlike that.

3. *Kinds.*] They are diſtinguiſhed by *Architects* into 5 Kinds, *viz. Tuſcan, Dorick, Ionick, Corinthian,* and *Compoſite,* according to the 5 Orders of Columns.

4. *Parts,* or *Members*] Exceed the number of the Kinds, becauſe that ſome Authors differ from others in their Form; of each of which I ſhall give you the following Account.

5. The

5. The *Tuſcan*] According to *Vitruvius*, muſt be $\frac{1}{2}$ a Module high; this croſs Member conſiſts of 3 Minuter Members, or Parts, *viz.* a *Plinth*, a *Thorus*, and a *Fillet*, and is thus divided, and ſubdivided; the whole Altitude being —- 30, is divided into 2 equal parts, the lower one of theſe is for the *Plinth*, and the upper part of the 2 is to be ſubdivided into 3 equal parts, the lower of theſe are for the *Thorus*, and the upper one for the *Fillet*.

Palladio alloweth this Baſe to be 30 Min. high alſo; the which he diſtributed amongſt 3 ſmaller Members, *viz.* a *Plinth*, or *Orlo*, a *Torus*, and a *Liſtella*. or *Ceincture*. The *Plinth* is 15 m. the *Torus* 12 $\frac{2}{1}$ m and the *Liſtella* 2 $\frac{2}{1}$ min. high

Scamozzi alſo alloweth this Baſe 30 m in heighth. but then he reckons but 2 Members, or Parts to it, and they are a *Plinth* of 18 m. and a *Thorus* of 12 m. altho' at the ſame time he places above the *Thorus* a *Liſt* of 3 m. which in all (I think) makes more than $\frac{3}{6}$ a Module by 3 m.

Vignola alſo makes the *Baſe* of 3 Parts. *viz.* a *Plinth* *Thorus*, and *Fillet*; all which he reckons 30 m. —— $\frac{2}{1}$ a Module

6. *Dorick.*] This *Baſe Vitruvius* reckons to conſiſt of 6 Parts. *viz.* a *Plinth* 2 *Thoruſes*, 1 *Scotia*, and 2 *Liſts*; the whole heighth of all theſe, he allows to be 30 m. which he thus divided, *viz.* *1ſt.* into 3 Parts, the lower one of theſe is for the *Plinth*, the 2 Parts remaining, he ſubdivides into the 4, the upper one of which 4, he allots to the upper *Thorus*, the 3 lower Parts of theſe 4, he divides into 2. The lower of which 2 is for the lower *Thorus*, then he ſubdivides the upper part of theſe 2 into 7 equal parts, the upper and lower of theſe 7, are for the 2 *Liſts*, and the 5 betwixt them is for the *Scotia*. But amongſt all theſe 6 Members, or Parts of the Baſe, there is one large *Fillet*, which is $\frac{1}{14}$ part of the Module, but this *Fillet* he reckons to be no part of the *Baſe*, but a part of the Body of the *Column*

Palladio aſſigneth 30 m. for the Altitude of this *Baſe*: According to his Scheme of this Member, it is compoſed of 7 Parts, *viz.* a *Plinth*, 2 *Thoruſes*, 3 *Annulets*, and a *Scotia*, or *Cavetto*; the which he thus proportions, *viz.* To the *Plinth*, (the which I think he may more properly call a *Scotia*, or *Caſement*) which is wrought hollow, he allots 10 m, to the lower *Thorus* 7 $\frac{1}{3}$ m, to the lower *Annulet* 1 $\frac{1}{4}$ m, to the *Cavetto* 4 $\frac{2}{3}$ m, to the middle *Annulet* 1 $\frac{1}{4}$, to the upper *Thorus*, 4 $\frac{1}{4}$, and to the upper *Annulet* 1 $\frac{1}{4}$.

Scamozzi makes the *Dorick baſe* 30 m, high, the which he ſubdivides amongſt 6 ſmall Members, *viz* (beginning below, and ſo aſcending) *1ſt.* A *Plinth*, to which he allows 10 $\frac{1}{2}$ m. 2 A *Thorus* of 8 m. 3. A *Liſt* of 1 m. 4. A *Scotia* of 4 m. 5. A *Liſt* of 1 m. And 6. A *Thorus* of 5 $\frac{1}{2}$ m. Above all theſe he places a *Liſt* of 2 m which he doth not reckon into to the *Base*, but to be part of the Body of the *Column.*

Vignola also reckons the heighth of the *Base* ½ the Diameter of the *Column* below, but he makes it to consist of but 4 parts, *viz.* A *Plinth*, a large and small *Thorus*, and a *List.*

7. *Ionick,*] According to *Vitruvius*'s Order, is ½ a Module high, he describes 2 sorts of Bases in this Order, one for the *Ionick* Column without a Pedestal, the other with; each of which *Bases* consist of smaller Members; but the *Bases* differ in the Dimensions of their Parts. The Members whereof they consist, are these following, *viz.* A *Plinth*, 4 *Fillets*, 2 *Scotia*'s, 2 *Astragals*, and a *Thorus* I shall 1*st.* describe the Dimensions of the Parts of the *Ionick Base*, without a Pedestal. This Base he thus divideth and subdivideth, *viz.* He divideth the whole heighth of the Base into 3 equal Parts; the lower one of which is the heighth of the *Plinth*, the 2 upper, and remaining Parts, he subdivideth into 7 equal Parts, the upper 3 of which. make the *Thorus*, the 4 7*th* remaining, he subdivideth into 8 equal parts, ½ of the lower 8*th* makes the lower *Fillet*, the other ½ and the 2 *d.* 8*th* and ½ the 3 *d* 8*th.* makes the 1*st Scotia*, and the upper ½ of the 3*d* 8*th.* makes the 2*d. Fillet*, the 4*th* and 5*th.* makes the 2 *Astragals*, ½ the 6*th.* 8*th* makes the 3 *d. Fillet*, the upper ½ of the 6*th* 8*th.*, and all the 7*th.* and ⅓ of the last, or upmost 8*th.* makes the second *Scotia*, the ⅔ of the last 8*th.* remaining, makes the upper *Fillet*, which subjoyns to the *Thorus*: Above the *Thorus* he places another *Fillet*, which he doth not reckon any Part of the *Base*, but part of the Body of the *Column*, which Fillet is $\frac{1}{12}$ of the Body of the *Column*, — 5 m.

The *Ionick Base* with a Pedestal, he thus divides into parts, *viz.* 1*st* Into 3 equal parts, the lower one of these is the heighth of the *Plinth*, the ⅔ remaining, he divides into 3 equal parts. the upmost of which he assigns for the *Thorus*, the ⅔ remaining, he subdivides into 12 equal parts, ½ the lower $\frac{1}{12}$ he assigns for the *Fillet*, above the *Plinth*, the remaining ½ of $\frac{1}{12}$, and the 3 next 12*ths.* make the first *Scotia*, the 5*th* 12*th.* make the 2*d. Fillet*, the 6 and 7*th.* makes the 2 *Astragals*, and ½ the 8*th.* makes the next *Fillet*, the other ½ of the 8*th* and the 9*th* 10*th.* and 11*th* make the 2*d.* *Scotia*, and the 12*th* and last part makes the upper *Fillet*, which is under the *Thorus*.

There is also a *Fillet* above the *Thorus* which is of the same heighth with that without a *Pedestal.*

Palladio assigneth 30 m. for the Altitude of this *Base*, and according to his Scheme of this Member into 6 smaller Members, 1*st.* A *Plinth*, (or rather as he delineates it a *Casement*) of 10 m. 2 A *Thorus* of 7 ½ m. 3. A *List* of 1 ¼ m 4. A *Scotia* of 4 ¾ m. 5. Another *List*, or *Ceincture* of 1 ¼ m.

6 A *Thorus*

6. A *Thorus* of 5 $\frac{1}{4}$ m. all which makes 30 m. which compleats his *Base*. Above which on the Foot of the Body of the *Column*, he places an *Astragal* of 2 $\frac{1}{4}$ m. and above that a *Ceincture* of 1 $\frac{1}{4}$ m. all which makes 33 $\frac{1}{2}$ m. *Scamozzi* makes the *Ionick Base* 30 m. high also, and of the same number of parts and form with *Palladio viz.* 1*st*. A *Plinth*, (which is concave) of 10 $\frac{1}{2}$ m A *Thorus* of 8 m. 3. A *List* of 1 Min. 4. A *Scotia* of 4 $\frac{1}{2}$ m. 5 A *List* of 1 m. 6. Another *Thorus* of 5 m. all which makes the *Base* of 30 m, above which on the *Column*, are 2 small Members more, *viz*. An *Astragal* of 2 $\frac{1}{2}$ m. and a *List* of 1 $\frac{1}{2}$ m. all which added together make 34 m. in heighth

Vignola Composes his *Ionick Base*, of the same number of small Members, and of the same form with *Vitruvius*.

8. The *Corinthian*,] According to *Vitruvius*, is $\frac{1}{2}$ a Module high, both in the *Corinthian Column*, with a *Pedestal*, and without a *Pedestal*; that without a *Pedestal*, he makes to consist of 11 smaller Members, *viz*.: A *Plinth*, 2 *Thoruses*, 4 *Fillets*, 2 *Scotia's*, and 2 *Astragals*: This *Base*, *viz*. 1*st* He divides the whole heighth into 4 equal parts, the lower one of these Divisions he assigns for the *Plinth*, the 3 parts remaining, he again divides into 5 equal parts, the upper one of these 5 he allots for the upper *Thorus* (which is the highest Member in the *Base*) the lower *Thorus* he maketh to contain 5 quarters of one of these 5th parts, *viz*. All the 1*st*. or lower 5th. part, and $\frac{1}{2}$ of the 2*d*. so that $\frac{1}{5}$ be taken for the upper *Thorus*, and one 5th. and a quarter of $\frac{1}{5}$ below for the lower *Thorus*, there remains but 2 of these 5ths. 3 of one 5th, the which he subdivides into 12 equal parts of $\frac{1}{2}$ of the lower 12th. he makes the 1*st*, or (lowest *Fillet*) then of the other $\frac{1}{2}$, and all the 2*d* 3*d*. 4th. and $\frac{1}{2}$ the 5th. he makes the lower *Scotia* of the remaining $\frac{1}{2}$ of the 5th, 12th. part, he makes the 2*d*. *Fillet*, of the 6 and 7th parts he makes the 2 *Astragals*, of $\frac{1}{2}$ the 8th. he makes the 3*d*. *Fillet*, of the other $\frac{1}{2}$ of the 8, and all the 9, 10, and 11 and $\frac{1}{2}$ the 12th, he makes the 2*d*. *Scotia*, and of the last $\frac{1}{2}$ of the 12th. part, he makes the 4th. or last *Fillet*, which subjoyns the under side of the upper *Thorus*. Above the *Base* he adds a *Fillet*, which is $\frac{1}{24}$ of the Diameter of the *Column* in heighth, which is — 2 $\frac{1}{2}$ m.

The *Base* for the *Corinthian Column*, with its *Pedestal*, is of the same heighth, and number of parts, and each part hath the same Dimensions, with that which hath no *Pedestal*.

Palladio makes this Base to contain 8 smaller Members, *viz*. 1 *Orlo*, 2 *Thoruses*, 2 *Astragals*, 2 *Ceinctures*, and 1 *Scotia*, 'Tis my Thoughts, that either the Author, or the Ingraver, have made a great blunder in the Division, and Subdivision of this *Base*, which I exhibit to you as I found it, tho' I do suppose it to be false: The *Orlo* he makes 9 $\frac{2}{7}$ m. the lower *Thorus* 7 m. the lower *Astragal* $\frac{3}{4}$ m. (the which I am confident is too

too little,) the lower *Cincture* $\frac{1}{4}$ m. the *Scotia* 3 $\frac{1}{4}$ m. the next *Ceincture* has nothing set to him, but he appears about the same size with the other *Ceincture*; then comes the next *Astragal* of $\frac{1}{2}$ m. and then the upper *Thorus* of 5 m. above all these 8 Members of the *Base*, he places another *Astragal* of 2 $\frac{1}{2}$ m. and above that a *Ceincture*: Thus I have given you a very lame account of this Member, but I may thank the Author, or Graver or both for it, that it is no better.

Scamozzi, according to his Portraicture of this *Base*, makes it 30 m. high, and he divides this grand Member into 8 petty Members of the same Form with *Palladio*, *viz.* 1. Orlo of 9 $\frac{1}{2}$ m then a *Thorus* of 7 m. next an *Astragal* of 2 m. next a *List* of 1 m. then a *Scotia* of 3 $\frac{1}{2}$ m next another *List* of 1 m and then another *Astragal* of 1 $\frac{1}{2}$ m. and last of all, another *Thorus* of 4 $\frac{1}{2}$ m. all which makes 30 m Above the Base he places two other Members on the Foot of the *Column*, *viz.* An *Astragal* of 2 $\frac{1}{2}$ m and a *List* of 1 m *Vignola* allows this Base to be 30 m. also; and as to the Form he makes it much the same with *Vitruvius*

9. *Composite*, *Compound*, or *Roman*. *Vitruvius* makes this *Base* to contain 30 m. in Altitude. This grand Member he divides into 10 smaller, *viz* A *Plinth*, 3 *Thoruses*, (one of which is in the middle, where the two *Astragals* are in the Corinthian Order) 4 *Fillets*, and 2 *Scotia's*. This Member he 1st divides into 4 Parts, the lower one of which is for the Heighth of the *Plinth*, the other 3 Parts he subdivides into 5, of the upper one of the 5 he makes the upper *Thorus*, the lower *Thorus* he makes of the lower 5th and $\frac{1}{4}$ of the 2d 5th. (so that the lower *Thorus* is $\frac{4}{5}$ high; the 2d 5th Parts, and $\frac{3}{4}$ remaining, he subdivides into 12 equal Parts, of $\frac{1}{2}$ the lower 12th, he makes the 1st *Fillet*, of the other $\frac{1}{2}$, and all the 2d, 3d, 4th, and $\frac{1}{2}$ the 5th he makes the 1st *Scotia*, of the remaining $\frac{1}{2}$ of the 5th, he makes the 2d *Fillet*, of the 6th and 7th, he makes the middle *Thorus*; then of $\frac{1}{2}$ the 8th he makes the 3d *Fillet*, of the remaining $\frac{1}{2}$ of the 8th, and all the 9th, 10th, and 11th, and $\frac{1}{2}$ the 12th, he makes the 2d *Scotia*, of the remaining $\frac{1}{2}$ of the 12th he makes the last *Fillet*, which is just under the upper *Thorus* Above the *Base*, on the Foot of the *Column*, he makes a *Fillet*, which is $\frac{1}{24}$ of of the Diameter of the Column below.

Palladio makes this Base 30 m. high, which he divides into 11 smaller Members, *viz*, An *Orlo*, 2 *Thoruses*, 4 *Lists*, 2 *Scotia's*, and 2 *Astragals*, to the 1*st*. Member, being an *Orlo*, (which is Concave) he allows 9 m. then follows 2 *Thoruses* of 7 m. then a *List* of $\frac{1}{2}$ m. next a Scotia of 3 m. then another *List* of $\frac{1}{2}$ m then the 2 *Astragals*, each of 1 m. a-piece, then a *Fillet*, or *List* of $\frac{1}{2}$ m. then a *Scotia* of 3 m. and then another *List* of $\frac{1}{2}$ m. and then the upper *Thorus* of 4 m. above which on the Foot of the *Column*, he places another

ther *Aſtragal* of 3 m. and above that a *Liſt* of 1 m.

Scamozzi makes the *Roman Baſe* 30 m high, the which he divides amongſt 7 ſmaller Members, *viz* 10 m. to a Concave Plinth, 7 m. to the 1ſt *Thorus*, 2 m. to an *Aſtragal*, 1 m to the 1ſt *Liſt*, 4 m. to the *Scotia*, 1 m. to the 2d *Liſt*, and 5 m. to the upper *Thorus*, which is the higheſt Member in the *Baſe*; but above the *Baſe*, he places 2 Members, *viz*. An Aſtragal of 2 $\frac{1}{2}$ m. and a *Liſt* of 1 $\frac{1}{4}$ m.

Vignola makes his *Roman Baſe* very much like *Vitruvius*'s, only he places 2 *Aſtragals* in the middle betwixt the 2 *Scotia*'s, where *Vitruvius* has a *Thorus*.

Baſilic.

This among the Ancients, was a large Hall, with Portico's, Iſles, Tribunes and Tribunal, where the Kings themſelves adminiſtred Juſtice; but the Name is ſomewhat differently applied now-a-days; being given to Churches and Temples; as alſo to certain ſpacious Halls in Princes Courts, where the People hold their Aſſemblies, and the Merchants meet and converſe together; as that for Inſtance of the Palace at *Paris*. *Baſilicos*, in Greek, ſignifies *Royal*.

Batement.

A Term uſed by ſome *Carpenters*, ſignifying thereby to abate, or waſte a piece of Stuff, by forming it to a deſign'd Purpoſe. Thus, inſtead of asking how much was cut off from ſuch a piece of Stuff, ſome *Carpenters* will ask what *Batement* that piece of Stuff had.

Batten.

Say ſome, is a Scantling of Stuff, of 2, 3, or 4 Inches broad, and but ſeldom above 1 Inch thick, and the Length unlimited. But I muſt tell you what Senſe I have obſerv'd Workmen to uſe it in, (by Workmen, I mean *Carpenters* and *Joyners*) which is this, *viz*. In Doors, and Windows of Shops, *&c*. which are not framed of whole Deal, or 1 $\frac{1}{4}$ Inch Oak, with *Stiles*, *Rails*, and *Pannels* (as Wainſcot is framed) and yet they are made to appear as if they were, by Means of Pieces which are bradded on (upon the plain Boards, which are joined together for) the Door, or Window, all round the Edges, and ſometimes croſs them, and up, and down, *&c*. According as how many *Pannels* the Workman deſigns the Door, or Window ſhall appear to have. Theſe Pieces which are thus bradded on to repreſent *Stiles*, *Rails*, and *Montans* are of different Breadths, according to the Deſign of the Artificer, as from 2, to 6 or 7 Inches, and commonly on one Edge of thoſe which repreſent the *Stiles*, and the upper and lower *Rails*, and on both Edges of thoſe which are to appear like *Montans*, and middle *Rails*, there is commonly ſome Moulding ſtruck; as a *Bead*, an O.-G. or the like.

Batten

Batten Doors.

1. *What.*] *Batten Doors* are (as I ſaid before) ſuch as ſeem to be Wainſcot ones, tho' they are not ſo ; for Wainſcot ones the *Pannels* are groved into the Framing ; but here they firſt joint and glue the Boards, which are cut to the full Length and Breadth of the Door-caſe, which Gluing being dry, they traverſe them over, both in Length, and Breadth, with a long Plane, and then ſmooth them, and then fit on the *Battens* on the Front-ſide. And this is what they call ſingle *Batten doors* ; for you muſt note, there are double *Batten doors*, viz. Such as are batten'd on both Sides, tho that is but ſeldom uſed.

But there are commonly uſed batten'd Doors, which are call'd double Doors, *viz.* Such as are front, or outer Doors ; they are commonly made of whole Deal, and then batten'd on the out-ſide, and Pieces of 4 or 5 Inches broad, miter'd round on the Edges, on the inſide of the Door, and then croſs the Door betwixt theſe Pieces, it is lined with ſlit Deal, which makes it level with the miter'd Pieces. I have ſeen ſome *Doors* that have been lin'd with Pieces put Bereling, and not at right Angles, but near Miter to the Sides of the Door, and when all is plained off level, it hath been divided out in Rhombuſes, and ſtruck with a Pencil, and at the Angles of the Rhombuſes, were round headed Nails driven, which added ſomething of Beauty to the Work : This Way of Lining upon the Doors, *viz.* Pointing from the lower Corner behind, toward the upper Corner before, I believe may be a good way to prevent a Door from ſagging, or ſinking at the Fore-corner, when-ever the Joints ſhall happen to unglue.

2. *Price.*] As to the Price of ſuch *Doors*, vid. *Doors*, N°. 4. where you will find Price of Materials, and Workmanſhip; but I ſhall here add, that for Workmanſhip of making *batten Doors* of ſlit Deal, about an Inch thick, (or of thin whole Deals) glued, and batten'd on one Side, 4 *s. per* Door, is a good moderate Price : But ſuch as are mentioned above, (which are for Front, and other outer Doors) *viz.* both Batten'd, and Lin'd, are worth 7 *s. per* Door Workmanſhip.

Batter.

A Term uſed by Workmen, to ſignifie that a Wall, a piece of Timber, or the like, doth not ſtand up-right, but leans from-you-ward, when you ſtand before it ; but when it leans toward you, they ſay it *over-hangs*, or *hangs-over*.

Bay.

The Word is uſed, to ſignifie (as it were) the Magnitude of a *Barn* ; for if a Barn conſiſt of a Floor, and two Heads, where they lay Corn, they ſay a Barn of two *Bays* ; theſe *Bays* are

are from 14 to 20 Foot long, and Floors from 10 (which is the ſmalleſt Size) to 12 broad, and uſually 20 long, which is the Breadth of the Barn: If a *Bay* be 20 Foot long, then there is commonly a pair of *Prick-poſts* in the middle, and a *Beam* to hold in the Rod from bending the Raiſons; but if the *Bays* are not above 16 Foot, and the Timber ſtout then there is no *Poſts*; but at the End of each *Bay* where there is always hanging *Braces*, framed into the *Beam*, and *Poſts*, and alſo a croſs *Cell* to hold in the ſide *Cells* from flying out when the *Barn* is fill'd; and 'tis common for large Barns to conſiſt of divers ſuch *Bays*.

Bay-window.

I underſtand to be ſuch a one which is compoſed of an Arch of a Circle; and ſo by Conſequence, ſuch a one will ſtand without the Streſs of the Building. By which Means Spectators may the better ſee what is acted in the Street.

Bead.

A Moulding, ſo call'd, which is commonly made upon the Edge of a piece of Stuff; as for Example, 'tis common to ſee Joyners make a Moulding which is about a quarter of a Circle, on the upper Edge of a *Skirting board*; alſo on the Lining of a Door-Caſe, or the like, alſo on the inner, or lower Edge of an *Architrave*. A *Bead* and a *Boultin*, differ very little, only in Magnitude; for when they are large, Workmen commonly call them *Boultins*. Sometimes a *Bead-plain* is ſet on, upon the Edge of each *Faſcia* of an *Architrave*, and ſometimes alſo this Moulding (eſpecially in the *Corinthian* and *Roman* Order) is cut, or carved, in ſhort Emboſsments, which reſembles Women's Beads in Semi-relief; and ſometimes likewiſe an Aſtragal is thus carved; in both which, theſe Carvings are call'd *Beads*.

Beam.

1. *What.*] In Building is a piece of Timber, which always lies croſs the Building, into which the Feet of the principal Rafters are framed; no Building hath leſs than 2 of theſe *Beams*, *viz.* one at each Head; into theſe *Beams* the *Girders* of the Garret-floor are framed; and if it be a Timber-Building, the *Teazle Tennons* of the Poſts are framed. The *Teazle Tennons* are made at right Angles to thoſe which are made on the Poſts to go into the *Raiſons*, and the *Reliſh*, or Cheats of theſe *Teazle Tennons* ſtand up within an Inch and a half of the top of the *Raiſon*; and the Beam is *cauked* down (which is the ſame as Dove-tailing a-Croſs) till the Cheeks of the Mortices in the Beam conjoin with thoſe of the *Teazle Tennon* on the Poſts.

2. *The Size.*] The *Beams*, according to an Act of Parliament, for the re-building of the City of *London*, after the dreadful Fire, were appointed to be of the following Scantlings, *viz.* Foot

	Foot.		Inch.		Inches.
In length	15	muſt be	7	and	5
	16	in that	8		6
	17	Square.	10		6

And ſo proportionably to their Lengths. But in the Country, where Timber is more plentiful; they generally make their *Beams* ſtouter.

Sir *Henry Wotton* adviſes, That all *Beams*, *Girders*, and *Summers*, ought to be of the ſtrongeſt, and moſt durable Timber.

Bear.

Timber is ſaid to *bear* at its whole Length, when neither a *Brick-wall*, or *Poſts*, &c. ſtand between the Ends of it. But if either a *Brick-wall*, or *Poſts* be trimm'd up to the Timber, then it is ſaid to *bear* only at the Diſtance between the *Brick-wall*, or *Poſt*, and either end of the Timber. Thus Carpenters uſually ask what

Bearing?

Such a Piece of Timber has the Anſwer to ſuch a Demand, or Queſtion, is 10, 12, 15 *&c.* Foot, according to the Length of the whole Timber, or elſe according to the Diſtance between either End of the Timber; and a

Bearer.

Viz. A Poſt, or Brick-wall, that is trimm'd up between the ends of a piece of Timber to ſhorten its *Bearing*.

Beam-filling.

1. *What.*] Is PlaiſterersWork, 'tis only filling up the vacant Space betwixt the *Raiſon* and the Roof, whether Tiling, Thatching, or any other Roof. 'Tis a ſort of Work that is very cuſtomary in the Country, where they do not Parge, or (which is all one) Plaiſter their Garrets, they thus perform this ſort of Work, *viz.* They take ſome pieces of Stones, or elſe Bricks, and lay them betwixt the Rafters upon the *Raiſon*, and then Plaſter upon it with *Loam*, or elſe they ſet ſome *Tiles*, with one edge upon the *Raiſon*, and the other leans againſt the Roof, and then upon theſe *Tiles* they Plaiſter with *Loam*.

2. *Price.*] The uſual Price for Workmanſhip only, in the Country, is $\frac{1}{2}$ *d. per* Foot, or 1 $\frac{1}{2}$ *d. per* Yard, Lineal Meaſure.

Bed.

Of a *Stone*, What. V. Stone, N. 4.

Bed-moulding,

Or as ſome will have it,

Bedding-moulding.

Is a Term uſed commonly amongſt Workmen (but I did never obſerve it in any one of the

the Treatiſes of the *Greek*, or *Italian Architects*. But however, our modern *Artificers* make uſe of this Compound Word, to ſignifie thoſe Members in a *Cornice*, which are below the *Coronet*, or Crown, E. G. 'Tis now common for *Joyners* to have their *Bed-moulding* to conſiſt of theſe 4 Members, *viz* 1. (below) an O. G. 2. A *Liſt* 3. A large *Boultin* And 4th and laſtly, under the *Coronet* another Liſt. This is what they frequently call a *Bed-moulding*.

Bevil.

Any Angle that is not ſquare, is call'd a Bevel Angle; whether it be more obtuſe, or more acute than a Right Angle; but if it be ½ as much as the Right Angle *viz*. 45 Degrees, then Workmen call it *Miter*; they have alſo a Term, *half Miter*, which is an Angle that is ¼ of a Quadrant, or Square, *viz*. An Angle of 22 ½ Degrees; this they call ½ *Miter*.

Bill.

All know a Bill to be an Account of Work done, Materials uſed, &c As for the Method of drawing up Tradesmen's Bills we ſhall give ſome Examples, under the following Heads, *viz*. *Bricklayers Bills*, *Carpenters Bills*, *Glaziers Bills*, and *Smiths Bills*; all which ſee in their proper Places.

Binding Joyſts.

1. *What.*] *Binding Joyſts*, are thoſe *Joyſts* in any Floor, into which the *Trimmers* of Staircaſe (or Well-hole for the Stairs) and Chimney-ways are framed; theſe *Joyſts* ought to be ſtouter than common *Joyſts*.

2. *Scantlin*, or *Size*.] The Size of theſe, as well as all other Timber Members, was ſettled by an Act of Parliament, before the Re building of *London*. According to which Act, *Binding-Joyſts*,

	Foot		Inches.	Inches.
Which contain in Length	7	muſt be in their Squares.	6	and 5
	9		7	
	11 or 12		8	

So large they were order'd to be, and no leſs. They might (I ſuppoſe) be as much bigger as they pleaſed.

Boarding of Walls.

See *Weather-boarding*.

Boat.

A Term uſed among Ironmongers, to ſignifie a certain ſort of *Nails*; for which, ſee *Nails*.

Bolts.

Of Iron, are of various Sorts. In *Navigation*, they have

have ſix Kinds, for which they have diſtinct Names, which we ſhall not here ſtay to deſcribe. Some Iron-mongers diſtinguiſh thoſe for Houſe-building, into but 3 Sorts, *viz.* Plate, Round, and Spring *Bolts.* Plate, and Spring-bolts are made uſe of, to faſten Doors and Windows; and theſe are of different Sizes, and Prizes. I have known ſmall *Spring bolts* ſold at 3 *d.* ½ *per* Piece, others at 9 *d* others at 14 *d.* and ſo likewiſe *Plate-bolts,* ſome are 9 *d.* 10 *d.* *&c.* *per* Piece. There are alſo *Braſs-knob'd Bolts, ſhort,* are about 10 *d. per* piece; *long,* for *Folding-doors,* about 18 *d. per* Piece. *Iron Balcony bolts,* about 1 *s* There are alſo *Braſs-plate Bolts,* at about 10 *d per* Piece. There are alſo *Round bolts* (or long Iron pins) with a Head at one End, and a Key-Hole at the other; theſe are commonly ſold by the lb. *viz.* 3 *d.* ½, or 4 *d. per* lb.

Bond.

A Term uſed amongſt Workmen; for when they ſay, Make good *Bond,* they mean, faſten the two, or more Pieces of Timber well together, either with Tenanting and Morticing, or Dove-tailing, *&c.*

Botham.

Is an Iron-monger's Term, which they uſe to ſignifie a certain ſort of *Nails*; for which ſee *Nails,* N. 2.

Boulder-Walls.

1. *What.*] That is Walls made of round Flints, or Pebbles, which are found where the Sea hath a Beach caſt up, and alſo at ſome other Places where there are plenty of Flints.

2 *The Method of building them.*] As I am inform'd by a Bricklayer that hath been uſed to ſuch Work; 'tis their uſual Way (if they can ſo fit it) for two to work upon it at a Time, one at one ſide of the Wall, the other at the other ſide, and one to be Right-handed, the other left; for two ſuch fit beſt to work together in this ſort of Work: They have a Hodd of Mortar pour'd down on their Work, and ſo they part it betwixt them, ſpreading it each toward himſelf, and then they lay their *Boulders,* or *Flints:* But he ſays, They always work with a very ſtiff Mortar, and had need to have a good Length of Work before them; for they work but one Courſe in Heighth at a Time; for, ſaith he, If we ſhould do more, it would be apt to ſwell out at the Sides, and run down; and therefore we are forced to work continually in Length: He ſaith likewiſe, That if it chance to be miſty Weather, 'tis very difficult to make the work ſtand.

3. *Price*] He tells me, That their Cuſtom is to work by the Square, or 100 Foot; for which their uſual Price is 12 *s.* for Workmanſhip only.

Boults.

Boults.

See *Bolts.*

Boultin.

In Architecture, is a Convex-moulding, that consists of an exact $\frac{1}{4}$ of a Circle; being the Member next below the Plinth in *Tuscan* and *Dorick Capital.* See *Quarter Round.*

Brace.

In a Building, is a piece of Timber, which is framed in with Bevel Joints. Its use is to keep the Building from swerving, either this, or that way; they are sometimes call'd *Strutts, viz.* When they are fram'd in the *King-piece*, and principal Rafters.

Brads.

1. *What.*] Are a sort of Nails without Heads, some Ironmongers distinguish them by six Names, as followeth, *viz.*

2. *Joyners.*] Plain for hard Wood-wainscot, from 1 Inch to 2 $\frac{1}{4}$ in Length.

3. *Batten.*] For soft Wood-wainscot, the Sorts are, 1 *d.* 2 *d.* 3 *d. Ditto*, large 4 *d. Ditto* large, 5 *d.* 6 *d.*

4. *Flooring.*] Plain for soft Wood, Joysts, the Sorts are 14, 15, 18, 19, 20, 21, 22, 23, 28, 32, and 36 lb. *per* M.

5. *Ditto strong.*] Fit for hard Joysts, the Sorts are 15, 18, 19, 24, and 32 lb. *per* M.

6. *Quarter-heads.*] For soft Wood the Sorts are 10, 13, 15, 18, 19, 20, 22, 23, 28, and 32 lb. *per* M.

7. *Ditto strong.*] For hard Wood-Joysts, the Sorts are 14, 20, 34, 44, and 54 lb. *per* M.

N. B. All *Bill-brads*, alias *Quarter-heads*, are very fit for shallow Joysts that are subject to warp, or for Floors laid in Haste, or by unskilful Persons, because the Bill to the Head will hinder the Boards from starting from the Joysts, but doth not make so smooth Work as the plain Brads.

Lastly, As to the *Prices* of *Brads*, I shall set down but a few at present, which shall be these following, *viz.*

1. Of *Batten-brads*; in the Country they vulgarly call 'em *Joyners-brads*, the usual Price

of a M. of	2	Inch	is	20 *d.*
	1$\frac{1}{2}$			15
	1			11

2. Of *Quarter-heads*, or *Bill-brads* for soft Wood-floors, the usual Price

of a M. of	15	lb.	is	4 *s.* 9 *d.*
	18			5 *s.* 6 *d.*

Break-in.

Is a Term used by Carpenters, when they cut, (or rather break) a Hole in Brick-walls with their Ripping-chissel.

Brest.

A Term in *Architecture*, made use of by some to signifie the same Member in a Column, that others call a *Thorus.*

A Brew-

A Brew-houſe.

None need be told what it is; but what I here mention it for, is becauſe it is a neceſſary Part in all Dwelling-houſes, eſpecially in the Country: Now Sir *Henry Wotton*, in his *Elem. Arch.* ſaith, That all Offices that require Heat; as *Brew-houſes*, *Bake houſes*, *Waſh-houſes*, *Kitchins*, and the like, ought to be placed in the Meridional Part of the Building, if the Poſition of the Houſe, in reſpect of the High-ſtreet or the like will admit of it; for it would be but an odd Contrivance, if a Houſe ſtood on the *North*-ſide of a High-ſtreet, to place all the Offices in the Front of it; and it would be very ridiculous to paſs thro' a Bake houſe, Brew-houſe, or Waſh-houſe, into Rooms of Entertainment, in a Noble-man's, or Gentleman's Houſe: And therefore we may ſee the old Proverb holds good ſtill, *viz.* That there is no general Rule, without ſome Exception.

Breſt ſummers.

In a Timber Building, are Pieces into which the Girders are framed, in all the Floors, but the Ground-floor (then they call it a Cell) and Garret-floor, (then it is call'd a Beam.) As to their Size, or Square, 'tis the ſame by the Act of Parliament with *Girders*, which ſee. You muſt note by the way, that I do not mean all the Pieces which have *Girders* framed into them, (and are not in the Garret, or Ground-floor.) But I mean all ſuch Pieces which are in the exterior Part of the Building, whether in the Front, Flanks, or Rear of the Building; for you muſt note the Pieces in the internal Part of the Building, into which the Girders are framed, are call'd *Summers*. The *Breſt-ſummers* in *London*, Mr. *Leybourn* ſaith, are uſed to be meaſured by the Foot. running Meaſure; but whether he means only for the Work, or Timber, or both, I do not know: Now *Com. Comer* ſaith, That *Breſt-ſummers*, in *London*, are valued by the ſolid Foot, if of Oak 3 *s. per* Foot; if Fir, 2 *s.*

Bricks.

I. *What.*] All know them to be a fictitious or artificial kind of Stone, of a reddiſh Colour; and as to their Form and Magnitude 'tis various, as alſo their Uſes, of both which, and likewiſe their Conſiſtence, we ſhall ſpeak in their due Places.

II. *Whereof made*.] I will 1ſt, tell you what *Pliny* ſaith of this Matter; he ſaith, That if you would have good *Bricks*, they muſt not be made of any Earth that is full of Sand or Gravel, nor of ſuch as is gritty and ſtony, but of a grayiſh Marl, or whitiſh chalky Clay; or at leaſt a reddiſh Earth. But in Caſe you are forced to uſe that which is ſandy, be ſure to make choice of that kind of Sand which is tough and ſtrong.

ſtrong. The beſt Seaſon (ſaith *Pliny)* is in the Spring, to make Bricks; for in the middle of the Summer, they are ſubject to crack and be full of Chinks. He further addeth, That the Lome, of which Bricks are made, ought to be well ſteeped, or ſoaked, and wrought with Water.

Mr. *Inco. in Mec. Ex.* ſaith, That *Bricks* are made of Earth, of which the whitiſh chalky Sort of Earth, and the reddiſh, are beſt.

At *Lunenburgh* in *Saxony* they make them of a fat Earth, full of Allom.

Alſo there are good Bricks made at *Patane* in *Aſia*, of a Pumice ſort of Earth, which being dried, will ſwim in Water, and not ſink.

The Ancients likewiſe made them of Earth which was ſandy.

But here, in *England*, they are made for the moſt part of a yellowiſh colour'd fat Earth, ſomewhat reddiſh, [vulgarly call'd Lome.] Mr. *Leybourn* ſaith, Bricks are made of a reddiſh Earth, which ought to be digged before Winter, but not made into Bricks till the Spring Seaſon.

III. *Of their Kinds, and Appellations.*] The Sorts or Kinds of *Bricks* are as various as their Appellations, and their Appellations are attributed to them for Diſtinction ſake (as the Names of all other Things are) thereby to be known. And theſe different *Names* were at firſt impoſed upon them, according as the firſt Donors of theſe Names thought they did ſee a Reaſon to beſtow ſuch Names on them, either from ſome Accident in their making, or from their Dimenſions, or from their Form or Figure, or from Cuſtom, or from Method in making or from the Place where or by whom made, or from their Uſe, *&c.* Now theſe which derive their Names from Accident, are *Clinkers*, *Samel*, or *Sandal*: Thoſe from their Dimenſions, are the *great* and *ſmall* (or Statute) and *Didoron*, *Tetradoron*, and *Pentadoron*: Thoſe from their Form and Figure, are *Compaſs*, *Concave*, *Feather-edgy*, and *Triangular*; thoſe from Cuſtom, *Statute*, and *Cogging*. Thoſe from the Method of making, are *Place*, and *Stock-bricks*. Thoſe from the Place where, or by whom, are *Dutch*, or *Flemiſh*; and thoſe from their Uſe, are *Buttereſs*, or *Pilaſter*, *Coping* and *Paving*. Of all which we ſhall treat in their Order. And,

1. *Compaſs-bricks*.] Theſe are of a circular Form; their Uſe is for Steening of Walls; the which I was told by an ancient experienced Workman, he uſed to perform thus, *viz.* He having firſt laid a good Bed of Clay, for the Bottom, they paved it with *Common* or *Statute bricks*, only laid down on it, and well ſettled thereon; and then they began their Compaſs-work with the *Compaſs bricks*, and as they carried up their Courſes, they ramm'd Clay in behind them (for they had Room left behind for the Purpoſe) which made all the Joynts of the Bricks pen cloſe and tight together. He ſaith,

He

He hath done ſuch Work, where the Walls have been but a little Depth in the Ground, and in a looſe open Mold (where the Water hath been brought in by Concave-bricks) and hath known ſome which he did betwixt 20 and 30 Years ago to do very well.

As to the Price of theſe Bricks, he could not certainly tell me; but he thought, not much dearer than *common* or *Statute bricks*; but then he ſaith, He that hath them made for his Uſe, is commonly at the Charge of a Mold made according to the Circumference of his Wall.

2. *Concave*, or *hollow Bricks.*] Theſe are like *Statute*, or *common Brick* on one Side, but on the other Side they have a Concavity, which is Semicylindrical. This Cavity is about ¾ n. deep, and 1 ½ n. broad; ſo that when 2 of theſe *Bricks* are placed with their Hollows together, they are like a Pipe of 1 ½ Bore; they are uſually about 12 n long 4 ½ broad, and 2 ¼ n. thick

As to their laying them in the Ground, they generally do it in Clay; but an ancient Workman did inform me, That there muſt be Care taken, that there do not grow any Trees, Buſhes or Brambles over theſe Bricks where they are laid to convey Water, nor yet very near them; for if there do, their Roots are apt to get in betwixt the Joynts of the Bricks, and there dilate themſelves with fibrous Roots, which meet together like a Ball of Hair, in the Concavity, which will endanger the ſtopping of it, and hinder the Current of Water. Now, if this Annoyance could be infallibly prevented, it would be the cheapeſt Way I know of to convey Water to a Houſe; for 6 or 8 *s* worth of Bricks would do about 6 *Rods*; and then ſuppoſe, that the Ditch digging, and laying the Bricks, the Charge of Clay, and ramming up again, ſhould be as much more, *viz.* 6 or 8 *s* for 6 Rods; according to this Propoſal, 1 Rod would coſt but 2 *s.* or 2 *s.* 8 *d.* and if the fore-cited Objection could be removed, this Work would laſt (I had almoſt ſaid) for ever: It would not be the 6th Part of the Price of Lead pipes, and every whit as ſerviceable, if not to be preferr'd before them; becauſe I do not ſuppoſe the Froſts would hurt this (tho' it often burſts Leaden-pipes), for put the Caſe the Water ſhould be frozen up in them, the Ice would then, I do believe, by its Expanſion, open the Joynts of the *Bricks*; but we may well ſuppoſe from the Nature of the Thing, that they will come together in their due Places when the Froſt is gone by the natural Gravity of the Earth; for then there will be no ſolid Body betwixt the Joynts, to hinder the cloſing again of the Bricks.

Altho' *Alder pipes* be much cheaper than *Lead*; theſe *Bricks*, will not be much above (if they are at all) half the Price of *Alder pipes*

As to the Price of theſe *Bricks*, I have known them ſold in *Kent* for 4 *s. per* Hundred

dred, and in *Suffex* for 3 *s.* 200 of thefe Bricks at a Foot long will lay 6 Rods.

3. *Cogging-bricks,*] Are a kind of Bricks which are in ufe in fome Parts of *Suffex* to make their *Toothing*, or Indented Work under the Copeing of Walls, built of *great Bricks*

They are about 10 n. long, 4 n. broad, and $2\frac{1}{4}$ n thick, and are commonly fold at the Price of common Bricks.

In ufing them, they lay them on the top of the Wall, juft under the *Coping bricks*, in an Oblique Pofition, fo that one Corner, or Angle projects over about $2\frac{1}{2}$ n. on one fide, and the Oppofite Diagonal Angle at the other, and projects as much over the other fide.

4. *Copeing-bricks,*] Are neceffary Concomitants to *great Bricks* for building Fence-walls, and are much ufed in fome Parts of *Suffex*.

The Size and Form of thefe *Copeing-bricks*, are as follow, *viz.* They are about 12 n. fquare, and $4\frac{1}{2}$ n. thick, having one flat or plain fide, and 2 flat ends, the 2 edges and upper fide, are all comprehended under one Curvelinear Surface, the 2 edges confifting of 2 *Boultins*, join'd by 2 *Cafements*, or Hollows, to an *Aftragal*, which is the top of the Brick, after this Form. 

Their ufual Price is from 12 to 16 *s. per* hundred.

5. *Dutch or Flemifh Bricks.*] I am informed by one that they are $6\frac{1}{4}$ n. $2\frac{1}{2}$ broad and $1\frac{1}{4}$ n thick; another tells me that they are 6 n. long 3 n. broad, and 1 n. thick, as for my own part, I never meafured any of them.

They are of a yellowifh Col.

The Paveing with thefe Bricks, is neater and ftronger than common. They muft be layed in Sand.

They are commonly ufed here in *England*, to pave Yards and Stables withal, and they make a good Pavement, and are very lafting, and being laid edge-ways, look handfomely, efpecially if laid Herring-bone fafhion.

They are alfo ufed in Soap-boilers Fats, and in making of Cifterns.

If we allow $\frac{1}{4}$ n. for the Joint, then 72 of thofe which are $6\frac{1}{4}$ n. long and $2\frac{1}{2}$ broad, will pave a Yard Square, but if they are fet on edge, it will require 113 to pave a fquare Yard.

But of the other Size 6 n. long, 3 n. broad, and 1 n. thick, being laid the flat way, 63 will pave a fquare Yard, but being fet edge-ways, it will require 165 to pave a Superficial Yard.

Thefe *Bricks* are ufually fold for 2 *s. per* Hundred at *London*.

6. *Clinkers*,] Are fuch *Bricks* as have much Nitre, or Salt-peter in them, which with the violence of the Fire runs and Glazes them.

7. *Didoron*] Were a fort of *Bricks* ufed by the Ancients, which were $1\frac{1}{2}$ Foot long, or 2 Spans, [the word *Doron* being Greek for a Span, or the fpace betwixt the top of the Thumb, and little Finger extended] and 1 Foot broad: Thefe were the fmalleft fort of

of Bricks uſed by the *Greeks* about their private Buildings; they having 2 larger Sizes for their publick Buildings, as you will find by the Sequel of the Diſcourſe.

8. *Feather-edge*,] Are a ſort of *Bricks* formerly uſed in ſome parts of *Kent* and *Suſſex*, they being of the ſame Size with *Statute-bricks*, but made thinner at one edge, than they are at the other, on purpoſe to pen up their *Brick-pannels* (as they call'd them) in Timber Buildings, and they were uſually ſold amongſt the *Statute-bricks* for that Purpoſe.

9. *Great Bricks*.] They are a ſort of *Bricks* that are 12 n long. 6 n. broad, and 3 n. thick. The weight of one of thoſe *Bricks* being examined, it was found to be about 15 ℔. ſo that 100 will weigh about 1500 ℔ and a Thouſand 15000 ℔. which is 6 Tun 13 c. 3 q. 20 ℔. So that about 150 will be a Tun weight.

The uſe of theſe *Bricks*, is to build Fence-walls, together with *Pilaſter*, or *Buttreſs-Bricks*, and *Copeing Bricks*: I know one Place in *Suſſex*, where they are much in uſe for that Purpoſe. Theſe Walls are but 6 n. thick, only at the *Pilaſters* they are 12 n. thick, and they uſually ſet a *Pilaſter* at every 10 Foot. I know a Wall of about 9 Foot high, of theſe ſort of Bricks that ſtands very well, which hath been built near 30 Years: I am informed they are much cheaper than Brick, and $\frac{1}{2}$ Walls, or 14 n. Walls of Statute-bricks; of which ſee Walls N. IV.

Theſe Bricks are uſually ſold at 2 *l. per* Thouſand, which is 4 *s. per* Hundred.

10. *Paving-bricks*.] They are by ſome call'd *Paving-Tiles*. Of theſe ſort of *Bricks*, there are various Sizes, according to the Fancy of Workmen, and the Cuſtom of Places. Theſe, ſaith Mr. *Leybourn* are of ſeveral Sizes, *viz*. 6. 8, 10. and 12 n. ſquare, in value from 6 to 20 *s. per* Hundred, and if you would know how many of either ſort will pave a Room, or the like,

Note that	36	Bricks of	6	Inches Square will pave a Square Yard.
	21		8	
	13		10	
	9		12	

In *Surry*, and ſeveral Counties of *England*, are made *Paving-Bricks* of 3 ſeveral Magnitudes, *viz*. 12 n. ſquare, and 1 $\frac{1}{2}$ n. thick, 10 n. ſquare, and 1 $\frac{1}{4}$ n thick) and 8 n. ſquare, and 1 n. thick, either of theſe ſorts being Poliſhed, or rubbed with ſharp Sand on the Surface, and well joyned, and the ſides made equal by hewing them with a Brick-ax, and rubbing them on a rubbing Stone with ſharp Sand, makes an an excellent Pavement, and very pleasing to the Eye, especially when laid Arras-ways.

I have seen Experiments made on some *Sussex Paving bricks*, which were 6 ½ n. square 1 ⅞ n. thick, 2 of them weighed 11 ℔. Tare, so that 100 of them would weigh 550 ℔ and a 1000 5500 ℔. and by consequence about 407 of them would weigh a Tun.

I have known some made of 9 n. square in *Sussex*, which use to be sold for about 8 *s. per* Hundred.

An experienced old Workman told me he had made *Paving Bricks* of Clay that were 15 n. square, which he was very much troubled to prevent their warping. These Bricks, when burnt, were of a pale red Colour, as were also some which he made 6 n. square of another sort of Clay, some Miles distant from the former.

He saith, that *Paving Bricks*, made of *Lome*, have the reddest Colour, when burnt: But they ought to be made of better Earth than *common Bricks*, tho' they seldom are, by those that make them for Sale.

He saith also, that beside the goodness of the Earth in *Paving Bricks*, there ought to be a great deal of care taken in the drying of them, to prevent their warping, and also when they are dry, to take them and dress them smooth and strait, on that which is to be the upper Surface, and also to pare the edges straight, and a little under, making an acute Angle with the upper side, and to see that they be exactly square, and then put them in the Kiln, and burn them.

The usual *Price* of 9, or 10 n. *Paving bricks* is from 8 to 12*s per* 100 in the Country. I have known 10 n. ones from *Surry* brought by Water to Sea-port-towns in *Kent* and *Sussex*, and sold for 10 *s per* 100.

11. *Pantadoron,*] Are a sort of *Bricks in use* formerly among the *Greeks*, being 3 f. 9 n long, and 1 f. broad, with these they did build their publick Edifices. See *Didoron*.

12. *Place bricks*] This is a general name for all sorts of *Bricks* that are made after the insuing Method, from whence they derive their Name. Now Workmen tell me they are forced to have above one Method in making of *Bricks*, not for Fancy sake, but out of pure Necessity; the Reason of which proceeds from certain different Qualities, inherent in different Earths But to proceed, *Place bricks, and Stock-bricks* are the 2 Kinds that receive their Names from the Method of their making.

Place.bricks are generally made in the *Eastern* part of *Sussex*; so call'd, because there is a *Place* just by where they Strike (or Mold) their *Bricks*, which is a level smooth piece of Ground, prepared for the *Bearer-off* (who carries the Bricks from the *Striker*) to lay them singly down in Rows (which they call *Ricks*) as soon as they are Molded, and there they are left till they are a little dried, *viz.* Till they are stiff enough to be turned on their Edges, and *Drest* (that is, cut

cut off their Inequalities, and Raggedness) and when they are dry, they carry them to the *Hacks* (or Places where they row them up, like a Wall of 2 Bricks thick, with some small Intervals betwixt them, to admit the Wind and Air to dry them) when the *Hack* is fill'd they are covered with Straw on the top, till they are dry enough to be carried to the Kiln to be burnt.

13. *Pilaster, or Buttress bricks.*] These Bricks are of the same length, breadth, and thickness with the *great Bricks*, 6. 9. they differ from them only in this; they have a notch at one end, which is half the breadth of the *Brick*, in breadth, and also in the length; they are made in the same Mold with the *great Bricks*, only when they make *Pilaster Bricks*, they put into one corner of the Mold, a Cube of Wood of 3 n. square; which piece causes the notch in the *Bricks* when they are Molded.

The use of these *Bricks* is to Bond the Work at the *Pilaster* of *Fence walls*, built of *great Bricks*. These Pilasters are made a Foot square, *viz.* A Brick in length, or 2 Bricks in breadth, alternately throughout the whole heighth of the *Pilaster*. So that the Pilaster stands out 3 n. beyond the Surface of the Wall on each side.

14. *Samel, or Sandal-bricks,*] Are those which lie out most in a Kiln, or Clamp, where the Salt-peter is not digested for want of Heat, and these are very soft and will soon moulder to dirt.

15. *Stock bricks,*] These differ not from *Place bricks* in Form; their difference lying conceal'd in the Quality of the Earth; they are made upon a Stock, *viz.* The Mould is put on a Stock, after the manner of Moulding, or Striking of Tiles, and when one Brick is Moulded, they lay it on a little piece of Board, a little longer than the Brick, and on that Brick they lay another piece of Board, like the first, and on that another Brick; after this manner, they lay 3 Bricks on one another, and so they continue to strike and place them on the Stage, as they do Tiles, till the Stage is full, and then they take each 3 successively, and carry them to the *Hacks*, and turn them down on their edges; so that there will be the thickness of a thin piece of Board betwixt each Brick. When the *Hack* is fill'd with one heighth of Bricks, from one end to the other, then they begin to set them up upon those, which were first laid on the Hack, by that time they will be little dried, and will bear the others; for they are Moulded of very stiff Earth; when they come to set a second or third, *&c.* Heighth, or Course they cater them a little, as they call it, to prevent their reeling: When the Hack is as high as they think fit, they cover them with Straw as they do *Place-Bricks* till they are dry enough to burn. This way Workmen tell me is more trouble than the other way, *viz.* of making Place bricks, and for making and burning (besides the

the digging of the Earth) they have 6*s*. *per* 1000, which is 1 *s*. *per* 1000 more than they ufually have for making of *Place-bricks* : But they are forced to make them fo, becaufe, if they lay them abroad in a Place to dry, as they do *Place-bricks*, the Nature of the Earth is fuch, that they will burft to pieces.

I very well remember an Inftance of this Kind, that was told me by an ancient experienced maker of Bricks and Tiles ; one that ufed to make Bricks about the Country (in *Kent* and *Suffex*) for Gentlemen : This Man was fent for to *Rumford* in *Effex* to make 100,000 of Bricks there for a Gentleman ; he having procur'd his Materials and Utenfils, went to Work (unadvifedly, not knowing the Quality of the Earth) and having ftruck about a 1000, when they had layen in the Place to dry, (according to the way of making *Place-bricks)* till about 10 a-Clock, when the Sun began to fhine very hot, his whole 1000 of Bricks burft to pieces, fo that he was forced to throw them all away, and then went to work again, and ftruck more ; and then, before the Sun fhone too hot, he *thackt* them, (*i. e.* cover'd them) over with Straw till the next Morning, and then he raked off the Straw with a Rake, and they did very well when they came to be fet on the *Hack* ; and when they were burnt, they were curious red Bricks, that would ring, when they were hit with any hard Thing. They did always ufe to make *Stock-bricks* at this Place, before he found the way of making *Place-bricks* of this fort of Earth.

16. *Statute*, *fmall*, or *common Bricks*] Their Dimenfions, *viz*. Of the Mold by the Statute, ought to be, as follows, *viz*. In length within 9 n. in Breadth 4 $\frac{1}{2}$ n. and in Thicknefs 2 $\frac{1}{4}$ n. Bricks made in fuch a Mold (the Earth being firft well temper'd) dried, and burnt, they will be lefs and lighter, yet they fhrink in Thicknefs but little; in Breadth lefs, and in their Length not difcernable : The Weight of Bricks is uncertain, the Gravity of Earths being very different alfo ; yet commonly one Brick will weigh about 5 ℔. faith Mr. *Leybourn*, and will contain 90 Cubick Inches, and from fome Molds 100. I once made an Obfervation on the Weight of *Statute Bricks*, I took 4, and meafured them, and weighed them, I found each Brick to be 9 n. long, 4 $\frac{1}{4}$ n. broad, and 2 $\frac{1}{3}$ n. thick, and I found that the 4 weigh'd 22 ℔. fo that one did weigh 5 $\frac{1}{2}$ ℔. and 100 of thefe would weigh 550 ℔. and a 1000, 5500 ℔. and about 407 will be a Tun Weight. Thefe were *Suffex* Bricks, of which they commonly reckon 500 to the Load, which Number of Bricks, according to this Proportion, will weigh about 24 Hundred and a half.

Thefe Bricks are frequently ufed in *Paving* of Cellars, Wafh-houfes, Sinks and Fire-hearths, and the like, 30 of thefe theſe made, according to the Statute, will pave a Yard ſquare, and 330 of them will pave a Square of 100 Foot. But you muſt Note, 'tis here meant, when the Bricks are laid the flat way, and not ſet on their Edges; for then it will take up near as many more.

But, I have it from Obſervation, That there muſt be 32 Bricks laid flat to pave a Yard, and 64 Bricks ſet an Edge to pave a Yard ſquare.

I find alſo by Computation, from an Obſervation, That there muſt be 4600 Statute-bricks to make a ſuperficial Statute-rod of Brick work, at a Brick and ½ thick, and by Conſequence 1700 to the Square, and 155 to the ſuperficial Yard; on a Wall of a 1 ½ Brick thick, V. P. Numb. VIII.

Mortar, the Quantity to a Rod of Brick-work.] Some allow more than others do, and the Truth is, they may well enough do ſo; for ſome Workmen have got a Habit of making larger Joints than others. Some uſually allow about a Load of Lime, and two Load and a half of Sand (at 36 Buſhels to the Load of Sand) to a Rod of Brick-work, or 4600 of *Statute-bricks*. And ſome others allow a Load and half a quarter of Lime, and 2 Load of Sand, *i. e.* two Buſhels of Sand to one of Lime; and others will allow but 1 ½ Load of Sand to a Load and ⅛ of Lime.

Price of theſe Statute or common Bricks.] This is various, for in different Parts of the Kingdom, they commonly have a different Price, which is not all neither; for Bricks in the ſame Kiln, ſhall have a different Price ſometimes, if the Maker of them be to lay them in at a greater Diſtance than uſual; and as Mr. *Leybourn* ſaith, Something ought to be conſidered, in reſpect to Workmens Wages, and the Price of Fuel to burn them with; but, ſaith he I never knew them cheaper than 9 *s.* nor dearer than 18 *s. per* 1000, deliver'd in any part of *London*.

In ſome Parts of *Suſſex* and *Kent* I have known *Statute*, or *Common bricks*, ſold for 16 *s. per* 1000, laid in a Mile or 2 diſtant from the Kiln, and at others for 20 *s.* At another Place in *Suſſex*, they ſell them at 25 *s per* 1000, if they lay them in about 2 or 3 Miles diſtant, that within this 10 or 12 Years, they did ſell them there for 20 *s. per* 1000; but ſince the late War began, the Iron-works in that part of the Country, have devour'd a great Quantity of their Wood, ſo that Fuel, of late Years, is got to be a fourth or more dearer than it did uſe to be; upon this Account they have now raiſed their *Bricks* to 25 *s. per* 1000.

Mr. *Wing* tells us, That in *Rutland* Bricks are but 12 *s. per* 1000 at the Kiln.

The Price of making Statute-bricks] In the Country, their uſual Price is 6 *d per* 1000 for the Molder, the Bearer off hath 4 *d.* and he that tempers the Earth ready for Uſe, hath 4 *d. per* 1000; and he that

that diggs it hath 6 *d per* 1000; for making the Earth ready (after it is digged, the digging being not reckon'd into the making), molding, bearing off, *&c* and burning, their usual Price is 5 *s. per* 1000.

Mr. *Leybourn* tells us, That about *London*, they allow the Molder 4 *d.* 5 *d.* or 6 *d. per* 1000; and that Bricks made at Home will stand the Maker of them in (besides the value of the Earth) betwixt 5 and 6 *s. per* 1000. But I am sensible it will be more in *Kent* and *Sussex*, at least some Parts of those Countries.

17. *Tetradoron.*] An ancient sort of *Greek* Bricks, which were 3 Foot, or 4 Spans long; and one Foot broad, being one of their larger Size, with which they built their publick Buildings. See *Didoron*.

18 *Triangular Bricks*] This Form of Bricks is described by *Daniel Barbaro*, Patriarch of *Aquileia*, in his largest Edition of his Comment upon *Vitruvius*. He would have these Triangular Bricks consist of an equilateral Triangle, each Side to be a Foot, and the Thickness but an Inch and a half. This sort of Bricks he highly commends to us for many good Properties; as 1st. That they are commodious in the Management. 2dly. Of less Expence. 3dly. Of fairer Shew, adding much Beauty and Strength to the mural Angles where they fall gracefully into an indented Work: So that Sir *Henry Wotton* wonders that we (in *England*) have not taken them into Use, being propounded by a Man of so good Authority in this Kind of Knowledge; but the Truth is, That all Nations are apt to start at Novelties, and are very apt to be wedded to their own Ways and Methods.

IV. *Of the Method of Making*] Of this see *Stock* and *Place-Bricks*. Mr. *Worlidge*, in his *Syst. Agricul.* is for exciting Brick-makers to try their Skill, in making a Composition of Clay and Sand, of which they may form in Molds, *Window frames* for Houses, of different Forms and Magnitudes, and also *Chimney-pieces*, and *Frames for Doors*, &c. in several Pieces made in Molds, that when they are burnt, they may be set together with a fine red Cement, and seem to be as one intire Piece; whereby may be imitated all manner of Stone-work now used in Building, and it will very well supply its Defect where Stones are wanting, or scarce and dear, and also save very much Timber, which is now used in Brick-Buildings, and appear much more compleat and beautiful, and be of greater Strength, and more durable for lasting, than Timber, or ordinary Brick; and one wou'd think it should be very feasible, as we may perceive by the *Earthen pipes* made fine, thin, and durable, to carry Water under Ground at *Portsmouth* in *Hampshire*, and by the *Earthen backs*, and *Grates* for Chimneys, made by Sir *John Winter*, formerly at *Charing-Cross*, of a great Bigness and Thickness, which are evident, and

and sufficient Demonstrations of the Possibility of making Work fine, thin, and light, for *Tiles*, either plain or curved, and for making of greater work in Moulds, and through burning of them, for *Doors*, *Windows*, and *Chimney-frames*, &c

This, saith he, is one of the most feasible, and beneficial Operations that I know in *England* to be neglected.

It is really my Thoughts, much might be done concerning making of *Chimney pieces*, *Stone-mouldings*, and *Architraves* Doors, and Windows, and *Architraves* or *Fascia's* for Fronts of Buildings. *&c* if that Men of this Profession would but set their Minds to work, to contrive some good Composition of Earth, and a way to manage it well in Molding, Burning, *&c*. But (the more is the Pity) Men of this Profession are like the Materials they work upon, *viz*. heavy and lumpish.

It might be made a Query whether a Composition of Earth, something like to common Crockers Earth, would not in some Measure answer the Design, since it is apparent, That whatever Form the Crockers are pleased to put their Earth into, it retains it after drying and burning, altho' Crocks, and such like Things are formed very thin: Now, suppose that *Chimney pieces*, or the like, were made in Molds, and dried and burnt, when they came to be set up, if they were not thought smooth enough, they might be polished with sharp Sand and Water, or a piece of sharp Stone, and Water. Or were there but Care taken of such Things as these (which are for Ornament, as well as Use) when they were half dry, or more in the Air, then to let them be polished over with an Instrument for the Purpose, either of Copper, or Iron, or some hard Body, and then leave them till they were dry enough to burn; 'tis my Thoughts such would not want much polishing afterwards.

And let me further add, I am very apt to think, that ingenious Men of this Profession, might make very handsome and beautiful *Chimney-pieces*, Stone *Moldings* for *Doors*, &c. fit for Noblemen's Houses. and all others that would be at the Charge.

What I would here propose is by way of Glazing, as Potters do their fine Earthen Ware, either white, or any other Colour, or it might be vein'd in Imitation of Marble, or be painted and anneal'd with Figures of various Colours, or some History, Perspective, or the like, which would be much cheaper, if not also as durable, and every whit as beautiful as Marble it self. And, had I Time, and Room in this small Treatise, I could here have added the Methods of Glazing, and painting of various Colours to have encouraged the Lovers of such Arts, to set their Heads to work at it. So that I am inclined to think, we rather want Art or Ingenuity, and Industry, than Materials

terials to satisfie our greatest Curiosity in Building. 'Tis not the Baseness of our *English* Materials, but want of Skill, and Diligence in managing them, that make our *English* Buildings in the least Measure inferiour to any Foreign ones: I well remember an Instance of this Nature, which was an Observation of an *English* Ambassadour, which was this *viz.* That we ought not to be discouraged with our ignoble Materials for Building, which we use in *England*, in comparison of the Marbles of *Asia*, and *Numidia*: For, saith he, I have often at *Venice* viewed with much Pleasure, an Antiporch, after the *Greek* Manner, erected by *Andreas Palladio* upon 8 Columns of the Roman Order, the Backs of Stone without Pedestals, the Shafts or Bodies of mere Brick, 3 ½ Foot in Diameter below, and consequently 35 Foot high, as himself hath described them in his second Book, than which, saith the Ambassadour, mine Eyes never yet beheld any Columns more stately of Stone, or Marble; for the Bricks were first formed in a Circular Mold, and were cut before they were burnt, into 4 Quarters, or Quadrants, or more than 4 Parts; for he could not certainly tell how many the sides were; afterwards in laying, jointed so close and nicely, and the Points concenter'd so exactly, that the Pillars appear one entire Piece.

And therefore I would not have *English* Men be disheartned, that we do here want those firm and solid Stones, which Nature hath furnished other Nations with; but rather to exercise their Ingenuity, to supply our selves by Art, with those things which Providence hath thought fit we should want, unless we would do so. And if we can but bring such things to Perfection which have been here hinted at, it may hereafter redound to the Honour of the *English* Nation. I mention these things here purely to stir up inquisitive Persons, to endeavour after an Improvement of such Arts, and that they might not be so stupid, as to suppose, That either they, or their Fore-fathers, were arrived at the *ne plus ultra* of this, or any other Art; and to perswade them, (if possible) to throw off that slothful and dangerous Principle, of resting contented with being possessed with the same degree of Knowledge, which our Predecessors had before us; and of thinking that they have skill enough, because the barbarous part of the World, doth not practice so much as they: But I would very fain (if it lay in my Power) prevail upon Mechanicks, to see what Improvements in their Professions they can bring forth.

V. *Of the Method of Burning Bricks*, &c.] All Bricks that are burnt, are burnt either in Kilns or Clamps

An experienced old Brick-burner, or Maker, tells me, That his (and some other experienced Brick-burners) Method in burning of *Bricks* and *Tiles*,

Tiles, was thus, *viz*. The Kiln being ſet, and cover'd with Pieces of Bricks, they firſt put in ſome Cord, (or great) Wood to dry the Ware, with a gentle, even Heat or Fire; which Fire they continue till the Ware is pretty dry; which they know by obſerving the Reek which aſcends out at the Top of the Kiln; for when it is changed from a thick Vapour, betwixt a whitiſh and darkiſh Colour, to a kind of a black Smoke, which is more tranſparent than the Vapour which firſt aroſe from the Kiln; after this blackiſh Smoke hath aſcended for ſome Time, they put in no more great Wood, but proceed to make ready for burning; which is performed, either with Spray, Buſh, Furz, Heath, Brake, or Fern-Faggots; but before they put in any Faggots, they damm up the Mouth of the Kiln (or Mouths; for ſome Kilns have more than one Mouth) with that *Shinlog*, as they call it (which is pieces of Bricks piled upon each other, with wet Brick-Earth, inſtead of Mortar.) This *Shinlog* they make ſo high, that there is but juſt room above it to thruſt in a Faggot, *viz*. Betwixt a ½ Foot and 2 Foot; for the whole Heighth of the Mouth is about 3 Foot. The Mouth being thus Shinlog'd, they proceed to put in Faggots, till they make the Kiln and its Arches look white with Heat, and the Fire begins to appear at the Top of the Kiln, and the Kiln and Arches below begin to change from white to a greyiſh Colour; then he ſaith, they ſlacken the Fire for ſome Time, *viz*. for about ½ an Hour, or an Hour, as they think fit, that the Fire, or Heat, may aſcend to the Top of the Kiln, by the Motion of the Air in at the Mouth, and alſo that the lower Ware may ſettle and cool, and not be burnt more than that above it. Thus they continue to do, heating and ſlacking alternately, till the Ware be thorough burnt, which it will be (he ſaith) in about 48 Hours: According to this Method, he ſaith he hath burnt many Kilns of Ware ſo equally, that thoſe on the Top were almoſt as hard as thoſe below (if not altogether.) He told me he had burnt ſeveral Kilns of *Tiles* and *Bricks* together, *viz*. About 3000 Bricks, and 10 or 11,000 of Tiles, and hath not had above 50 waſte, broken, and *Sandal* Tiles in all (which I will aſſure you is very Rare); whereas, ſaith he, ſuch Brick-burners as continue their Fire without any Intermiſſion, make their lower Ware extream hard; and that on the Top of Samel-bricks, or Tiles; nay, and which is worſe, they make the lower Ones run ſo with the exceſſive Heat, that they are almoſt united in one entire Body; ſo that they are forced to get them out with Wringers (or Iron-bars) and each *Bolt* of *Tiles* ſhall be one entire Maſs, (which I have obſerved my ſelf.)

And then, as to cooling of Kilns of Ware, ſome unwiſe Burners, as ſoon as the Ware is is burnt, they immediately ſtop up the reſt of the Mouth of the Kiln, which was left open above the *Shinlog*, by which Means it is long in cooling; ſo that they (*viz.* Such indiſcreet Burners) are commonly a Fortnight, or almoſt three Weeks, in Setting, Burning, and Cooling, and drawing of a Kiln of Ware. Whereas, ſaith he, I have Set, Burnt, Cool'd, and Drawn a Kiln a Week for ſeveral Weeks together: But then I never ſtopped up the Reſt of the Kilns Mouths above the *Shinlog*, but left it open for the Air to paſs in and cool the Ware.

He alſo told me, That 600 of Faggots would burn a Kiln of 10 or 11,000 of *Statute-bricks*. Mr. *Wing* informs us, That a Chaldron of Coals will burn about 4,200 of Bricks.

I have been informed, that their Method of burning *Bricks* in *Clamps*, is ſomething after this Manner, *viz.* They build their *Clamps* of the *Bricks* that are to be burnt ſomething like the Method of Building the Arches in Kilns, *viz.* with a Vacancy betwixt each Bricks Breadth, *&c* for the Fire to aſcend by; but with this Difference, that inſtead of Arching, they truſs, or ſpan it over, by making the Bricks project over, one beyond the other, on both ſides the Place, for the Wood and Coal to lie in, till they meet, and are bonded by the Bricks at the Top which cloſes up the Arch this Place for the Fuel, they carry up ſtrait at both Sides, or which is the ſame Thing, upright at both Sides, till it is about 3 Foot high, and then they begin to lay the *Bricks*, projecting over inwards, till they meet in the middle, which they will do in about 3 or 4 Courſe of Bricks in Heighth, the Width of the Mouth being but about two and ½ half Foot. Above this Arch they lay the Bricks in the Order they do in a Kiln, to 8 or 10 Foot in Heighth, according as the *Clamp* is to be in Bigneſs; for they uſually burn a great many Thouſands in a *Clamp* at a Time, ſo that they build them 8 or 10 Foot above the Arching.

But you muſt further Note, That after they have begun to make the Place (or Places) to receive the Fuel, before it is cloſed up at the Top, it is almoſt filled with Wood, and on that they lay a Thickneſs of Sea-Coal, and then they over ſpan the Arch; but they ſtrew Sea-coal all over the *Clamp*, from bottom to top, *viz.* betwixt all the Rows of *Bricks*; for they are not laid Contingent in their Vertical Rows; and one Courſe of Bricks is laid one way, and the other another; ſo that there is ſmall Interſtices betwixt all the Bricks, for the Coal to be ſtrewed into, from the bottom to the top: This being done, they fire the Wood, and that fires the Coal; the which, when 'tis all burnt out, they conclude the Clamp of Bricks to be burnt.

VI. *Of the Quantity of Earth to make a Thouſand of Bricks*, &c]

I am

I am inform'd that 1 Load of Lome (a Load being 12 Bushels) will make about 200 of *Statute-Bricks*, and then by Consequence, 5 Load will make a 1000. Also that 19 Load of Lome will make 1600 of great Bricks, and 12 will be sufficient for a 1000 of the same.

VII. *Of the Choice of Bricks*, &c.] *Pliny* advises in making choice of Bricks for Building, to be sure (if possible) to procure such as are two Years old at least. There are commonly, and general in all Kilns and Clamps, three Degrees of Bricks, in Goodness, *viz.* The first and best sort are those which lie next the Fire, (*viz* Those are best for lasting) and have, as it were, a Gloss on them, which proceeds from the Salt-peter, which is inherent in them, and which by the Violence of the Fire, runs and glazes them; these are call'd *Clinkers*.

The second and most general sort for Building, are those which lie next in the Kiln, or Clamp, to those before mentioned.

The third and worst sort, are those which lie on the out sides of the Kilns and Clamps, where the Salt-peter is not digested for want of due Heat; and these, when they come to be exposed to the Weather for some Time will moulder away like Dirt; and these Workmen call Samel or Sandal-bricks. 'Tis an Observation, That whilst Bricks are burning, those on the windy Side of a Clamp, are the worst of all.

VIII. *Of Observables in Buying and Laying Bricks*, &c.] And first, of Buying; the last Number will direct any Master or Workman (that doth not understand it) how to choose good Bricks; and in the 16 Section of Bricks, *viz.* Under the Head Statute-bricks, you have some Directions, as to the Number of Bricks; but you must Note, 'Tis impossible to be certain (to know) how many will be wanting exactly; because in such Cases there can be no infallible way discovered; and that for several Reasons, *viz.* (altho' the *Bricks* were all made in the same Mould, and burnt in the same Clamp, or Kiln) the Bricklayers Hand may vary in laying his Mortar. Secondly, Many Bricks warp in burning, (and the Seller will bring you some such, in Spight of all your Care in chusing) Thirdly, Some miscarry, and are spoiled in every Carriage. Fourthly, The Tally, or Tale, is, for the most part, too little, if not well looked to. And besides all these Uncertainties, when Bricks are dear, and Lime cheap (which sometimes happens so) if you put your Work out by the Great, or by Measure, and he is to find Materials that doth the Work, the Workman, without good looking after, will certainly use the more Mortar, and make very great Joints; which is a Defect in any Building.

Secondly, Of laying Bricks, which is a Thing of no small Consequence in a Building; for the well working, and bonding

bonding of Brick-work (or as ſome Workmen call it, breaking of Joint,) conduces very much to its Strength; I think therefore it may not be amiſs to add ſome particular Notes about it, which experienced Workmen have thought convenient to commend to the Publick, as well worth their Obſervation.

Firſt, Let me commend to your Care, to be ſure to procure good ſtrong Mortar; of which ſee *Mortar*.

Secondly, If your *Bricks* are laid in Winter, let them be kept and laid as dry as poſſible; if they are laid in Summer-time, it will quit coſt to imploy Boys to wet them; for they will unite with the Mortar much better, than if they were laid dry, and will make the Work much ſtronger. But perhaps it may be well objected, That it will be too much trouble to wet all the *Bricks* (by dipping them in Water) if the Building be large; and beſides, it makes the Workmens Fingers ſore. To prevent theſe Inconveniencies, there may be Water throw'd on each Courſe of *Bricks* after they are laid, as I am inform'd was done at the Building of *Phyſicians College* in *Warwick Lane*, by Order of the Surveyor, the ingenious Mr. *Robert Hooke*.

Thirdly, If your *Bricks* are laid in the Summer-time, be ſure to cover them, to prevent their drying too faſt; for if the Mortar dry too haſtily, it doth not cement ſo firmly to the *Bricks*, as when it dries gradually.

Fourthly, If *Bricks* are laid in Winter, be ſure to cover them very well, to protect them from Rain, Snow, and Froſt, which laſt is a mortal Enemy to all Mortar, eſpecially to all ſuch as have taken Wet but juſt before the Froſt aſſaults it.

Fifthly, Let Care be taken that *Bricks* be not laid Joynt on Joynt, in the middle of *Walls*, as ſeldom as may be, but let there be good Bond made there, as well as on the Out-ſides; for ſome Workmen, in working Brick and ½ Wall, lay the Header on one ſide of the *Wall*, perpendicular on the Header on the other ſide of the *VVall*, and ſo all along thro' the whole Courſe, which indeed neceſſarily follows, from the unadviſed ſetting up of the Quoin at a Toothing; for 'tis common to Tooth in the Stretching courſe two Inches, with the Stretcher only, and the Header on the other Side to be perpendicular over the Header on this Side, which cauſes the Headers to lie Joynt in Joynt in the middle of the *VVork*.

Whereas, if the Header on one Side of the *VVall* were toothed as much as the Stretcher on the other Side, it would be a ſtronger Toothing, and the Joynts of the Headers of one Side, would be in the middle of the Headers of the Courſe they lie upon on the other Side.

All that can be pretended to excuſe this ill Cuſtom of working thus, is this, That the Header will not hang 2 n. over the *Bricks* underneath it. This

This indeed I do grant to be an Objection, but not so great, but that it may be removed, and that without much Difficulty, *viz.* thus, By having a piece of Wood of the Thickness of a Course of *Bricks*, and 2 n. broad, and lay it on the last Toothing Course to bear it, or a Brick-bat put upon the last Toothing, will bear it till the next Quoin is set upon it, and then the Bat may be taken away.

Sixthly. The same Inconveniency happens at an upright Quoin in a Brick and ½ *Wall*, where 'tis usual to lay a Closer next the Header, on both sides of the *Wall*, and in so doing, 'tis Joynt in Joynt all the Length of the *Wall*, except by chance a 3 quarters Bat hap pen to be laid.

To prevent which Inconveniency, and thereby make the *VVall* much firmer, lay a Closer on one Side, and none on the other Side; but lay a 3 quarter Bat on the Quoin in the Stretching-course, and in the Heading-course adjoyn an Header next to the Header at the Quoin.

Also in two *Brick-walls*, it is the best way in Stretching Courses, wherein they lay stretching on both Sides the *VValls*, next the Line so also to lay stretch ng in the middle of the *VVall*, and Closers next to each Stretching-course that lies next the Line.

A Bricklayer and his Labourer (having all his Materials ready) will lay in a Day about 1000 *Bricks*, in whole Work on a solid Plain, and some very expeditious Bricklayers, will lay 12 or 1500.

IX. *Of Facing Timber-buildings with Bricks.*] In some Places this Method of Facing Timber-building is in Use, but I think it should be call'd Caseing; for 'tis covered all over on the Out-side with Brick, so that no Timber is to be seen. The which is performed after this Manner *viz.* All betwixt the Timber the Wall is a *Brick* a Length thick (or 9 n. Wall of *Brick*, but against the Timber, the Wall of *Bricks* is but 4 ½ n. or half a Brick, or the Breadth of a *Brick* thick (beside the Timber.)

But this Method is not approved of by able Workmen, because the Mortar doth so extreamly corrode and decay the Timber.

For I remember an experienced Bricklayer told me, That he did pull down such Work at *Eridge-place* (which is one of my Lord of *Abergaveny*'s Country-Seats) and the Timber was extreamly corroded, and eaten with the Mortar.

Bricklayers.

1. *VVork*] The *Bricklayers-work* in the City is of various Kinds, *viz. Tyling*, *VValling*, *Chimney-work*, and *Paving* with *Bricks* and *Tiles*. But in the Country 'tis common for the *Bricklayer's* Trade to comprehend the *Masons* and *Plaisterers* also. All which Particulars will render it too large to be comprehended under the general Head of *Bricklayer's work*; I therefore think it will be more

more convenient to rank it under its particular Branches, or Parts, *viz.* *VValling*, *Tyling*, *Chimney-work*, *Paving*, &c.

Bill of making.] A Bricklayer's Bill may be Composed after this Method.

Mr. Robert Rich, *of* Rochester, *his Bill of Materials, had of, and Work done by* Benjamin Bennet, *Bricklayer*, October 5. 1724.

	l.	s	d.
For 12 Thousand of Bricks at 15 *s. per* M.	09	00	00
For 8 Thousand of Tiles at 20 *s per* M.	08	00	00
For 17 Hundred of Lime at 14 *s. per* C.	11	18	00
For 15 Load of Sand at 2 *s.* 6 *d. per* L.	02	05	00
For 10 Hundred of 9 n. Paving-tiles at 10 *s.* 6 *d.* *per* Hundred	05	05	00
For 40 Ridge-tiles at 1 ¾ *d. per* Piece.	00	05	10
For 3 Weeks and 3 Days Work for my self, at 3 *s. per diem.*	03	03	00
For 27 ½ Days for my Man at 2 *s* 6 *d. per* Day.	03	08	09
For a Labourer 27 ½ Days at 1 *s.* 8 *d. per* Day.	02	05	10
Sum Total is	45	11	05

But Note, if Bricklayers do not work by the Day, then they use a different Method in Writing their *Bills*; for then they either take their Work by the Great, *viz* To do all, and find all belonging to Bricklayers Work, or else he is to do it by Measure, and to find all Materials and Work, at such a Price by the Rod for *Walling*, by the Square for *Tiling*, and by the Yard for *Paving*, *&c.* But if he finds no Materials, he may also work by Measure, and then the Bill must be made after this manner, *viz.* For so many Rods of *Walling*, at so much (according to their Agreement) *per* Rod, *&c.*

Note also, That in some Buildings *Chimneys* are put out to the Bricklayer by the Hearth, either only to build, or to find Materials also, and then the Bill is made according to the Agreement

There are some other things which come into a *Bricklayer's Bill*, viz. All kind of ornamental Work in Brick, which is commonly set down, or rated at so much *per* Foot, or so much *per* Piece, except a good Rate be allowed by the Rod, *&c.* Or there be a Sum of Money over and above the Price, or Value of the Rod-work allowed, and so the Ornamental Work be included in it. By Ornamental Work, is to be understood, streight, or circular *Arches*, over Windows, or Doors; *Fascia's*, with, or without Moldings, *Architraves*, round Windows, or rubbed *Returns* *Friezes*, *Cornices* of all sorts, *Water Tables* wrought and

and *Water courſes*: All which are valued by the Foot running Meaſure; to which I muſt add *Baſe-mouldings*, and *Plinths*, and the *Splaying* of the *Jambs* of Windows and Doors on the inſide of Buildings. Alſo *Pilaſters*, *Peers*, *Pediments Grotto's*, and *Ruſtick Quoins*. Theſe five laſt mentioned, are valued at ſo much *per* Piece, according to the largeneſs, and goodneſs of the Work and Materials; and thus all *Ornamental Work*, ought to be valued. By the word *Ornamental Work*, is to be underſtood in *Bricklayers Work*; all kind of *Brick Work*, that is hewed with an Ax, or rubbed on a Rubbing-ſtone, or of Stone wrought with *Chiſſels*, or rubbed with Stones, or Cards, all ſuch is *Ornamental Work* and ought to be paid for, beſides the Rod-work, *&c.* I ſhall now proceed to ſpeak of that Part of Bricklayer's work, which is called

Brick Work.

1. *Some Notes about Meaſuring*, &c.] Sometimes *Brick-walls* are wrought 2 n. thicker than the reſt of the Work, part of the way, which 2 n. ſerveth for a *Water-table* to the Wall which is uſually ſet off about two Foot above the Ground; and therefore the Brick-work may be meaſured at the ſame Thickneſs that is above the *Water-table*, and then the 2 n. Work may be thus added to it.

Suppoſe a Wall 20 Foot in length, and 2 Bricks thick above the *Water-table*.

After the Dimenſions of the Wall is taken (from the bottom, to the heighth it is to be taken at 2 Bricks) then add 20 Foot in length by the heighth of the 2 n. Work *viz.* From the bottom to the ſetting off, or *Water table*, which being halfed, is ſo much 4 n. Work, and then reduce it to a Brick and a ½ Work.

As for *Ornamental Work*, we need not to mention that here, it being mentioned above.

2. The Meaſuring of *Gable-ends* in *Brick work*, is done after the ſame Method that Carpenters meaſure Gables, (only this is reduced into Rod-work) See *Gable end*, N. 2.

3. Be ſure to obſerve, in taking Dimenſions of Walls that joyn to an Angle, that the length of one Wall be taken at the out-ſide of the Angle, and the others length to the in-ſide of the Angle.

4. If there be a *Gable end* to meaſure, and the width of the Houſe be given (or known) which is the Baſe of the *Gable-end*, and the length of the Perpendicular is required, there is a brief way uſed amongſt Meaſurers to find it. To make it the plainer, I will propoſe an Example, *viz.* Suppoſe the Baſe of the Gable be 24 f and the length of the Perpendicular is required; take the length of the Rafter (which will be) 18 f. to which add ½ it ſelf, *viz* 9 Foot, it makes 27, the ½ of it is 13 f. 6 n the length of the Perpendicular But tho' this way be commonly practiſed, it is not exact, for it makes the Perpendicular a little too much: This you muſt note is practiſed

practiſed for Roofs, that are $\frac{3}{4}$ pitch; and therefore I would not adviſe any to make uſe of this Method in any other Pitch. Now I am upon diſcourſing of *Gable-ends*, I will here add two exact Ways of finding the Perpendicular; the firſt ſhall be by Proportion. thus, *viz.* As 30 to 22, 35, ſo is the length of the Rafter to the Perpendicular required; or ſubſtract the Square of $\frac{1}{2}$ the Baſe, or $\frac{1}{2}$ width of the Houſe, from the Square of the Rafters length, there will remain a Number, whoſe ſquare Root is the length of the Perpendicular.

5. In taking out the Deductions for the Doors, and Windows, *&c.* if any happen in Brick-work, of 2 $\frac{1}{2}$ Bricks thick, or in 2 Bricks thick then add $\frac{2}{3}$ to the length, for thoſe in the 2 $\frac{1}{2}$, Brick work, and $\frac{1}{3}$ to the lengths of Doors, or Windows, in 2 Br. (or it may be $\frac{2}{3}$, or $\frac{1}{3}$ to the breadth, and not the length, according as which will be ſooneſt divided) and then the lengths and breadths being multiplied one into the other, the Product is the proper Deductions in Brick $\frac{1}{2}$ work, without any further trouble; and it will neither wrong Maſter nor Workman.

6. Our ſixth Note ſhould have been on *Chimneys*, but of that ſee *Chimneys.*

I did alſo intend (when I began this Head of *Brick-work*) to have inſerted here ſeveral other things appertaining to *Brick-work*,) *viz.* The Method of Meaſuring, Reducing to Standard-thickneſs, finding the value of any odd Foot, Price of this Work, in diverſe Parts of the Kingdom, of laying Foundations of Walls, *&c.* But finding of B. will be a very copious Letter, I ſhall refer it to Walls of Brick, which ſee N. IV.

Brick-Walls.

See *Walls*, N. IV.

Bridge.

Of Timber to Build over any Brook, Gill, or ſmall River, if it do not exceed 40 or 50 Foot in length, and that without ſetting any of the Timber down in the Water, it is a cheap and ſafe way of building a *Bridge* of that length.

To perform this piece of Art, the Timber muſt be ſo joynted, as to reſemble (in ſome meaſure) an Arch of Stone, or Brick, the Joynts ought to be well made, and ſhut together ſtrongly with Cramps and Dogs of Iron. This Bridge muſt be made to reſt upon two ſtrong firm Pillars of Wood, at either end of the Bridge, both being well propped with Spurs or Braces; there muſt be two good Butreſſes of Brick for theſe wooden Pillars, and Spurs to ſtand in, that they may not give way, or ſlip; this being done, the Bridge may be Planked over, and Graveled, and it will laſt a long time. This hath been already practiſed, ſaith Sir *Hugh Plat.*

Bring up.

Bring-up.

'Tis a Term uſed among Workmen, eſpecially *Carpenters*, when they diſcourſe with *Bricklayers*, and then they ſay *Bring-up* the Foundation ſo high, *Bring up* ſuch a Wall, *Bring up* the *Chimneys*, *&c.* Which is as much as to ſay, build the Foundation ſo high, build the Wall, build the Chimneys, *&c.*

Broad-ſtone.

1. *What.*] 'Tis the ſame with Free-ſtone, only this is ſo called, becauſe they are raiſed broad and thin out of the Quarries, *viz.* not above 2 or 3 Inches in thickneſs.

2. *Uſe.*] The Uſe of theſe ſort of Free ſtones, which are called *Broad ſtones*, is for Paving of Yards, and Paſſages, and before Shop-doors and Stalls, *&c.*

3. *Price.*] If they are promiſcuous breadths and lengths, then the uſual Price for the Stone fitting, and laying in Mortar, from 6 to 8 *d. per* Foot ſquare, or from 4 *s.* to 6 *s. per* Superficial Yard.

But ſome of theſe Stones are cut into perfect Squares, as Paving-tiles are, but much larger, as 18, 20, and 24 Inches ſquare, or more, but thoſe, as they are neater, ſo they are dearer; ſome Paving with theſe, being worth 1 *s. per* Foot, but 'tis worth 15 or 16 *d. per* Foot, if the Stones be good and well poliſhed, as they ought to be, for Kitchins, Dairy-houſes, and neat private Places.

Building.

I. *Conſiderations about it.*] Every Man that is diſpoſed to *Build*, either out of Choice, or thro' Neceſſity, ſhould firſt ſit down, and ſeriouſly conſider of the whole Deſign, *viz.* Both of the Manner and Method, as well as the Charge and Expence. And I am ſatisfied, that Premeditation is a very neceſſary Maxim, or Preliminary to Building, becauſe we have it from no leſs Perſon than our Bleſſed SAVIOUR himſelf, who ſaith in *Luke* 14. 18. *Which of you intending to build a Tower, ſitteth not down firſt, and counteth the coſt, whether he have ſufficient to finiſh it?* The Reaſon and Neceſſity of it follows, *v.* 29. and 30. *Leſt haply after he hath laid the Foundation, and is not able to finiſh it, all that behold it begin to mock him, ſaying, This Man began to build, and was not able to finiſh.*

And let me perſwade all *Builders*, to make choice of ſuch *Surveyors*, and Workmen, as underſtand what they are going about, before they begin the Work, *viz.* Such as be Maſters of what they pretend to, as a Surveyor that underſtands how to give the Draught, or Model of a Deſign; ſo as that when it is erected, it may anſwer to the end, which is to Build well; and a Building is ſaid to be well done, when it is ſo contrived, and perfected, that it is poſſeſt of the following Qualifications, *viz.* Accommodation, or Uſefulneſs, Proportion, (Beauty, or Handſomeneſs

ſomeneſs) and Uniformity in its Parts; Firmneſs with Duration. For that Fabrick cannot be accounted perfect, which is uſeful only for but a ſhort ſpace of Time, or not convenient for a longer; and hath not alſo Decency and Beauty, which is derived from Proportion and Uniformity: I would therefore (if it lay in my Power) endeavour to perſwade all Builders to procure ſuch Surveyors, and Workmen (if poſſible to be procured in the Country, where the Builder lives) as underſtood the Theory and Practice of Architecture, and alſo of Arithmetick, (which is the Ground of all Arts) without the Knowledge of theſe two, the beſt Mechanick, or Handicraft Man will be but an imperfect Builder, and ſubject to fall into many Errors, and be guilty of committing many Faults, and making many Miſtakes. For Gentlemen, and others that are Builders, are too often prevail led upon, and perſwaded by ſuch Workmen, as are wedded to their own Wits (tho' they were never verſt in the Grounds of Architecture, and were wholly ignorant of the meaning of Proportion, Uniformity, and Accommodation in Building) and tied to their own odd ſort of irregular old way; which is no better than a deformed Cuſtom, and ſuch Men will not (for the moſt part) be prevailed on, or perſwaded to a more compleat way, though it be much more beautiful, and regular, and alſo with leſs Materials, and cheaper, and more convenient than the other; and all the Reaſon they will, or can render for it, is, becauſe it is New to them, and they were never accuſtomed to ſuch a way of working; neither do they underſtand it: For ſay they, Our Fore-fathers did it not before us, which is a very prevailing Argument with ſome that know no better; yet perhaps the Maſter-Builder is willing to beſtow Expence enough on his Houſe, &c. to Enrich, and Adorn it; but his Workmen, thro' ignorance perform it with very little Skill or Art. But I do think none are ſo Senſeleſs and Stupid, as to deny, that it is better to erect ſuch a Fabrick, that ſhall be more uſeful and neceſſary, and alſo more pleaſing, both to wiſe Men, and Fools, than that (tho' done by the ſame Coſt and Expences) which will only pleaſe an ignorant Workman or two, (who are poſſeſt with an over-weening Affectation of their own Skill, which at beſt is but conceited Blundering or Unskilfulneſs) and perhaps it may alſo ſatisfie ſome few others, that do not underſtand the Methods, and Maxims of Architecture, and ſo were eaſily prevailed upon by theſe abſurd Workmen, and Proſelyted to be of their Opinion, be it right or wrong. Having thus laid down ſome Conſiderations about *Building*, I ſhall next proceed to ſpeak of certain

II. *Aphoriſms neceſſary to be known, and obſerved in Building.*] Dr. *Fuller*, *Prebend* of *Sarum*, ſaith,

ſaith, He that alters an old Houſe, is tyed as a Tranſlator to the Original, and is confined to the Fancy of the firſt Builder. Such a Man were unwiſe to pull down a good old Building, to erect (purchance) a worſe new one. But thoſe that raiſe a new Houſe from the Ground, are Blameworthy, if they make it not handſome and uſeful, ſeeing to them Method and Confuſion are both of a Price. In Building, ſaith he, we muſt reſpect *Situation*, *Contrivance*, *Receipt*, *Strength* and *Beauty*, to which I will add *Form* or *Figure*.

1. Of *Situation*] The Precepts belonging to *Situation*, ſaith Sir *H. W.* do either concern the total Poſture, or Poſition (as I may term it) or *placing* of the *Parts*: The firſt of theſe is uſually reckon'd by *Architects*, as part of their Profeſſion, but the Truth is, it is borrowed from other Parts of *Learning*, there being betwixt *Arts* and *Sciences* (as well as betwixt Men) a kind of Society, and Communication of *Principles*.

For ſome of them are purely *Phyſical*, touching the Quality and Temper of the *Air*, *viz.* That it be a good Healthy Air, not ſubject to Foggy Noiſomeneſs, from Fens, or Marſhes, that are adjacent; that it be alſo free from Noxious, Mineral Exhalations. And let not the Place want the ſweet Influence of the Sun-beams, nor be wholly deſtitute of the Breezes of Wind, which will Fan and Purge the Air; the want of which would make it like a ſtagnated Pool, or ſtanding Lake of Air, (which is very unhealthy,) as ſaith *Alberti*, the Florentine *Architect*. He alſo warneth us to avoid ſuch Places, as are ſubject to *Earthquakes*, *Contagions*, *Prodigious Births*, and the like.

Dr *F's.* Phyſical Advice is, *viz. chiefly to chuſe a wholſome Air*; For Air, ſaith he, is a Diſh one feeds on every Minute, and therefore it had need to be Salubrious. Wherefore great Men (who may Build where they pleaſe, and poor Men where they can,) if herein they prefer their Profit above their Health, I refer them to their Phyſicians to make them pay for it accordingly. *Cato* ſaith, Let your Country-houſe have a good Air, and not be open to Tempeſts, ſeated in a good Soil; let it therein exceed, if you can, and let it ſtand under a Hill, and behold the *South*, in a healthy Place.

Pliny adviſeth not to ſet a Country-houſe too near a Fen, or ſtanding Water, nor yet over againſt the Stream and Courſe of a River; for ſaith he, (as *Homer* ſaith to this Purpoſe) The Fogs, and Miſts that ariſe from a great River, betimes in the Morning before Day-light, cannot chuſe but be very cold and unwholſome.

Oeconomical.] Saith Sir *H. W.* let the Houſe, or Seat be well Watered, and well Fuelled; let not the way to it be too ſteep, and of an incommodious Acceſs, which will be a Trouble to both Friends, and the Family. And ſee that it be not Seated too far from ſome Navigable River, or Arm of the Sea, which

which will conduce to the Eaſe of the Family, in procuring Proviſions, and other Domeſtick Neceſſaries.

Dr. *F.* ſaith, *That Wood and VVater are two ſtaple Commodities*, where they may be had. The former I confeſs hath made ſo much Iron, that it muſt be bought with the more Silver, and grows daily dearer. But it is as well Pleaſant as Profitable, to ſee a Houſe Caſed with Trees, like that of *Anchiſes* in *Troy*.

The worſt is, where a Place is bald of Wood, no Art can make it a Perriwig in haſte.

And as for Water, the want of it is a great Inconveniency, the Miſchief of many Houſes, where Servants muſt bring the Well upon their Shoulders.

Optical Precepts or Maxims.] Such I mean, (ſaith Sir *H. W.*) as concern the Properties of a well choſen *Proſpect*, which may be ſtiled the Royalty of Sight: For as there is a Lordſhip (as it were) of the Feet, wherein a Man walketh with much Pleaſure about the Limits of his own Poſſeſſions; ſo there is a Lordſhip likewiſe of the Eye, which being a Ranging, and Imperious (I had almoſt ſaid) Uſurping Senſe, cannot indure to be Circumſcribed within a ſmall Space, but muſt be ſatisfied both with Extent, and variety; yet on the other ſide, I find vaſt and indefinite Proſpects, which drown all Appreheniſons of very remote Objects condemned by good Authors, as if thereby ſome Part of the Pleaſure (whereof we were ſpeaking) did periſh.

A Pleaſant Proſpect is to be reſpected.] Saith Dr. *F.* A medly View (ſuch as of Water and Land at *Greenwich*) beſt entertains the Eyes, refreſhing the weary Beholder with exchange of Objects. Yet ſaith he, I know a more profitable Proſpect, where the owner can only ſee his own Land round about him. To this Head of *Situation* he adds what follows, *viz.*

A fair Entrance, with an eaſie Aſcent, gives a great Grace to a Building.] where the Hall is a Preferment out of the Court, Parlour out of the Hall, (not as in ſome old *Buildings*) where the *Doors* are ſo low, Pigmies muſt ſtoop, and the Rooms ſo high, that Giants may ſtand a Tip-toe.

A *Political Precept.*] I remember (ſaith that great Architect, Sir *H. VV.*) One private Caution, which I know not well how to Rank amongſt the reſt of the Precepts, unleſs I call it *Political*, which is this, *viz.* By no means to *Build* too near a great Neighbour, which were to be as unfortunately Seated on the Earth, as *Mercury* is in the Heavens, for the moſt part ever in Cumbuſtion, or Obſcurity, under brighter Beams than his own. We are next to come to

2. *Contrivance.*] When the Situation is reſolved upon, the next, in Order is *Contrivance.* The which being a thing of great Moment, in this Affair of Building, I cannot enter upon it, before I have given ſome few general Precautions.

And

And First, I would by no means have any one that intends to Build a Structure(that shall be either useful or ornamental) set to work without Advice or Assistance of a *Surveyor*, or a Master-workman, that understands the Theory of Architecture, and is capable of Designing a *Draught*, or *Model*, according to the Rules of Art. If a Draught be resolved upon (which may serve indifferent well for small ordinary Buildings) there ought to be the *Ichnography* of each Floor, and also the *Orthography* of each Face of the Building, *viz.* The Front, the Flanks, and the Rear. But if the Workman be skill'd in Perspective, than more than one Face may be represented in one Diagram *Scenographically.*

In the Contrivance of these Designs, whether for *Draught*, or *Model*, the Quality of the Persons, for whom the Building is erected, must be considered in respect of the Ichnographical Plots especially. For Noble-men have occasion for more Rooms of Office, than others of a meaner Degree; all which much be design'd according to their most convenient Occasions, with the lengths and breadths according to Proportion; also the Ichnography of all Chimneys, both in length and breadth of the Hearths, and Jambs, Bed-places, Stairs, and the Latitude of all Doors and Windows, in each Contignation, or Floor. And if it were required in Timber Buildings, the Longitude, Latitude, and Crassitude of Ground-plates, or Sells, Brest-summers, and in all (whether Timber, Brick, or Stone, Buildings,) the Dimensions of Summers, Girders, Trimmers, and Joysts. Also in the upper Floor, the Scantling of the Draggons-beams, Raisons, or Raising-pieces, or Wall-plats, *&c.* And also the Crassitude of Partitions, Walls, *&c.* in Brick, or Stone-fabricks.

All which, and all other Parts (whether in the *Ichnography*, or *Orthography*) of Buildings, ought to be represented (as also *Ovens, Stoves, Broilers, Furnaces, Coolers, Fats* for *Brewing, &c*) with their just Measures, for the best Advantage, as to Commodiousness, Health, Strength, and Ornament. All which Dimensions I would advise to be set in the proper Places to which they belong in the Diagrams, in Characters; because unless the Schemes be very large, it will be very difficult to take the Dimensions nicely, of the smaller parts, if not of the great ones likewise; it will scarce be practicable to take either of them to an Inch, nor perhaps, to two, three, nor four, according as the Diagram may be in Amplitude.

In the *Orthographical Schemes*, there must be the true Delineations, and Dimensions of each Face, and all its Concomitants, as *Doors, Windows, Balconies, Turrets* or *Cupolaes, Chimney-shafts, Fascia's, Rustick Quoins, Architraves, Friezes, Cornishes, Pediments, Pilasters, Columns, Shells* over *Doors, Lantherns*, and all other *Ornaments*. And if it be a a Timber-building, then all the Members in that Face ought to have their several Sizes, in Characters, and true Positions by the Scale. As for Example, the Ground-plates, or Cells, Interduces, Brest-summers, Beams, Principal-posts, or Braces, Quarters, Prick-posts, or Window-posts; Jambs, or Door-posts, or Puncheons, King-pieces, or Joggle-pieces, Struts, Collar-beams, Door-heads, Principal-rafters, Shreedings, *&c.* The Ichnography, Orthography, and Scenography of the Stair-case, may be also delineated, and all its Parts, as Hand-rail, Risers, Noseing of the cover, or top, String-board, and Mouldings on it, or Cartouses, Ballisters, Pendents, *&c.* with their true Positions, Forms, and Dimensions, all which being carefully done by an ingenious Surveyor, I think 'tis almost impossible for a Workman to mistake, or to commit any Blunders; tho' to my knowledge they are too subject to do it. More of this see in *Draughts.* You shall next hear what Sir *H. VV.* saith of this Matter; his Precautions are as follow, *viz.*

First, (saith he) Let no Man that intends to Build, settle his Fancy on a *Draught* in Paper (or Vellum) of the Work or Design, how exactly soever delineated, or set off in Perspective, without a *Model,* or Type of the whole Structure, and of every Parcel, and Partition, either in Past-board, or Wainscot.

Secondly, Let the *Model* be as plain as may be, without Colours, or other Beautifying, lest the Pleasure of the Eye, preoccupate the Judgment.

Lastly, The bigger this Type is, it is so much the better; not that I would perswade any Man to such an Enormity, as that *Model* made by *Antonio Labaco,* of St. *Peter's* Church in *Rome,* containing 22 Foot in length, 16 in breadth, and 13 in heighth, which cost 4184 Crowns, the Price of a reasonable Chapel, yet in a Fabrick of 40 or 50000 Pounds, there may be very well expended 30 *l.* at least to procure an exact *Model,* for a little Penury in the Premises, may easily create some Absurdity, or Error, of a far greater Charge in the Conclusion.

What Sir *H. VVotton* doth here Caution, is very proper and requisite, in large and sumptuous Buildings, whether publick or private; as for Noblemen's Mansion-houses, and the like: but it is not worth the while, to be at the Trouble, and Cost to procure a Model for every little Dwelling-house that Men build for their own Conveniency.

Having thus given sufficient Caveats, I will next proceed to discourse of the Compartition, or *Contrivance,* whereby to distribute the whole Ground-plot, *&c.* into Rooms of Office, or Entertainment, as far as the Capacity of the Building, and the Nature of the Climate will correspond, yea so far as it may be both decent and useful. But in the mean while we are to consider, whether the Building be to be erected in a City, or or great Town of Trade; and whether for a Gentleman, or a Shop-keeper, which is the chief Thing to be considered of by the *Surveyor*, or Master-workman, before he makes his Draught. For a Gentleman's House must not be contrived like a Shop-keeper's, neither must all Shop-keepers Houses be a like; for some Trades require a deeper, others may dispense with a shallower Shop, and so an Inconveniency may arise in both; for if the Shop be hollow, the Front Rooms upward ought to be shallow also; because, by the strict Rules of Architecture all Partitions of Rooms ought to stand directly one over the other: For if the Shop stands in an eminent Street, the front Rooms are commonly more Airy than the back Rooms, and always more commodious for observing publick Passages in the Street; and in that Respect it will be inconvenient to make the front Rooms shallow; but if there be a fair Prospect backwards, of Gardens and Fields, *&c.* (which seldom happens in Cities) then it may be convenient to make the back Rooms the larger for Entertainment, *&c.*

'Tis observed by some, That in building of Houses long, the Use of some Rooms will be lost, and it takes up more for Entries and Passages, and requires more Doors: And if a Building consist of a Geometrical Square, if the House be any Thing large, there will be want of Light to the middle Rooms, more than if it be built like an H, or some other such like Figure (unless it have a Court in the middle of it, which was the Method of building great Houses formerly.) This Way, like a Roman Capital H, is much applauded by some; for say they, This Form maketh it stand better, and firmer against the Winds, and Light, and Air comes every Way to it, and every Room is near the one to the other. Some affect this Figure very much, because the Offices may be remote from the Parlour, and Rooms of Entertainment, and yet in the same House, which may serve very well for a Country Gentleman's House. Now, the Method which some propose for such Buildings, is thus, In the Front of one of the long Parts of the H is the *Kitchin*, and the *Bake-house*, *Brew house*, and *Dairy-house*; in the same part behind it, the Hall in the middle of the H, which seperates the Parlours (which are in the other long Part) and Rooms of Entertainment from the Offices.

I shall here add a cheap Contrivance in *Building*, approved of by some, and then proceed to Sir *Henry VVotton*'s Method of contriving Noble Buildings.

Now this cheap way is thus, *viz.* Where Bricks may be had, the Walls of a Building may be best, and most securely raised with them, and with little Cost, if there be firm and strong Quoins, or Columns raised at the Corners of the House, of sufficient

Strength

Strength, to ſupport the Floors and Roof or the main Beams of it ; they may be built Square, and between them the Walls may be raiſed of the ſame Materials, and they may be worked up together with the Quoins, leaving the one half of the extraordinary Breadth of the Quoins without, and the other within the Wall, whereby there will be much Charge ſaved, both in Materials, and Workmanſhip, and yet the *Building* be firm and ſtrong.

According to Sir *Henry VVottons* Definition of *Contrivance*, it conſiſts of theſe two Heads, or Principles. Gracefulneſs, or Decency, and Uſefulneſs.

Decency or *Gracefulneſs*, he alſo ſaith, conſiſts in a double Analogy, or Correſpondency Firſt, Between the Parts, and the whole, whereby a great Fabrick ſhould have great *Apartments*, great *Lights*, or Windows, great *Entrances*, or *Doors* great *Stair-Caſes*, great *Pillars* or *Pilaſters* ; in fine, all the Members and Parts great, proportionable to the Building.

The ſecond Analogy, is between the Parts themſelves, not only conſidering their Breadths, and Lengths, as where we ſpeak of Doors and Windows, which ſee ; but here, ſaith Sir *Henry*, enters a third reſpect of *Heighth*, a Point ſaith he, I muſt confeſs) hardly reduceable to any general Precept. The Truth is, The Ancients did determine the *Longitude* of all Rooms which were longer than broad, by the Double of their *Latitude*. *Vitruvius*, Lib. 6. Cap. 5. And the Heighth by half the Breadth and Length added together ; but when the Room was a Geometrical Square, they made the Heighth half as much more as the Latitude, which Dimenſions the modern Architects have taken leave to vary upon Diſcretion : Sometimes ſquaring the Latitude, and doubling that ſquare Number, the ſquare Root of that Number is the Heighth, and ſometimes more, but ſeldom lower than the Breadth.

But what is here mentioned, I think is not now practiſed neither, unleſs it be in ſome Nobleman's Houſe, who will have a Hall, or the like, higher pitch'd than the reſt of the Rooms in the Building, and ſometimes a Dining-room ; or elſe, for the moſt part, all the Rooms of a Floor are of an equal Heighth ; and in my Judgment, 'tis by far the moſt commodious Method ; becauſe then there is no Room loſt, (as there muſt be where one Room is open almoſt to the top of the Houſe, as I have obſerved it in ſome old Buildings.) And then the Floor of the ſecond Story will lie level and even, and not in the odd old Method of Steps, out of one Room into the other.

As to the Heighth of *Rooms*, they are various amongſt us, according as what Perſons they are built for, and Cuſtom of the Place in the Country ; ordinary Timber-buildings, are about 7 $\frac{1}{2}$, or 8 Foot at moſt, betwixt Floors : The ſecond ſort of Houſes in the Country,

try, is about nine Foot betwixt the Floors, which for the most part is the Pitch of their Rooms at *Tunbridge wells*.

The third sort in the Country, (*viz.* in *Kent* and *Sussex*) are Gentlemen's Seats, which for the most part are 10 or 12 Foot high, such as are New Buildings: But 'tis common in Old Stone-buildings to be much higher, *viz.* 14 or 16 Foot.

By ACT *of* Parliament *for the Building of* London, *there was reckoned* 4 *Rates of Houses*, viz

The	Rate		
1	Rate	2	Stories, Cellars and Garrets.
2		3	
3		4	
4		5	

The	Rates, Cellars in heighth betwixt Floor and Ceiling.		
1		$6\frac{1}{2}$	Foot.
2		$6\frac{1}{3}$	
3		$6\frac{2}{2}$	
4			at the Discretion of the Builder.

The	Rate 1 Story		
1		9	Foot.
2		10	
3		10	
4			at Discretion, *&c.*

The	Rate 2 Story		
1		9	Foot.
2		10	
3		$10\frac{1}{2}$	
4			at Discretion, *&c.*

T	Rate 3 Story		
1		9	Foot.
2		9	
3		$9\frac{1}{2}$	
4			at Discretion, *&c.*

The	Rate 4 Story		
3		$8\frac{1}{2}$	Foot high.
4			at Discretion, *&c.*

As to Sir *Henry*'s 2d. Point of *Contrivance*, viz *Usefulness*, which will consist in a sufficient Number of Rooms, of all sorts, and in their due and apt Coherence without Distraction or Confusion, so as the Spectator may not only call it *Una Fabrica ben racolta*, (as the *Italians* used to say of well united pieces of Work) but likewise that it may appear Airy and Spiritous, and fit to welcome chearful Guests, about which the principal Difficulty will be in Contriving of the *Lights*, and *Stair-cases*, whereof I will give you a Note or two: For the

the firſt, I obſerve that the Ancient Architects were at much eaſe ; for both *Greeks* and *Romans*, (of whoſe private Dwellings *Vitruvius* hath left us ſome Deſcription) had commonly two *Cloyſtered* open *Courts* one for the Womens ſide and the other for the Men; who perhaps would now take ſuch a Separation unkindly However, by this means they had a good Conveniency, to admit Light into the Body of the Building, both from without, and from within, which we muſt now ſupply by ſome open Form of the Fabrick , or (among other graceful Refuges) by Terraſing any Story, which is in danger of being too dark; laſtly, by Perpendicular Lights from the Roof which are the moſt natural of all others. As to the ſecond Difficulty, *viz* Contriving of the *Stair caſes*, which is no hard Point in it ſelf, the only thing in contriving them, is to make them handſome, convenient, and in as little Room as may be, that they be no hindrance to any other Room, or Rooms I have, (ſaith Sir *H. W.)* obſerved that the *Italian* Architects, are inclined to place the *Kitchin*, *Bake houſe*, *Pantry*, *Waſhing-rooms*, and the Buttery likewiſe under Ground, level with the Cellar-floor, raiſing the firſt Aſcent 15 Foot, or more up into the Houſe; by which Method, beſides removing of Annoyances out of ſight, and having thereby much more room above, it doth alſo by the Elevation of the Front, add Majeſty to the whole Aſpect, and with ſuch a Diſpoſition of the Principal *Stair-caſe*, which commonly doth deliver us into the Plain of the ſecond Story, where Wonders may be done with a little Room : (I have obſerved, that they commonly place all their Rooms for Office, about five Foot under Ground at *Tunbridge-wells*, the firſt Stories being about 8 Foot, and then the Lights or Windows of them, be juſt above the Groud without ; but then you muſt note that thoſe Houſes always ſtand upon an Aſcent, that they may have good Sewers to keep theſe lower Rooms dreín'd dry from Water.) But the petty Offices, (ſaith Sir *Henry*) may be well enough ſo remote in *Italy*, yet by the natural Hoſpitality of *England*, the *Buttery* muſt be more viſible, and we have occaſion for larger Ranges, or Chimneys, and more ample *Kitchins* than the *Italians*, or than perhaps the aforeſaid Compartition will bear ; and likewiſe not ſo remote from the Dining-room, or elſe (ſaith he) beſides other Inconveniences, perhaps ſome of the Diſhes may ſtraggle by the way.

Here (ſaith Sir *Henry*) let me note a common Defect that we have in our *Engliſh Buildings*, *viz.* The want, or neglect of a very uſeful Room, call'd by the *Italians*, *Il' Tinello*, 'tis very frequent, nay almoſt eſſential in all their great Families.

'Tis a place properly appointed for a Conſervatory of the Meats that are taken from the Table, 'till the Waiters are ready to eat, which with us us is (according to an old fashion) more unseemly set by, in the mean time.

Now touching the Distribution of *Lodging-chambers*, I must here presume to reprove an odd Custom they have in *Italy*, without any ancient Precedent, as far as I can learn by *Vitruvius*.

Namely, That they so contrive their Partitions, as when all the Doors are open on a Floor, one may see through the whole House, which doth necessarily put an intolerable Servitude upon all the Chambers, except the inmost, where none can arrive but through the rest; or else the Walls must be extream thick for secret Passages, and yet this will not serve the turn without three Doors to every Room; a thing not to be born with in cold and windy Regions, and every way no small weakning to the Work: This Custom I suppose to be grounded upon a fond Ambition of displaying to Strangers all their Furniture at one View.

There is likewise another Defect, (for Absurdities are seldom solitary) which will follow by consequence, upon such a servile disposing of the *Inner chambers*. That they must be forced to make as many common great Rooms, as there shall be several Stories, which (besides that they are usually dark, a thing hardly to be avoided, running as they do quite through the House) do likewise devour so much Place, that thereby they want other Galleries and Rooms of Recreation, which I have often consider'd amongst them.

Having thus given some general Hints and Directions, and detected some Faults, the rest must be committed to the Sagacity of the Architect, who will be often put to diverse ingenious Shifts, when he is to wrestle with Scarcity of Ground.

As sometimes to dam one Room (the *Italians* call it *Una Stanza dannata*, as when a *Buttery* is cast under a Stair-case, or the like;) altho' of great use for the Beauty and Benefit of all the rest; at another time to make those fairest which are most in sight, and to leave the other (like a cunning Painter) shadowed. I will close this part (saith Sir *Henry*) of Compartition, with a short Description of a *Feasting or Entertaining room after the Egyptian manner*, who seem (at least till the time of *Vitruvius*) from the ancient *Hebrews*, and *Phœnicians* (whence all Knowledge did flow) to have retain'd with other Sciences in a high Degree, also the Principles and Practices of this magnificent Art. For as far as I can learn, and conjecture by *Vitruvius*, *lib. 6. cap 5.* there was no Form for such a Royal Use, comparably imagined like that of the aforesaid Nation; which I shall now proceed to explain.

Let us conceive a Floor, or Area of a good length (e. g at least 120 Foot) with the Latitude somewhat more than $\frac{1}{2}$ the Longitude (the Reason whereof shall be in its due place given) along the two sides

ſides and head of the ſaid Room ſhall run an Order of Columns, or Pillars, which *Palladio* doth ſuppoſe Corinthian ones, ſupplying that Point out of *Greece*, becauſe we know no Order proper to *Egypt*.

The other Head, or fourth Side, I will leave free for Enterance: on the aforeſaid Pillars was laid an Architrave, which is only mentioned by *Vitruvius*; *Palladio* adds thereto (and not without Reaſon) both Frieſe, and Cornice, over which went up a continued Wall, and therein $\frac{1}{2}$ or $\frac{3}{4}$ Pillars, anſwering directly to the Order below, but $\frac{1}{4}$ part leſs, between theſe $\frac{1}{2}$ Columns above, the whole Room was Windowed round about.

Now from the loweſt Pillars, there was laid over a Contignation, or Floor, born up on the outward Wall, and the Head of the Columns with Terrace and Pavement, *ſub dio*, ſaith *Vitruvius*, and ſo indeed he might ſafely determine the Matter in *Egypt*, where they fear no Clouds; therefore *Palladio* (who leaveth this Terrace uncovered in the middle, and Balliſter'd about) did perhaps conſtrue him rightly, tho' therein differing from others. We muſt underſtand a ſufficient breadth of Pavement left between the open part and the Windows, for the Pleaſure of the Spectators that look down into the Room. The Latitude I have ſuppoſed a little more than $\frac{1}{2}$ the length, becauſe the Pillars ſtanding at a competent diſtance from the outer wall, will by Interception of the Sight, ſomewhat diminiſh the breadth in Appearance; in which Caſe Diſcretion may be more Licentious than Art. This is the Deſcription of an *Egyptian* Room for Feaſts, and other Jollities. About the Walls whereof, we muſt imagine intire Sratues placed below, and illuminated by the deſcending Light from the Terraſs, and likewiſe from the Windows between the half Pillars above; ſo that this Room had abundance of Light, and beſides other Garniſhings, it muſt needs ſeem very ſtately to the heighth of the Roof that lay over two Orders of Columns.

Having thus far conſider'd of the Lower parts of the Building, the Houſe may now have its Hat put on; which point, tho' it be the laſt in this Art in Execution, (of any part of the bare Shell of the Houſe) yet it is always the firſt in Intention; for none would build but for Shelter: I ſhall now only deliver a few of the propereſt, and naturalleſt Conſiderations belonging to the *Roof*.

There are two Extreams to be avoided in the *Cover* or *Roof* of a Houſe, *viz.* That it be not too heavy, nor too light; the firſt will be objected againſt, for preſſing too much the under Work; the other contains a more ſecret Inconveniency; for the *Cover* or *Roof*, is not only a bare Defence, but likewiſe a kind of Band, or Ligature to the whole Fabrick, and therefore will require ſome reaſonble weight; but of the two a Houſe top heavy is the worſt: Next,

Next, there muſt be Care taken to contrive an Equality of the Preſſure of the *Roof* upon all the parts of the Edifice, *viz.* As much on one ſide, as it doth on the other. And here *Palladio*'s Advice is very good, which is this, *viz.* That the inward Walls may take their ſhare of the Burden, and the outer ones be the leſs charged with it. *Thirdly*, The *Italians* are very careful in giving the *Roof* a graceful Pendency, or Slopeneſs, dividing the whole breadth of the Building into nine parts, two of theſe Diviſions ſhall be the Perpendicular to the *Roof*.

But in this Point the Quality of the Region is to be the Rule to walk by, as *Vitruvius* obſerveth; that thoſe Climates that are ſubject to great Snows, ought to have ſharper *Roofs* than other places, where they are not ſubject to the like Accidents; and in all places Comelineſs muſt yield to Neceſſity.

I will now add Dr. *T. F*'s general Maxims [for Contrivance in Building, which are as followeth, *viz.*

Let not the common Rooms be ſeveral, nor the ſeveral Rooms be common] (by which, I ſuppoſe he means (by what follows) that the common Rooms ſhould not be private or retired, nor the private Rooms common) The *Hall* (which is a Pandochæum) ought to lie open, and ſo ought Galleries and Stairs (provided the whole Houſe be not ſpent in *Paths*,) *Chambers*, and *Cloſets* ought to be private and retired.

Light (God's Eldeſt Daughter) is a principal Beauty in Building] Yet it ſhines not alike from all parts of the Heavens. An *Eaſt-window* gives the infant Beams of the *Sun* before they are of Strength to do any harm, and is offenſive to none but a Sluggard. A *South-window*, in Summer, is a *Chimney* with a Fire in it, and needs to be ſkreen'd by a Curtain. In a *Weſt-window*, in Summer-time, towards Night, the *Sun* grows low, and over familiar, with more Light than Delight. A *North window*, is beſt for Butteries, and Cellars, where the Beer will be ſowre, becauſe the *Sun* ſmiles on it. *Thorow Lights* are beſt for *Rooms of Entertainment*, and Windows on the ſide for Dormitories.

3. *Receipt.*] As for *Receipt, a Houſe had better be too little for a Day, than to great for a Year.* And 'tis eaſier borrowing of thy Neighbour a brace of Chambers for a Night, than a Bag of Money for a Twelve-Month. It is Vanity therefore to proportion the *Receit* to an extraordinary Occaſion; as thoſe, who by over building their Houſes, have dilapidated their Lands, and their Eſtates have been preſſed to Death under the weight of their Houſe.

Strength.] As for Strengh, *Country-houſes muſt be Subſtantives, able to ſtand of themſelves.*] Not like City buildings ſupported by their Neighbours, on each Side. By *Strength*, I mean ſuch as may reſiſt Weather and Time, and not Invaſion

ſion, Caſtles being out of Date in *England*, only on the Sea-coaſt. As to making of Motes round about a Houſe, 'tis a Queſtion whether the Fogs that ariſe from the Water, be not more unhealthful than the Fiſh brings Profit, or the Water Defence.

In working up the Walls of a *Building*, do not let any Wall be worked up above 3 Foot high, before the next adjoyning Wall be brought up to it, that ſo they may be joyn'd together, and make good Bond in the Work. For there is an ill Cuſtom uſed among ſome Bricklayers, to carry, or work up a whole Story of the Party Wall (meaning in *London*) before they work up the Fronts, or other Work adjoyning, that ſhould be bonded, or worked up together with them, which occaſions Cracks, and Settlings in the Walls of the *Building*, which weaken it very much.

Sometimes the Strength of a Building is much impair'd, in the erecting of it, by Reaſon the Maſter did not procure ſufficient Stuff, or Materials, and Money before he began to build; for when Buildings are erected by Fits and Pauſes, now a piece and then another, the Work dries, and ſinks unequally, whereby the *Walls* grow full of Chinks and Crevices; this pauſing Humour is condemned by all Authors.

5. *Beauty.*] *Let not the Front look a-ſquint on a Stranger, but accoſt him right at his Entrance.*] Uniformity and Proportion, much pleaſeth the Eye, and 'tis obſerved that Free-ſtone, like a fair Complexion, ſooneſt waxeth old, whilſt Bricks keep their Beauty longeſt.

Let the Offices (ſaith Dr. *T. F.*) *keep their due Diſtance from the Manſion Houſe.*] Thoſe are too familiar which preſume to be of the ſame Pile with it. The ſame may be ſaid of *Stables* and *Barns*, without which a Houſe is like a City without Works, it can never hold out long.

'Tis very inconvenient (and rather a Blemiſh, than a Beauty to a Building) to ſee the Barns and Stables too near a Houſe, becauſe Cattle, Poultry, and ſuch like muſt be kept near them, which are an Annoyance to a Houſe.

Gardens alſo are to attend in their Place. When God (*Gen.* 2. 9. planted a Garden *Eaſtward*, he made to grow out of the Ground every Tree, pleaſant to the Sight, and good for Food. Sure (ſaith the Dr.) He knew better what was proper to a Garden than thoſe, who now-a-days therein only feed the Eyes, and ſtarve both Taſte and Smell. Let the Garden (ſaith Mr. *Worlidge*) joyn to one, if not more Sides of the Houſe; for what can be more pleaſant and beautiful for the moſt Part of the Year, than to look out of the Parlour, and Chamber-Windows into Gardens? For Beauty, alſo let there be Courts or Yards kept from Cattle and Poultry, *&c.* and planted with Trees, to ſhade, defend, and refreſh your Houſe, and the Walls alſo planted with Vines, and other Wall-

Wall-fruit, all which will add Pleasure and Beauty to your Habitation.

6. *Form or Figure.*] *Figures* are either simple or mixt; simple *Figures* are either Circular or Angular; and of Circular, either Compleat or Deficient, as Oval: The Circle is an unprofitable Figure in private Buildings, being the most chargeable, and much Room is lost in the bending of the Walls, besides an ill Distribution of the Light, except from the Center of the Roof, so as it is not used, only in Temples and Amphitheatres. The Oval, and other imperfect Circular *Forms* are subject to the same Exceptions, and are less Capacious.

Touching the Angular *Forms* or *Figures*, it is a true Observation, That this Art loves neither many Angles, nor few; first, The Triangle, which hath of all the others, the fewest Sides and Corners, is of all others the most condemned, being indeed both incapable and infirm, and not easily reduced into any other Form, but that of it self in the inward Partitions: As for *Figures* of 5, 6, 7, or more Sides and Angles, they are fitter for Military than Civil Architecture; tho' there is a famous Piece at *Caprarola*, belonging to the House of *Farnese*, contrived by *Baraccio*, in the Form of a Pentagon, with a Circle inscribed, where the Architect did ingeniously wrestle with divers Inconveniencies in disposing of the Lights, and in saving the Vacuities, But such Designs as these aim more at Rarity than Commodity, and are rather to be admired than commended. And therefore, by the Precepts and Practice of the best Builders, we resolve upon Rectangular Squares, as a mean betwixt too few, and too many Angles, and thro the equal Inclination of the Sides (which make the Right Angle) stronger than the Rhomb, or any other Quadrilateral Figure; but whether the Quadrat, or Rectangle Parallellogram, be the better, is not yet well determined, tho' I prefer the latter, provided the Length do not exceed the Latitude above $\frac{1}{3}$, which would much diminish the Aspect.

Of mixt Figures, partly Circular and partly Angular; there is a proper Objection against them, *viz.* That they offend *Uniformity*. Of which (having here mentioned it) I will add something concerning *Uniformity*.

In *Architecture* there seems to be Two opposite Affectations *Uniformity* and *Variety*; yet these seeming Opposites may be very well reconciled; as we may observe in our own Bodies, the great Pattern of Nature; which is very Uniform in the whole Figuration, each Side agreeing with the other both in Number, Quality, and Measure of the Parts: And yet some are round as the Arms, others flat, as the Hands, some prominent and others indented or retir'd; so the Limbs of a noble Fabrick may be correspondent enough, tho' they

they be various, provided we do not run out into extravagant Fancies, when we are contriving how to part and caſt the whole Work. We ought likewiſe to avoid enormous Heights of ſix or ſeven Stories, as well as irregular *Forms*; and on the contrary, are low diſtended Fronts, as unſeemly; or again, When the Face of a Building is narrow, and the Flanks deep.

III. *Of the modern Way of Building in* England, *compared with the ancient.*] When I compare the modern *Engliſh* way of Building with the old way, I cannot but wonder at the Genius of old Times. Nothing is, or can be more delightful, and convenient than Height, and Nothing more agreeable to Health than free Air. And yet of old, they uſed to dwell in Houſes, moſt of them with a blind *Staircaſe*, low Ceilings, and dark Windows; the Rooms built at Random, (without any thing of Contrivance) and often with Steps from one to another. So that one would think the People of former Ages, were afraid of Light, and good Air; or loved to play at hide and ſeek. Whereas the Genius of our Times is altogether for light *Stair caſes* fine *Saſh-windows*, and lofty *Ceilings*. And ſuch has been, of late, our Builders Induſtry, in point of *Compactneſs* and *Uniformity*, That a Houſe after the new way will afford upon the ſame Quantity of Ground as many more Conveniencies.

The Contrivance of *Cloſets* in moſt Rooms, and *painted Wainſcot*, now ſo much uſed, are alſo two great Improvements, the one for Conveniency, the other for Cleanlineſs and Health: And indeed, for ſo damp a Country as *England* is, nothing could be better contrived, than *VVainſcot*, to keep off the ill Impreſſion of damp Walls. In ſhort, for handſome Accommodations, and neatneſs of Lodgings, *London* undoubtedly has got the Preheminence.

The greateſt Objection againſt *London-Houſes* (being for the moſt part Brick) is their Slightneſs, occaſioned by the Fines exacted by the Landlords. So that few Houſes, at the common Rate of Building, laſt longer than the Ground-Leaſe, and that is about 50 or 60 Years. In the mean Time, if there happens to be a long Fit of exceſſive *Heat* in *Summer*, or *Cold* in *VVinter*, the Walls being but thin, become at laſt ſo penetrated with the Air, that the Tenant muſt needs be uneaſie with it; but thoſe Extreams happen but ſeldom. And this way of Building is wonderful beneficial to Trades relating to it, for they never want Work in ſo great a City, where Houſes here and there are always repairing, or building up again.

The *Plaiſtered Ceilings*, ſo much uſed in *England*, beyond other Countries, make by their Whiteneſs the Rooms much Lightſomer, and are excellent againſt raging Fires. They ſtop the Paſſage of the Duſt, and

and lessen the Noise over Head; and in Summer-time the Air of a Room is something the cooler for them, and in the Winter something the warmer, because it keeps out cold Air then, better than the Board-floors alone can do.

IV. *Some general Rules to be observed in Building.*] These following *Rules* were establi-shed by *Act of Parliament*, before the Re-building of *London*

First, In every *Foundation* within the Ground, add one *Brick* in Thickness to the Thickness of the Wall, next above the Foundation, to be set off in three Courses, equally on both Sides.

Secondly, That no *Timber* be laid within 12 Inches of the fore-side of the *Chimney-jambs*.

Thirdly, That all *Joysts* on the back of any *Chimney*, be laid with a *Trimmer*, at 6 Inches Distance from the back.

Fourthly, That no *Timber* be laid within the *Funnel* of any *Chimney*, upon Penalty to the Workman for every Default 10 *s*. and 10 *s*, every Week it continues un-reformed.

Fifthly That no *Joysts*, or *Rafters* be laid at greater Distances from one to the other, than 12 Inches; and no *Quarters* at greater Distance than 14 Inches.

Sixthly, That no *Joysts* bear at longer Length than 10 Foot; and no single *Rafters* at more in length than 9 Foot.

Seventhly That all *Roofs*, *VVindow-frames* and Cellar-floors be made of Oak.

Eighthly, That *Tile pins* be of Oak

Ninthly, That no *Summers* or *Girders* in Brick-Buildings, do lie over the Heads of *Doors* or *VVindows*.

Tenthly, That no *Summers* or *Girders*, do lie less than 10 Inches into the Brick-work; nor no *Joysts* less than 8 Inches, and that they be laid in Lome.

Some also advise, That all *Tarsels* for Mantle trees to lie on, or *Lintels* over Windows, or *Templers* under Girders, or any other Timber that must lie in the Wall to lay them in Lome, which is a great preserver of Timber; but *Mortar* eats and corrodes it. Some Workmen pitch the Ends of Timber that lie in the Walls to preserve them from the Mortar.

V. *Of Surveying of a Building*] I will here briefly touch upon the Method of *Surveying* of *Buildings*; by which the Manner and Form of taking Dimensions may be seen, which take as followeth.

A

A Survey *of a* Building *Erected by* R. M. *for* R. S. *the Thickneſs of the* Walls (*as by Agreemenet*) *Brick and* ½, *at* 3 l. per *Rod, for Mortar and Workmanſhip ; the Dimenſions were taken as followeth.*

	Foot. Parts.	
1. The length of one ſide. From the Foundation of the Raiſing.	40—50 16— 0	648—
2. The breadth at one end. The heighth to the Croſs-beam.	17—16 16—50	283—14
3. A Partition-wall within. Height to the firſt Story.	17—16 10—50	180—18
4. The length of the other ſide. From an old Wall to the Raiſing.	39—33 7—00	275—31
5. The breadth at the other end. From the Floor to the Croſs-beam.	17—00 4—83	82—11
6. A Water Table 30 Foot reduced to From the Foundation to the Table.	7—50 3—16	23—70
7. A Setting off on the other ſide of the Houſe.	16—83	16—83
8. A Gable-end.	66—00	66—00
The Total Area, or Content of theſe Dimenſions.		1575—27

Parti-

Particulars to be deducted.

		F.	P.	
1. One Door-case,	Broad	8	66	81—58
	High	9	42	
2. Another Door-case,	Broad	4	33	32—13
	High	7	42	
3. A Third Door-case,	Broad	4	33	22—34
	High	5	16	
4. A Window-case,	Broad	4	50	20—25
	Deep	4	50	
5. Another Window-case,	Broad	4	5	20—25
	Deep	4	5	
The Total of these Deductions.				176—55
Taken from the whole Content.				1575—27
There rests due to the Bricklayer.				1398—72

Which reduced into square Rods, is 5 Rods 38 Foot.

And then according to the Contract, there will due to the Bricklayer 15 *l*—8 *s.*—3 *d*

Thus far Mr. *Leybourn*: We will now see Mr. *Ven. Maudey*'s Method of Surveying Buildings, and taking Dimensions; and setting them down in a Pocket-book.

2. *Note*, Before you begin to set down your Dimensions, it is convenient to divide the breadth of the Page or Feet, into so many several Columns as you think you shall have Occasion for; either with Lines drawn with Ink, or a Pencil; your Pocket book being about 4 n. broad, (which is one of the broader sized Pocket books) you may then divide a Leaf into four Columns.

3. Before any Dimensions are set down, the Work-master's and Work-men's Names ought to be expressed; also the place where, the Day of the Month, and Date when you measure. I will suppose, for Example, That you were to measure Glazing; then observe if it were Glazed with Square-glass, you must write Squares above the Dimensions, and over those Dimensions which are appertaining to Quarry-glass (if there be any) you must write Quarries, that when you come to make the *Bill of Measurement*, you may express them severally, because they are of several Prizes.

4. The

4. The better to explain the Method, I will here insert a *Bill of Measurement of Glazing.*

Glazing *done for* C. D. *in* Long-acre, *by* T. G. *of St.* Martin's *in the Fields; Measured* October *the 19th.* 1725.

Quarries. F. I. P.		Products. F. I. P.	Squares. F. I. P.		Products. F. I. P.
5—8—6 5—7—3		31—11—30	04—03—00 01—02—00		04—11—06
5—3—6 2—4—6		12—06—09	02—00—00 01—06—00		03—00—00
2—6—0 1—2—0	(3)	08—09—00	06—00—09 05—00—03		30—05—03
2—1—0 1—8—6	(2)	07—02—04	01—02—00 03—00—00	(2)	07—00—00
		60—5—11			45—04—09

An Explanation of the Columns.

In the *first* Column towards the Left-hand, are the Dimensions of Glazings done with Squares; which you are taught to cast up in *Cross Multiplication* which see N. 2

In the *second* Column you have the Product of each Dimension just against it.

In the *third* Column you have the four Dimensions of Glazing done with Squares.

In the *last* you have the Product of each Dimension just against it also.

At the bottom of the *second* Column, you have the Sum Total of the Products of the Dimensions done with Quarries, which is 60 Feet, 5 Inches, and 11 Parts.

Also at the bottom of the *last* Column, there is the Total Sum of the Products of those Dimensions of the Glazing that was done with Squares, being 45 foot, 4 n. 9 p. As for the odd parts it signifies but little if they are left out in the Sum Totals of a *Bill of Measurement*, for it will amount to but very little in value.

5. *N. B.* When you are a taking of Dimensions, and setting them down in your Pocket-book, whether it be Glazing, or any other Tradesman's Work; you must observe to leave every other Column vacant, that so having set down all your Dimensions in your Book

Book (which muſt be generally done, before any is caſt up) when you come to caſt them up) which muſt be in another Book, or a Sheet of Paper) you may enter the Product of each pair of Dimenſions, juſt againſt them, as you ſee before.

6. If there be another to meaſure againſt you, and there ſhould be a Miſtake in either of your caſtings up of the Dimenſions (as it often happens thro' Security or Negligence) then one by reading over the Dimenſions in his Book with the Product to each Dimenſion as he goes on, and the other looking in his own Book, the Miſtake will ſoon be diſcovered, which muſt be rectified between you.

Therefore to be certain in caſting up your Dimenſions, you ought to caſt them up twice, if not three times, *viz.* After you have caſt them all over once, begin and caſt them over again, and ſee whether it agrees with your firſt caſting up, if not, then caſt up again. When you make your Bill of Meaſurement, your muſt ſet your Name to it at the lower end of the Bill.

7. An Example of a *Bill of Meaſurement.*

Glaziers *Work done for* G. D. *in* Long-acre, *by* T. G. *of St.* Martin's *in the Fields; Meaſured* October *the* 19*th*, 1725.

	l.	s.	d.
For 60 Foot, 5 Inches of Glazing, done with Quarries, at 5 *d. per* Foot.	01	05	1¼
For 45 Foot, and 4 Inches of Glazing, done with Squares at 7 *d. per* Foot.	01	06	2¼
Sum Total	2	11	3½

Meaſured the Day and Year above written, by T. S.

For the Satisfaction of the Curious, I will ſhew the Method of taking the Dimenſions of *Bricklayers work* which is the moſt troubleſome of any Mechanick's Work (relating to Building) to meaſure

8. Altho' I ſaid before, that you might divide a Page or Leaf of your Meaſuring-book into four Parts, or Columns; yet in meaſuring of *Briklayers-work*, it will be neceſſary to divide a Page into but three Columns, one large one for the Apellations, two ſmaller, one for the Dimenſions, the other for the Products.

As

As in this Example.

Appellations.	*Dimensions.*		*Products.*
Basis of the Front and Rear	3 B. f. n. 25—00 00—06	(2)	3 B. f. n. 25—00
Front and Rear	2 ½ B. 25—00 11—00	(2)	2 ½ B. 550—00
Basis of both the Flank-walls	2 ½ B. 36—02 00—06	(2)	2 ½ B. 36—02
Both the Flanks	2 B. 36—02 11—00	(2)	2 B. 795—08
The Wall between the Chimney	1 ½ B. 11—06 09—10	(2)	1 ½ B. 113—01
The falling back of both Chim	1 B. 05—00 04—00	(2)	1 B. 40—00
The four Jambs	2 B 14—00 11—00	(2)	2 B. 161—00
The fore part, or Brests of both Chimneys	2 B. 11—06 5—00	(2)	2 B. 115—00

9. The

9. The *Dimensions* with their *Products*, being set down, in the next place the *Deductions* of the Windows and Doors must be put down, and their *Products*. V. P. of Dimensions in *Brick-work*.

Deductions.

	Deduct. $2\frac{1}{2}$ B. f. n.		Prod. $2\frac{1}{2}$ B. f. n.
The 4 Windows	{06—06 / 04—00}	(4)	104—00
	$2\frac{1}{2}$ B.		
The 2 Doors	{09—00 / 04—00}	(2)	72—00

10. The next Thing in order to be done is, to add the *Products* of each several thickness into one Sum.

The Products of the several Thicknesses.

3 B.	$2\frac{1}{2}$ B.	2 B.	$1\frac{1}{2}$ B.	1 B.
25—00	550—00	795—08	113—01	40—00
	36—02	161—00		
		115—00		
	586—02	1071-08		

The several Products of each Thickness being added.

In the first Column on the Left-hand there is 25 Feet of 3 B.

In the second 586,2. of $2\frac{1}{2}$ B. *&c.*

How to find these Products, see cross Multiplication. N. 2.

Having found the total Sum of the Products of the Deductions; each total Sum must be substracted from the total Sum of the Products of the Dimensions that are of the same Crassitude.

E. G. the *Products* of the *Deductions* in

$2\frac{1}{2}$ B.
104—00
73—00

The Total Product in $2\frac{1}{2}$ B. is } 176—00

Which 176 Feet of $2\frac{2}{1}$ B. Work, being contain'd in the Windows and Doors; must be Subtracted from the 586 Feet, and 2 Inches, being the Total Product of all the Dimensions of that Crassitude, *viz.* $2\frac{1}{2}$ B: *viz.* $2\frac{1}{2}$ B. Work.

This

This is manifeſt to Reaſon. Becauſe when the *Dimenſions* of the Front and Rear were taken, the whole Length and Breadth was taken over the *Doors* and *Windows*, allowing Nothing of Abatement for them.

12. *N. B* That whatſoever *Doors*, or *VVindows*, or any other Vacancies, are meaſured over when the *Dimenſions* are taken; you muſt remember to make Deduction for them out of the Total Product of the *Dimenſions* of the ſame Craſſitude wherein they were ſituated.

To make it the plainer, take the following

Example.

The *Doors* and *VVindows*, being in 2 ½ Brick-work, I ſet down the Total Product of all the *Dimenſions* of that Craſſitude which is 586—02

which is	586—02
The Total Product of all the *Deductions* of that Thickneſs, which are to be ſubtracted, is	179—00
The Remainder is	410—02

The like Method muſt have been practiſed, if there had been any other *Deductions* in any other Craſſitude: Subtraction muſt have been made of all ſuch *Deductions*, out of the *Products* of the *Dimenſions*, before you went to reduce your ſeveral Craſſitudes to the Standard-thickneſs of 1 ½ B.

But ſeeing we have no other *Deductions* in our preſent Example of a Survey; the next Thing to be put in Practice, will be to reduce the ſeveral Craſſitudes to the cuſtomary Thickneſs of 1 ½ Brick. But I think it neceſſary to refer it to Walls, under the Head of meaſuring them.

More of this Nature, *viz.* Of Surveying *Buildings*, or taking *Dimenſions*, &c. may be ſeen under the different Heads of *Carpenters* Work, *Joyners*, *Bricklayers*, *Plaiſterers*, *Maſons*, *Painters*, *Thatchers*.

VI. *Of meaſuring Buildings.*] Having briefly treated of taking *Dimenſions*, &c. I ſhall here add but little under this Head of *Meaſuring*; only mention the Artificers relating to *Building* that uſually work by Meaſure; which are firſt *Bricklayers* ſecond *Carpenters*, third *Plaiſterers*, fourth *Painters*, fifth *Glaziers*, ſixth *Joiners*, and ſeventh *Maſons*. Some of theſe work by the ſuperficial Yard, ſome by the Rod, ſome by the Square, and ſome by the Foot: Of all which Works the *Dimenſions* are taken either with a 10 Foot Rod, or a 5 Foot one, or with a 2 Foot Rule, and ſometimes with a Line. But let the Dimenſions be taken how they will, they are uſually ſet down in Feet, Inches, and parts of Inches; or elſe in Feet and Centeſimal Parts of Feet, which laſt way is eaſieſt caſt up; the following Table will ſhew the Centeſimals.

A Table

A TABLE *of Centeſimal Numbers, for every Inch, and quarter of an Inch in a Foot.*

Inches.	100 P. Foot	1 quarter of an Inch. 100 P. Foot.	2 quarters of an Inch. 100 P. Foot.	3 quarters of an Inch. 100 P. Foot.
0	. 00	. 02	. 04	. 06
1	. 08	. 10	. 12	. 14
2	. 16	. 18	. 2	. 22
3	. 25	. 27	. 29	. 31
4	. 33	. 35	. 37	. 39
	. 42	. 44	. 45	. 47
6	. 50	. 52	. 53	. 55
7	. 56	. 60	. 62	. 64
8	. 66	. 68	. 7	. 72
9	. 75	. 77	. 79	. 81
10	. 83	. 85	87	. 9
11	. 92	. 94	. 96	. 98
1 Foot.	100			

To ſet any number of Feet, Inches and Parts, as ſuppoſe 30 Feet, 8 Inches and 2 Quarters, you muſt firſt ſet down 30 Feet with a Period, or Comma after it, thus, 30, and then look in the firſt Column of the Table for 8 Inches, and at the Head of the Table for 2 quarters, and then againſt 8 Inches, and under 2 quarters you will find 7; which ſet down beyond the 30 to the Right-hand, and it will ſtand thus, 30. 7.

I ſhall not here ſtay to treat of the Method of Meaſuring all theſe Artificers Works, becauſe they will be all treated of under their proper Heads. But ſhall now proceed to ſpeak ſomething.

VII. Of *Valuing Buildings*.] To eſtimate the Charge of Erecting any Houſe near the Truth, or to value one that is already built, ſo that you come ſomething near the Matter, (provided it be built of Brick and Timber, as they uſually build in *London*, and Gentlemen in the Country) there muſt be given,

(1.) The Dimenſions thereof, not only in length and breadth, but alſo in heighth, in reſpect of the Number of *Stories*: For,

(2.) By the length and breadth, the quantity of Squares upon each Floor may be found, and alſo the Squares of Roofing in the Carpenter's-work, and alſo Tyling in the Healers, or Bricklayer's-work. And,

(3.) By the heighth, one may give a near Eſtimate of the Rows of *Brick work*, contain'd in the Walls round about, and in the *Partition-walls*, if there be be any; and alſo in the *Chimneys*: Then,

(4.) Conſider how many pairs of *Stairs*, and of what ſort.

(5.) What Partitions of Timber with Doors.

(6.) What Timber Front.

(7.) What Number of Window-frames, and Lights.

(8.) What Iron-work; and

(9.) What Lead, *&c*

Of all which, ſee the particular Heads.

Now ſaith Mr. *Leybourn*, what will be the Charge of Erecting a Fabrick of Brick-walls and Timber, which ſhall be 20 Foot in Front, and 44 deep, (which is the Method in *London* (and I believe in moſt Cities, *&c.*) for the Front to be ſhorter than the Flanks) and to conſiſt of Cellars, three Stories, and Garrets; which is one of the ſecond Rate Houſes. We will ſuppoſe the Price of Materials to be as followeth, (in *London*,) viz.

		l. *s.* *d.*
For	Bricks *per* Thouſand	00—16—00
	Tiles *per* Thouſand	01—05—00
	Lime *per* Hundred	00—10—00
	Sand *per* Load	00—03—00
	Oak, or Fir Timber *per* Load.	02—15—00
	Deal-boards *per* Hundred	07—10—00
	Laths *per* Bundle.	00—01—06

As for the Plaiſterers Work.

For	Lathing, Plaiſtering, Rendring and Waſhing with White and Size, *per* Yard.	00—01—02
	Lathing and Plaiſtering, *per* Yard.	00—00—10
	Plaiſtering and Sizing, *per* Yard.	00—00—60

Smiths Work.

For	Iron Balconies, *per* lb.	00—00—05
	Folding Caſements, *per* Pair.	00—16—00
	Ordinary Caſements, *per* Piece.	00—04—06

For Painting.

Window Lights	00—00—06
Shop Windows, Doors, Pales, *per* Yard.	00—01—00

Now ſaith he, from theſe Rates of Materials for Building, and for Workmanſhip: Such a Houſe as is here propoſed, will amount to about 360 Pounds, which is near 41 *l.* *per* Square.

Now

Now Mr *H Phillips*, proposes the following Method to find the value of a *Building*, viz. Suppose, saith he, a House, which is 1 Rod or 16½ f. in Front, and 2 Rods deep back in the Flanks; the Compass of this House will be 6 Rods, and if this House stands in a high Street, having a Cellar 4 Stories, and a Garret, (which is one of the third Rate Houses) the heighth thereof will be 50 Foot, or 3 Rods; so that there will be 18 Rod of *Brick-work* in the Walls; which may be all reduced to a Brick and ½ thick, and suppose each Rod of *Brick-work* to contain 4500 of Bricks, and will cost about 7 *l.* the Building, *viz.* Bricks, Mortar, and Workmanship; then the whole 18 Rods of *Brick-work* will cost about 126 *l.* The Timber-work for Floors, Windows, Roofs, *&c.* about as much more; the Tiling, Plaistering, Lead, Glazing, and Painting, will be about as much more so that the whole will amount to 378 *l.* the Allowance for the *Party walls* will very well pay for the *Chimneys.* So that this House cannot amount to above 400 *l.* the Building, which is not full 3 *l. per* Square; but this is a very great Price in comparison of Mr. *Leybourn's*, but he saith that it will be worth more or less, according to the Market-price of the Materials.

The *Friendly Society* of *London*, for Insuring of Houses, have two Rules by which they value *Houses, viz.* either by the Rent, or number of Squares contain'd on the Ground plot. This last is the general Rule by which they value all *Buildings*, which is grounded on an Act of Parliament for the Rebuilding the City of *London*, made about *Anno* 18 *Car.* 2.

The *Buildings* of the City of *London* are valued according to their *Rates*, of which Rates there are four, *viz.*

1st. Rate	2	Stories, Cellars, and Garrets.
2 ——	3	
3 ——	4	
4 ——	5	

And the naked *Building*, or Shell of a Brick-house (the Floors being finished) is thus valued by the Square or 100 Foot, if in high Streets, *viz.*

1st. Rate at	25 *l*	*per* Square.
2 ——	35 *l*	
3 ——	45 *l*	
4 ——	50 *l.*	

But these Rates may be augmented at the Discretion of the Surveyor, or according to the finishing of the *House.* I shall not now insist any longer on this Head, but defer it till a better opportunity; see some thing more of this nature: *House.*

VIII. Of

VIII. *Of Cenſuring Buildings*] I ſhall here draw towards a Concluſion of this Head, with Sir *Henry Wotton*'s Methodical Directions, how to Cenſure, or paſs ones Judgment on Fabricks already erected; for without ſome way to contract ones Judgment, it will be confounded amongſt ſo many Particulars as a Building conſiſts of. I ſhould therefore think it to be (almoſt) a harder Task, to be a good *Cenſurer*, than a good Architect: Becauſe the *working* part may be helped by Deliberation, but the *Cenſuring* part muſt flow from an Extemporal Habit. Therefore (not to leave this laſt Part or Head without ſome Light) I could wiſh him that comes to examine any noble (or great) Work; firſt of all to examine himſelf, whether the ſight of many fine Objects, which he hath ſeen before, (which remain like impreſſed Idea's on the Brain) hath not made him apt to think that nothing is good, but that which is the beſt: to be thus affected, would reliſh too much of a Cynical Humour

Next, before he comes to give his Opinion concerning the *Edifice* let him endeavour to inform himſelf preciſely of the Age of the *Fabrick*, upon which he is to give his Sentiments or Opinion. And if he ſhall find the viſible Decays to exceed the Proportion of time it hath been built, he may then ſafely conclude, without making farther inquiry, that the *Materials* were bad, and too ſlight, or the Seat is nought, *viz*. Poſited on a bad Soil for Foundation, or expoſed too much to a Tempeſtuous Corroſive Air, *&c.*

Now after theſe Premiſes, if the Houſe be found to bear his Years well, (which is always a Sign of a good Conſtitution.) Then let him ſuddenly run backwards (for the Method of *Cenſuring* is contrary to the Method of *Compoſing*) from the *Ornaments* (which firſt allure the Eye) to the more Eſſential *Members*, till at laſt he be able to form this Concluſion, That the Work is *Commodious*, *Firm* and *Delightful*: which are the three capital Qualities in good *Buildings*. And this (as I may term it) is the moſt Scientifical way of *Cenſuring*.

But there are two other things which I muſt not forget. The firſt, (out of *Georgio Vaſſari*'s Preface before his Laborious Work of the Lives of *Architects*) is to paſs a running Examination over the whole Edifice, according to the Properties of a well ſhapen Man; as whether the *Walls* ſtand upright, upon a good Foundation; whether the Fabrick be of a comely *Stature*; whether for the *breadth* it appear well *burniſhed*; whether the principal *Entrance* be in the middle of the *Front*, like our Mouths; whether the *Windows*, as our Eyes, be ſet in equal number, and diſtance on both ſides the *Entrance*, *&c.* Whether the *Offices*, like the Veins in our Bodies be conveniently diſtributed, *&c.* For this Allegorical Review, may be driven as far as any Wit will, that is at leiſure.

The

The ſecond way, is in *Vitruvius, Lib.* 1. *Cap.* 2. Where he briefly determines ſix *Conſiderations*, which accompliſh this whole Art *viz.* 1. *Ordinatio* 2. *Diſpoſitio*. 3. *Eurythmia* 4. *Symmetria*. 5. *Decor*. And 6 *Diſtributio*.

The two firſt of theſe may be very well ſpared or omitted; for as far as I can perceive, either by his Interpreters, or by his own Text, (which in that very Place, where perhaps he ought to have been cleareſt, he is moſt obſcure) he meaneth by *Ordination*, nothing but a well ſettling of the *Model* or *Scheme* of the whole Work. Nor by *Diſpoſition*, he meaneth no more than a neat and full Expreſſion of the firſt Idea or Deſign thereof, which perhaps does more immediately belong to the Artificer, rather than to the Cenſurer. The other 4 are ſufficient to condemn or approve (or abſolve) any Fabrick whatever. Wherefore

Eurythmia is that agreeable Harmony between the Breadth, Length, and Heighth of all the Rooms of the *Fabrick*, which is very pleaſing to all Beholders; which is always ſo to all, by a ſecret Power, which is in Proportion: where, let me Note this, That tho' the leaſt Error or Offence that can be committed againſt the Sight, is Exceſs of Heighth, yet that Fault is no where of ſmall Importance, becauſe it is the greateſt Offence againſt the Purſe.

Symmetria, is a due Proportion of each Part in reſpect of the whole; whereby a great *Fabrick* ſhould have great *Apartments* or *Rooms*, great *Lights* or *VVindows* great *Entrances* or *Doors*, great *Stair-caſes*, great *Pillars* and *Pilaſters* &c. In fine, all the Members and Parts large, proportionable to the *Building*. For as it would be but an odd Sight to ſee a great Man with little Legs, Feet, Arms, Hands, *&c.* So alſo it would be undecent to ſee a great *Fabrick* conſiſt of little *Apartments*, *Lights*, *Stair-caſes*, *Entrances*, &c. So on the other hand, as it is ſtrange to ſee a little Man poſſeſt of great Members, and Limbs, it is alſo very unſeemly to ſee a *Fabrick* that is but ſmall to be contrived into great *Rooms* to have great *Stair caſes*, large *Entrances*, *Lights* &c But again, as it is an unbeſeeming Sight, to ſee either a great or little Man to have ſome of his Members proportionable to his Body, and other ſome ſo large, as if they did belong to a Monſter or Giant, and not to him to whom they are affixt: So it is equally as ugly to ſee a little *Houſe* or *Edifice* to have ſome of its Parts monſtrous alſo, *viz.* great in ſome Parts of the *Apartments*, and by Conſequence others muſt be as little, or ſome muſt be annihilated; and ſo by Conſequence will be wanting; or great *Stair caſes*, great *VVindows*, great *Doors*, or any other Members larger than ought to be, in reſpect of the *Symmetry* of the Parts with the whole. It's likewiſe unſeemly to ſee ſome of the Parts too little, (that they are not proportiportionable to the whole Structure) as to see a Man have one very small Leg, and the other proportionable to his Body, or to have one little Eye like a Bird's, and the other like (what it should be) a Man's, *&c.* Thus we see there are many Errors which may be committed, (for want of a due Consideration, and Premeditation) in the Proportion of the Parts, *&c.* of a Building, and this is too often too well known to some Gentlemen that are Builders, when they light on Workmen that are unskilful in the Rules of *Proportion*, and the Theory of Architecture; and such Men very often run themselves into a grand Error; so that they are almost confounded in their Business, and know not well which way to wind themselves out of it again; and being thus wandered from the right Road (for want of this necessary Knowledge,) they still blunder on in the dark, not knowing a better way than their old inartificial one; till they have finished their Course, *viz.* Compleated, or rather (which is more proper) concluded, their *Building*; according to their way of working by Guess; and these guessing Workmen too often guess wrong, and commit many Faults, which are easily detected, by any one, that has but a little Skill in judging of a *Building's Symmetry*, *&c.* Besides their conceal'd Crimes, which such Men are very subject to commit, *viz* (I mean) their cutting, mangging, and spoiling their Master's Materials, when they work upon his Stuff, as they generally do in the Country: Sometimes you may observe in your Course of *Censuring*, a Door, or Chimney so misplaced, either to the right or left, so as to spoil the intended use of a Room, and tho' it be not totally spoiled, yet it seems as if the Artificers, (or Surveyor) wanted Instinct as well as Reason, that they could not contrive so well as *Beavers*, and other Brutes which build their own Houses, convenient for their Occasions. Perhaps sometimes you may observe a Chimney so situated in the Angle of a Room, (tho' design'd for Conveniency, because it could not well be carried up otherwise from the Chimney below it) yet this Chimney shall spoil the intended use (in some Measure) of two Rooms, (*viz.* That in which the Chimney is, and the next adjoining to the Chimney jamb) thus I knew one that did observe two Chambers in a great Measure thus spoiled; the Chimney was set in the Angle of the inner one, and the Door came into it from the Chamber without, just by one Jamb, the which Jamb (and by Consequence, that whole Chimney) was carried a Foot too far out in the Room, (which they might as well have carried farther the other way) and by this Means the Door was placed too far toward the other Wall of the Rooms, so that the Partition Wall by this Means was made so short betwixt the other

other Wall and the Door (at the Chimney-jamb) that it was thereby rendred unfit to place any Beds against it; although it was the most becoming Place for Beds in the Rooms: Sometimes little sneaking ill-contrived *Stair-cases* are built in a good comely large *Structure*, and then again on the contrary, some small (or at most but midling) Houses, shall have a *Stair case* so large in 'em, that if you were first to have seen the *Stair-case*, before you had seen any of the other Apartments you might well conjecture, That the Rooms of that Fabrick (to be in proportion with the *Stair-case*) to be twice, or three times as large again as they are. Nay, perhaps, this is not all the Error neither; for these random Workmen, do so manage the Matter, that they spoil the Conveniencies of Closets under them (or any other Conveniency) tho' it be now A-la mode to make some little Conveniences under a *Stair case*; for *Closets* are accounted an Improvement in our modern Way of *Building*: Sometimes you may observe an ill Position of *Lights* (or Windows) to a *Stair case*, not out of Necessity, but thro' want of Skill and Precaution. Again, as to Lights (or *Windows*) you may sometimes in your Viewings observe an ill Position (as well as an irregular Disposition in them), *viz.* Either in respect of Uniformity, or to secure them from the Weather (I mean by this last, when they are placed too near the Surface of the Building, that the Walls do not project beyond them, the better to carry the wet from them which run down by the Walls in stormy Weather. And then as to Uniformity in *placing* them, it sometimes so happens, that they cannot place the Windows on the Garrets exactly over those in the Stories below; and therefore, when they will not be brought into Uniformity with those below them, they ought to be placed as uniform as may be within themselves. What I am now speaking of, hath been observed in a *Fabrick*, which stood in the Form of a Roman Capital L, having two Fronts on the out-side of the L, confronting two Streets which cross'd one another at Right Angles; the Foot, or short part of the L in the *Fabrick*, was not so wide but it might be spann'd with one Roof, but the long part was too wide to be spann'd with one Roof (unless it had been carried up a vast deal higher than the other part, which would have then rendered it very unseemly) and therefore three Roofs were set on the long part of the L, parallel with that on the short part, so that there was three Gutters, and four Gable-heads on that Front which was the long part of the L, and in each of these Gable heads there was a Window; now, according to the Divisions of the Apartments in the Stories below, the Windows in them would not fit to be placed (none of them) perpendicular under the middle of these Gables, the Artificer think-

thinking to make it ſomething nearer to Uniformity, places ¾ of theſe Windows all towards (nay, very near) one ſide of the Gables, pretending they were nearer directly over the others, no doubt; and therefore, in his Judgment, it's nearer to Uniformity; whereas, in Truth, it's farther from it; for by this Means they are not in a uniform Poſition, neither in reſpect of thoſe in the Stories below them, nor yet within themſelves, which laſt they would have been, had they been placed in the middle of each Gable, and would have been more decent and comely, in reſpect of the Front without, and the Rooms within: For Lights being thus diſorderly poſited, makes it look all a-ſquint, and as deformed (almoſt) as to ſee a Man have an Eye in his Temples. Nay, I could inſtance in other Things, concerning placing of Lights, or Windows, *viz.* of miſplacing them, in reſpect of Altitude, like the Eyes of ſome Monſters (mentioned by *Peter Lampagneus*) which were ſeated in the Shoulders, and Breaſts, nay, I think, almoſt in the Knees, or Feet, as it hath been obſerved by ſome in ſome parts of ſome *Fabricks*. But theſe are the Effects and Blunders of working without the Aſſiſtance of forecaſting, and contriving according to Art. And laſtly, (for I will not trouble you any longer at preſent, concerning *Symmetry*) you ought to obſerve whether Doors have their due *Symmetry*, as to their Dimenſions, as well as Poſitions, *viz.* That they be not too high, as if they were for a Barn, nor too low, as if they were made for Houſes in *Sophia* in *Bulgaria*, where both *Jews* and Chriſtians have their Doors of their Houſes but a little above three Foot high, which is ſo contrived, becauſe the *Turks* ſhould not bring in their Horſes, which they would do, and make uſe of them for Stables in their Travels, if it was not for this Contrivance. In fine, my Meaning is, That theſe, as well as all other Parts of a *Building*, ought to be Analogous to the reſt of the *Fabrick*. I ſhall now proceed to ſpeak of

Decor.] It is the keeping of a due Reſpect between the *Inhabitant*, and *Habitation*. Whence *Palladio* did conclude, That the principal *Entrance* was not to be regulated by any certain Dimenſions, but by the Dignity of the Maſter; yet to exceed rather in the more, than in the leſs, is a Mark of Generoſity, and may always be excuſed with ſome noble Emblem, or Inſcription, as that of the *Contè di Bevilacqua* over his Gate at *Verona*, where 'tis likely there had been committed a little Diſproportion,

Patet Janua: Cor magis.

Diſtributio, is that uſeful Caſting (or Contriving) of all Rooms for *Office*, *Entertainment*, or *Pleaſure*, of which I have ſufficiently treated before under this Head of Building, *Numb.* II. §. 2.

Theſe

These are the four general *Heads* which every Man ought to run over before he pretends to paſs his Judgment upon the *Building*, or undertake to *cenſure* the Work that he ſhall view.

I ſhall conclude this laſt Head, touching *Ornaments*, againſt which (methinks) I hear an Objection, even from ſome well meaning Man; that theſe delightful *Crafts*, or *Arts*, may be various ways ill applied in a Nation. I muſt confeſs, indeed, that there may be a laſcivious, and likewiſe a ſuperſtitious Uſe, both of *Pictures*, and *Sculpture*: To which Poſſibility of Miſapplication, not only theſe *ſemi-liberal Arts* are ſubject, but even the higheſt Perfections and Endowments of Nature. As *Beauty* in a light Woman, *Eloquence* in a mutinous Man, *Reſolution* in a Ruffian, prudent *Obſervation of Hours*, and *Humours*, in a corrupt Courtier, *Sharpneſs of Wit* and *Argument* in a ſeducing Scholar, and the like. Nay, finally, let me but ask, What Art can be more pernicious, than even *Religion* it ſelf, if it be converted into an Inſtrument of Art. Therefore, *Ab abuti ad non uti, negatur conſequentia.*

IX. I ſhall finiſh this Head of *Building* with that Concluſion of Dr. *T. F.* In *Building* (ſaith he) rather believe any Man than an Artificer, in his own Art for Matter of *Charges*, eſpecially if either he, or any particular Friend of his, be like to be concerned about that *Fabrick* which you deſign to erect; not but that they can tell nearly the Charge, when they know the *Deſign* viz. Some ingenious Workmen, but there is but few of them in the Country, that truly underſtand the Method of valuing a *Building* and thoſe that do, if they are like to have any hand in it, it is very rare, if they will give a juſt Eſtimate of it, according to their Judgment; becauſe they think, if they ſhould acquaint the *Builder* with the full Coſt at firſt, it would blaſt a young Builder in the Bud, and therefore they ſooth him up till it hath coſt him ſomething to confute them

The *Spirit* of *Building* firſt poſſeſſed People after the Flood, which then cauſed the Confuſion of Languages, and ſince, of the Eſtate of many a Man. I have known ſome Perſons that would curſe thoſe with whom they were angry (if they were Men capable of it) by wiſhing, That the *pirit of Building* might poſſeſs them.

Butment.

Is a Term uſed by Maſons and Bricklayers, by which they mean the Supporters, or Props, on or againſt, which the Feet of Arches reſt.

Buttery.

1. *What.*] In Noble men's, and Gentlemen's Houſes, 'tis the Room belonging to the Butler, in which he diſpoſes all his Utenſils belonging to his Office, as his Napkin-preſs,

press, Table-cloth, and Napkins, Pots, Glasses, Tankards, Monteth, Cistern, Cruets, Salvers, Pepper-boxes, Sugar-box, Mustard-pot, Spoons, Knives, Forks, Voider, or Basket, and all other Necessaries appertaining to his Office.

2. *Of its Position.*] According to Sir *Henry Wotton*'s Rule, it ought to be placed on the *North* side of the Building, that is design'd for the Offices. It is generally with us in *England* placed near the Cellar, *viz.* the Room commonly just on the Top of the Cellar-stairs.

Bullen-nails.

1. *What.*] Are a sort of Nails with round Heads, and but short Shanks, Tin'd, and Lacker'd; I think there is about three Sizes of them.

2. *Their Use.*] They are used in hanging of Rooms, and fitting up of Beds, covering of ordinary Stools, Chairs, Couches, Desks, Coffins, *&c.*

Buttress.

1. *What.*] A Term in *Architecture*, used to signifie a Prop or Support, either of Brick, or Stone, set to keep the Work the firmer in its Position, as against Brick, or Stone-walls that are high, or have any considerable Weight against them on the other side, as a Bank of Earth, or the like; they are also used against the Angles of Steeples, Churches, and other Buildings of Stone, *&c.* on their out-side, and along the Walls of such Buildings, as have great and heavy Roofs, which would be subject to thrust the Walls out, if they were not thick, if there be no *Buttresses* placed against them; *Buttresses* are also commonly placed for a Support, and *Butment* against the Feet of Arches, that are turned cross some great Halls, in old Palaces, Abbeys, *&c.* And generally at the Head of Stone-buildings, where there are great Crocket-windows, there are for the most part, *Buttresses*, placed for *Butments* to the Arches of these Windows. As to the Theory of *Buttresses*, or *Butments*, I never yet could see any thing of it. 'Tis my Thoughts, An ingenious, sagacious, and prying Architect, that is well grounded in the Mathematicks (and would but take the Trouble upon him, throughly to examine this Matter) he might bring it within the Bounds of Reason, and Rules; whereby to know nearly of what Size, and (by Consequence) what Weight a *Buttress*, or *Butment* ought to be (which you must Note must be various,) according to the Dimensions, and Form of the Arch, and the Weight which is superincumbent on it. As to the Weight of the Materials, both on the Arch, and in the *Buttress*, or *Butment*, 'tis no Difficulty to calculate. But perhaps it may be here objected, there is a sensible Difference as to the Strength and Goodness of the Morter, which may in a great Measure compensate for

for the Weight of the *Buttrefs*, or *Butment*; for where there is a ftrong firm Mortar made ufe of, lefs Weight (or Magnitude) of Brick, or Stone, fhall be capable to refift the Preffure of an Arch, with its Superincumbent Materials, than where the Mortar is bad and feeble. To which I anfwer. I could very well experiment the Strength of Mortar, both as to a direct, or oblique Force, *viz.* Either fhoving it out of its Pofition, or pulling it the fhorteft way from its Adherents. by which I mean lifting directly up I think it may alfo be feafible (and I am fure it would be ufeful) to make Experiments, to try what *Butment* would be requifite for Arches of any Dimenfion, or Form whether Straight, Semicircular, Skeen, or Scheam or of the third or 4th Point, or Eliptical, *&c.*

The ingenious Dr. *Hook*, Reader of Geometry in *Grefham College*, hath promifed, in his Treatife of Heliofcopes, to publifh fomething of this Nature; but if ever he did do it; I have not been fo fortunate as ever to fee it, or hear of it, but only in that Treatife; what he there promifed, was as follows, *viz.* A true mathematical and mechanical Form of all manner of Arches, with the true *Butment* neceffary to each of them A Problem (faith he) which no Architectonick Writer hath ever yet attempted, much lefs performed.

A Treatife of this Nature would be extream ufeful to the publick: For the want of a certain Rule in Arching, with its neceffary *Butment*, hath too often proved the Ruin of fome *Structures*, which have been of no fmall Charge; of which I could give an Inftance from my own Obfervation but very lately; which is of a large Stone-bridge, which was no fmall Charge to two Counties to erect, not above ten or a dozen Years ago, which is already fo intolerably gone to Decay, that it is almoft ready to fall, and muft be rebuilt again in a little Time; for fome of the Arches are forced to be propped with many pieces of Timber; the chiefeft Fault that I could ever yet obferve (but I did never make any ftrict Search into it, but only as I rode over it) feems to me, to be want of a good and firm *Butment*, for the Materials did not feem to me to be the Caufe of the Defect in the Work. I have only one Thing more to add concerning *Buttreffes*, and that is fomething of the Value of fuch Work.

2. *Of the Price of building Buttreffes.*] As to the Method of putting out fuch Work, if it be not done by the Day, it is commonly done by the Cubick Foot. A Gentleman that had an Occafion for two Buttreffes to be built againft an old Stone-building, defired me to difcourfe with his Workman about it, and to put it out by Meafure to him; when he and I came to treat of the Matter, I found he knew not well what to fay about it; at laft he told me he would do it for

5 *d.*

5 *d per* Foot Cubick, *viz.* For Workmanship only, it being a sort of Work, that neither he nor I ever knew put out by the Foot; and therefore, after a little considering the Work, I told him I did think less than ½ the Price would be sufficient, and therefore we came to no Conclusion at that Time; and before I should see him again, I knew how to inform my self from a new *Buttress* which was built by the Day (by two good Workmen) of five Foot square, and twelve high: When I had found the Solidity of this *Butteress*, I found, that according to their Days Works, it came to about 2 *d.* ½ *per* Foot, and this I reckon to be a top Price; because the Workmen are taken to be torpid Operators, and the Work was also very well done; upon which Account for 1 *d* ½ or 2 *d per* Foot, as I find by Observation (from the Quarry-stones at 5 *s. per* Load, Lime 25 *s* for 32 Bushels, Sand at 1 *s* 6 *d* for 12 Bushels to a Load) that such Work, Materials, and Workmanship, may be done for 6 *d.* or 7 *d. per* Foot Cubick.

Cabinet,

STrictly taken, is the most retir'd Place in a House. But a Cabinet in Palaces and great Houses, consists of an Outer-chamber, an Anti-chamber, and a Cabinet with a Gallery on the Side.

Caliducts,

That is, Conveyers of Heat. The Ancients used to warm their Rooms with certain (secret) Pipes (call'd *Caliducts*) that were convey'd in the Walls, transporting Heat to sundry Parts of the House, from one common Furnace.

Camber beam.

A Piece of Timber cut Arching (or with an obtuse Angle) in the middle. Camber beams are commonly us'd in Platforms, as Church leads, *&c.* And in other Cases, where there is occasion for long Beams, a Camber beam being much stronger than another of the same Size; for he being lay'd (as they generally are) with the hollow side downwards, and having good Butments at the Ends, is a kind of an Arch.

Cames.

The small slender Rods of Cast-lead, of which the Glaziers make their turn'd Lead. For their Lead being cast into slender Rods, of some 12 or 14 Inches long each, is call'd the *Came* (and sometimes they call each of those Rods a *Came*) which being afterwards drawn through their Vice, makes their turn'd Lead. See *Lead*, N. 10.

Camerated.

Vaulted or Arched.

Cant.

Cant.

A Term uſed by ſome Carpenters, when a piece of Timber comes the wrong way in their Work, they ſay *cant* it, *i. e.* turn it about.

Cantalivers.

1. *What.*] Several Ancient and Experienced Workmen tell me, That they are the ſame as *Modilions*, only thoſe are plain, but theſe are carv'd. They are both a kind of Cartouzes, ſet (at equal diſtance) under the *Corona* of the *Corniſh* of a Building.

2. *Price of making.*] Mr. *Leybourn* ſays, They are commonly made by the Piece, at different Rates, according to the Curioſity of the Work. And experienced Workmen tell me, They have commonly 2 *s.* 6 *d.* for making and carving of each. But in *London* they will carve them for 1 *s.* 8 *d* each.

3. *Price of Painting.*] Mr. *Leybourn* ſays, They are commonly Painted by Tale, or ſo much *per* Piece according to the Colour they are laid in.

Cantaliver-corniſh.

1. *What.*] Is ſuch a Corniſh as has *Cantalivers* under it.

2. *Price.*] Mr. *Leybourn* ſays, They are commonly made by the Foot, running Meaſure [*i. e.* by the number of Feet, in length only] at different Rates according to the Curioſity of the Work. And experienced Workmen tell me, That they commonly have 2 *s per* Foot for the Corniſh, it being plain without any Carving in it, and with the *Cantalivers*, about 3 *s.* 6 *d. per* Foot.

Canting-ſtairs.

See Stairs.

Capital.

Is the upper part of a Column. Such of theſe as have no Ornaments, as the *Tuſcan* and *Dorick*, we call Capitals with *Mouldings*; and the reſt which have *Leaves*, and other Ornaments, Capitals with Sculptures. The word is borrow'd from the Latin *Capitellum*, the Head or Top of any Thing.

Tuſcan.] According to *Vitruvius*, the height of the Tuſcan Capital (by the Aſtragal at the bottom) muſt be half the Diameter of the Body of the Column below. And this height being divided into 3 Parts, the firſt, and uppermoſt part goes to the *Abacus* [which is a Square, or flat *Moulding*] the ſecond Parts goes to the *Boultin*, and *Fillet* under it, [the Boultin is a quarter of a Circle, the Fillet a narrow flat Moulding] and this part is ſubdivided into 4 Parts, of which 3 go to the Boultin, and 1 to the Fillet, and the third and laſt Part goes to the Neck, which is flat and ſtraight. Again, the Neck is divided into two parts, one of which is the breadth of the *Aſtragal* under it (which conſiſts of a Semicircle, and a Fillet under it) The Aſtragal is is again divided into 3 parts, of which 2 go to the Semicircle, and 1 to the Fillet. The Projecture of the Capital ſhall be $\frac{1}{8}$ part of the Diameter of the Body of the Column below. The Aſtragal projecteth in a Square.

According to *Scammozzi*, the heighth of the Capital (by the Aſtragal at the bottom) muſt (alſo) be $\frac{1}{2}$ the Diameter of the Column below. And this heighth being divided into 60 parts, 20 of them ſhall go to the *Abacus* (or *Plinth*, as he calls it) 15 to the *Echinus* or half round (which *Vitruvius* calls the *Boultin*) 5 to the Rondel, or *Bead-moulding* [which is a Semicircle] 3 to the *Liſt* (which *Vitruvius* calls a *Fillet*) and 17 to the *Neck*, or *Frieſe*, (as he calls it.) Again, 7 ſuch Parts muſt go to the *Rondel* of the Aſtragal, and 3 to its *Liſt*.

According to *Palladio*, the heighth of the Capital is (alſo) half the Diameter of the Body of the Column below (*viz.* by the Aſtragal, which none of them reckon a part of the Capital, tho' in propriety of Speech, it ought to be ſo eſteem'd.) And this heighth is divided into three equal Parts, the uppermoſt of which goes to the *Abacus* (which he alſo calls the *Dado*, or *Dye*) the next part goes to the *Ovolo*, or *Echinus* (*Vitruvius* calls it the *Boultin*) the other part is divided into ſeven, of one of which is made the *Liſtella* (*Vitruvius* calls it the *Fillet*) under the *Ovolo*, and the other ſix Parts go to the *Collorino*, or Neck (he alſo calls it the *Hypotrachelium*, or *Frieze* of the Capital.

3. *Dorick.*] According to *Vitruvius*, the heighth of the Dorick Capital (by the Aſtragal at the bottom) is equal to $\frac{1}{2}$ the Diameter of the Body of the Column below. And this heighth being divided into three Parts, the firſt and lowermoſt goes to the Neck, the next Part goes to the *Boultin*, [by which Term he here comprehends ſeveral Members and deſcribes this Part in 2 Forms. 1. Of which is a *Boultin* (as 'tis deſcribed N. 2.) and 3 *Fillets* under it, and the other is a *Boultin*, and an *Aſtragal* under it] and this part is divided into 3 parts, 2 of which go to the *Boultin*, properly ſo call'd, and the other to the 3 *Fillets*, or to the *Aſtragal*; the Fillets are all of an equal ſize, in the Aſtragal, the Fillet is $\frac{1}{3}$ of the whole, the third and uppermoſt part of the Capital is again divided into 3, the 2 lowermoſt of which go the Square, and the other to the *Cimatium* [which is an O-G. (with the hollow downwards) and a *Fillet* over it. An O-G. is a Moulding, ſomewhat reſembling an S, which *Vitruvius* makes of two quarter Circles join'd together; and this *Cimatium* being alſo divided into 3 parts, 2 of 'em go to the O-G, and 1 to the *Fillet*. The Aſtragal under the Capital is equal to $\frac{1}{2}$ the Neck, and made as in N. 2.

Scammozzi makes the Capital of the ſame heighth, which he divides into 60 parts, of which 3 go to the Fillet of the Cimatium, 5 to the O-G of the Cimatium, 12 to the Square, 14 to the Boultin, 5 to the Rondel,

del, and 2 to the Fillet of the Aſtragal under the Boultin, and 19 to the Neck. The Aſtragal under the Neck contains 10 ſuch parts, of which 6 $\frac{1}{2}$ goes to the Rondel, and 3 $\frac{1}{2}$ to the Fillet. I have deſcrib'd it by *Vitruvius*'s Terms, for *Scammozzi* mentions none of them.

Palladio (alſo) makes the Capital of the ſame heighth with *Vitruvius*, which he divides into 3 parts, the uppermoſt of which he ſubdivides into 5 parts, 2 of which go to the *Cimatium* [which *Vitruvius* calls the *Cimatum*] and is again ſubdivided into 3 parts, one of which goes to the *Liſtella*, or *Annulet* [which *Vitruvius* calls a *Fillet*] and the other 2 to the *Scima recta*, [which is an *O--G*, as 'tis deſcribed in this Number] and the other 3 of the firſt Sub-diviſions of this part goes to the *Abacus* [which *Vitruvius* in this Number calls the Square.] The ſecond of the 3 grand Diviſions of the Capital, is ſubdivided into 3 parts, 2 of which go to the *Ovolo*, or *Echinus* [which *Vitruvius* (N. 2.) calls a *Boultin*] and the other to the *Annulets* under it, which are 3, and are equal. The third principal *Hypotrachelium*, or *Frize* [which *Vitruvius* calls the Neck] The *Aſtragal* under the Neck, is as high as all the 3 *Annulets*

4. *Ionick*] According to *Vitruvius* is made thus, divide the Semidiameter of the Body of the Column below into 18 parts, take 19 ſuch parts, of which three muſt go to the *Cimatium*, 1 to the *Fillet*, and 2 to the *Cima* [or *O--G*] under it. Then take 4 parts for the *Trochilus* of the *Volute*, or *Scroll* [the *Trochilus* is that Member from whence the Scroll begins] then take 4 parts from the *Boultin* [which is $\frac{1}{4}$ of a Circle] which muſt be carv'd with *Eggs* and *Anchors*. Then take 2 parts for the *Aſtragal* under the *Boultin*, the *Aſtragal* is carved with *Beads*, and has a Fillet on each ſide of it, each $\frac{1}{4}$ of the whole. Then the 6 remaining parts muſt go to the half of the Volute below. Then take 8 more ſuch parts, which muſt go to make the remainder of the Frieze, or Neck of the Capital, and 3 more ſuch parts for the Aſtragal under the Neck, of which one part goes to the Fillet.

[*Scammozzi*'s Deſcription of the *Ionick* Capital is ſo blind, that I think none is ever the wiſer for it. And *Palladio*'s Deſcription agrees with *Vitruvius*'s.

5. *Corinthian*.] According to *Vitruvius* the heighth of this Capital (by the Aſtragal at the bottom) is equal to the Diameter of the Body of the Column below, $\frac{1}{7}$ part of which goes to the *Abacus* (which conſiſts of a *Boultin*, a *Fillet*, and *Plinth*, [which is but a larger Fillet] the *Abacus* being ſubdivided into 3 parts, 1 of 'em goes to the *Boultin*, and a third part of the next goes to the *Fillet*, and the reſt to the *Plinth*. The heighth of the *Aſtragal* below the Capital, is $\frac{1}{12}$ part of the Diameter of the Body of the Column below, and is divided into 3 parts whereof the *Fillet* contains 1 part, and the *Boultin* 2.

Scam-

Scammozzi makes this Capital 1 $\frac{1}{7}$ Diameter of the Column high, which divided into 75 parts, 4 of 'em go to the *Boultin*, 1 to the *Fillet*, and 9 to the *Plinth*, and the reſt to the *Neck*.

Palladio alſo makes the heighth of this Capital equal to the whole Diameter of the Body of the Column below and $\frac{1}{8}$ part more, which is allowed to the *Abacus*; by which I underſtand he means all the Mouldings above the Acanthus Leaves.

6. *Roman*, or *Compoſite*.] *Vitruvius* makes, and divides this Capital like the *Corinthian*; and ſo does *Scammozzi*, and *Palladio*; only the Carving of this is ſomewhat different from that.

Carcaſs.

1. *What*.] The Timber-work (as it were the Skeleton) of a Houſe, before it is Lathed and Plaiſter'd.

2. *Price of Framing*.] The Price of Framing the Carcaſs of a Houſe (in the Country) as Workmen inform me, is about 8 *s. per* Square, if the Workman pay for the Sawing; if not but about 4 *s.* 6 *d. per* Square.

Cariatides.

From the *Greek*, *Kariatydes*, a People of *Caria*. By theſe, in Architecture, are meant certain Figures of Captive-Woman dreſt after the manner of that Nation, and ſerving inſtead of Columns to ſupport the Entablements.

Theſe are called, by *M. le Clerc*, very properly, Symbolical Columns. The occaſion of 'em (as he tells us, and all Authors agree it was) take, as follows. The ancient *Greeks*, to preſerve the Memory of their Victories, had a Cuſtom in the Columns of their Publick Buildings, to add Figures and Repreſentations of the Enemies they had ſubdu'd. The Wives of the rebellious *Carians*, when reduc'd to Obedience, and the *Perſians* vanquiſht by the *Lacedæmonians* at *Platæa*, were the firſt Subjects of theſe Columns; which have preſerv'd, to late Poſterity, both the Glory of the Victors, and the Diſhonour of the Vanquiſhed. Hence, originally, came the Names *Cariatides* and *Perſian* Columns; which have been ſince applied to all Columns made in Human Figures, tho' with Characters very different from one another.

We don't now repreſent the *Cariatides* (or *Cariates*) as formerly, with the Marks of Servitude and Slavery: Such Characters were injurious to the Fair-Sex, and for that reaſon we give them others entirely oppoſite; never uſing them in Building, but as ſingular Beauties, and ſuch as make the greateſt Ornament thereof. They never make their Appearance, now, but under the noble Symbols of Prudence, Wiſdom, Juſtice, Temperance, *&c.* When the *Cariates* are Inſulate, they ſhould not have any Weights to ſupport, greater than thoſe of Balconies, little Galleries, or or slight Crownings; and their Entablature may be *Ionick*.

The *Cariates* should aways have their Legs pretty close, the one a little a thwart the other, with their Arms either join'd to the Body or to the Head, or at least but very little asunder; that, as they do the Office of Columns, they may, as much as possible, bear the Figures of 'em

There is a particular Defect in the *Cariates*, that being the Figures of Women, they don't seem altogether proper to do the Office of Columns: But this is easily amended when they join to a Wall, there being nothing to do in that Case but to place a Console over them, which shall appear to bear all the Weight of the Entablature. This will have a good Effect; and the *Cariates* will serve for Columns; without appearing over-burden'd.

If the *Cariates* have a Projecture beyond the Wall in the manner of Pilasters, they may be used in the Architecture of a Gallery or Salon; provided they ben't made to sustain any thing but the Entablature; the Weight of the Vault being born by the Wall behind, which serves them as a Ground or Bottom.

The *Cariates* should never be made of an immoderate Stature; lest being too big, they become frightful to the Ladies. For this reason one would sometimes chuse to confine them under the Impost of a Portico; such Imposts serving them for an Entablature. Further, on occasion, one may raise 'em on Pedestals, which, however, must not have less than one third of their Height. And if besides this, one place Consoles over their Head, the Figures may be made of a reasonable Size.

The *Cariates* and the common Columns should never be used together, under the same Entablature; for besides that there can never be a just Symmetry between them; the Figures of Women as high as common Columns would be monstrous, and make all the rest of the Architecture appear mean and pitiful.

There are some *Cariates* that have their Arms cut off; as those, for Instance, in the Hall of the Swiss-Guards in the *Old Louvre*. But these kind of Mutilations; which are only used to make the Figures more light and delicate, or rather to make them more conformable to the other Columns, are only proper for Termini or Forms, which are a kind of half human Figures seeming to proceed out of a Vagina or Sheath (as is taken notice of under that Word.)

The *Cariates* must always appear in Characters proper to the Places they are used in. Those for instance, which support the Crowning of a Throne ought to be Symbols, or Representations of Heroick Vertues. Those that serve as Columns in a Place of Devotion, should bear the Characters of Religion, and those again in Halls and Banquetting-Rooms, carry the Marks of Gladness and Rejoycing.

'Tis

'Tis not proper to use *Cariates* in the Figures of Angels excepting at Baldaquins and Altars; and such as do appear under that Holy Form, ought, in my opinion, to support the Entablature with their Heads, as bearing it easily and without trouble.

The Entablature supported by Angels may be *Corinthian*, and the Virtues *Ionick*; And both the one and the other somewhat less massive than the ordinary. Thus far *Sebastian le Clerc*, concerning the *Cariatick Order*, in his excellent *Treatise of Architecture.*

Carpenter's.

1. *Work.*] The several kinds of Work done by Carpenters, (in relation to Building) with their Prizes, and Methods of Measuring them, *&c.* are too many to be comprehended under this so general a word of *Carpenter's work*; and therefore I shall refer them to their *Particulars*, (as Framing, Flooring, Roofing, *&c.*) where they will much more readily be found.

2. *Bill to make.*] a Carpenter's Bill should be made after this manner.

Mr. William Liberal *of* London, *his Bill of Materials had of, and Work done by* Thomas Trueman, June *the 24th* 1725.

	l s. d.
For 17 Load of Oaken-timber, at 22 *s.* the Load	18—14—00
For 28 Load of Fir-timber, at 35 *s.* the Load	49—00—00
For 180 Feet of Oaken-plank, 2 Inches thick, at 3 *d* the Foot	02—05—00
For 17 M. of 10 *d.* Nails, at 6 *s.* the M.	05—02—00
For 7½ C. of Deals, at 6 *l.* 5 *s.* the C.	46—17—06
For 28 ℔. of large Spikes, at 4 *d.* the ℔.	00—09—04
For 8 Weeks Work for my self, at 3 *s.* the Day	07—04—00
For 8 Weeks 2 Days Work for my Man, at 2 *s.* 6 *d.* the Day	06—05—00
Sum is	135—16—10

But, *Note*, If the Carpenter do not work by the Day, then he writes, For so many Square of Roofing (at what Price they agree upon *per* Square) so much Money Likewise for so many Square of Flooring, at so much *per* Square, so much Money. Also for so many Square of Partitioning, at so much *per* Square, so much Money. And for so many Square of Cieling Joysts, *&c.* The Windows they set down either at so much *per* Light, or so much *per* Window The Door-cases at so much a piece, either with, or without Doors. The Mantle-

Mantle-trees, Taſſels, *&c.* at ſo much a piece. The Lintelling, Guttering, Corniſh, Winder-boards, *&c.* at ſo much *per* Foot. Stairs, at ſo much *per* Step, or ſo much a Pair, *&c.*

Car-tooſes, touzes, -touches.

In Architecture are much the ſame as *Modilions*; only theſe are ſet under the Corniſh in Wainſcotting, and thoſe under the Corniſh at the Eves of a Houſe. Some Workmen call them *Dentils*, or *Teeth*.

Cartouche.

Perrault ſays, It is an Ornament of *Carv'd Work*, of no determinate Form, whoſe Uſe is to receive a Motto, or Inſcription, the word being borrowed from the *Italian Cartoccio*.

Cartridges.

In Architecture are the ſame as Cartoozes.

Cart-nails.

See Nails.

Caſeing of Timber-work.

1. *What.*] Is a Plaiſtering of a Houſe all over on the outſide with Mortar; and then ſtriking it (by a Ruler) with the corner of a Trowel, or the like, to make it reſemble the Joynts of Stone; that ſo the whole houſe may ſeem to be Built of Stone.

2 *Of the beſt way of doing it.*] Experienced Workmen tell me, That it ought to be done upon Heart-laths, becauſe the Mortar will decay the ſap ones in a little time: And tho' it be more work to Lath it with Heart, than with Sap-laths; yet 'tis better for the Mortar to hang to, becauſe Heart-laths are narroweſt, and they ought to be cloſer together for Mortar than for Lome. They alſo tell me, That they commonly uſe to lay it on in two Thickneſſes, *viz.* the laſt before the firſt is dry.

3. *Of its Price.*] I have put out this kind of Work to Plaiſterers for 3 *d.* or 4 *d.* the Yard, including Doors and Windows [*i. e.* meaſuring it as if there were none] and for 6 *d. per* Yard, excluding Doors and Windows [*i. e.* deducting them from the whole.]

Caſe of Glaſs.

1. *Crown.*] A Caſe of Crown-glaſs (as a Glazier in *London* informs me) contains 24 Tables, each Table being Circular, or nearly ſo, and (as I obſerved) about 3 Foot 6 Inches, or 3 Foot 8 Inches Diameter. See P. Glaſs, N. 3.

2. *New-caſtle.*] The ſame Perſon tells me, That they have 35 Tables of *New-caſtle-glaſs* to the Caſe; and that there ſhould be 6 Foot of Glaſs in a Table: I am ſure there is as much (and more) in one of them I ſaw in his Ware-houſe; for I obſerved them to be much of the Form of a Corner-tile ſuppos'd to be preſt out flat; and by meaſuring one of them, I found it to be

be about 3 ½ Foot on the upper or circular End, about 18 or 20 Inches on the lower and opposite End, and the Perpendicular heighth about 3 Foot. Mr. *Leybourn* (and Mr. *Wing* from him) says, That a Table of *New castle-glass* contains about 5 Foot, and that 45 of these Tables go to a Case.

3. *Normandy.*] Mr. *Wing* says, That 25 Tables make a Case of *Normandy* Glass.

Casement,

In Architecture is a hollow Moulding. Some Architects make it ⅙ of a Circle, others ¼.

Casements,

1. *What*] In Architecture are Windows to open.

2. *Price.*] Mr. *Leybourn* says, They are valu'd (according to their largeness, and the goodness of their Workmanship in their Locks and Hinges,) from 3 *s* to 20 *s*. a Casement. Casements about 2 ½ Foot long, are about 4 *s*. or 4 *s*. 6 *d* a piece. Folding-casements of the like size, with Bolts, Hinges, *&c* about 12, or 13 *s* the pair. Mr. *Wing* says, They are worth 7 *d*. or 8 *d*. the Pound, some 9 *d*. *viz*. Folding-casements Some Smiths in *London* ask'd me 6 *d*. *per* Pound for Casements, others said they were worth more, if they had Locks to them; but 6 *d* was their Price, if they made them with Turn-bouts (or Turn-buckles, as some call'd 'em) or Cock-spurs, and Pull-backs at the Hind side to pull them to with. One Smith told me, he would make them for 5 *d*. *per* Pound. I know some Smiths in the Country, make 'em by the Foot (measuring the whole Circumference round by the outer Edge of the Casement; thus, if a Casement be 2 Foot long, and 1 ½ Foot broad, they reckon him 7 Foot) A Smith at *Rye* ask'd me 9 *d*. *per* Foot for ordinary Casements, which I think is dear; for in other Parts of *Sussex*, they proffer'd me to make 'em for 6 *d*. *per* Foot, if ordinary; but if something extraordinary (as Folding-casements, *&c*.) then 8 *d*. *per* Foot.

3. *Of Painting.*] Casements are commonly Painted by the Piece, at three Half-pence, 2 *d*. or 3 *d* a piece, according as they are of bigness.

4. *Of Hanging.*] Glaziers in the Country tell me, That 'tis the Smith's Work to hang up the Casements; and if they don't do it themselves, they pay the Glaziers for doing it; who have 2 *d*. a piece for hanging of small Casements, and 3 *d*. a piece for large ones.

5. *Of Pining.*] See Glazing. N. IV. §. 1.

Cast,

In Architecture a piece of *Timber*, or a *Board*, or *the like*, is said *to Cast*, or *to be Cast*, when (by its own Drought, or Moisture, or by the Drought, or Moisture of the Air, or other Accident) it alters its flatness, and straightness, and becomes crooked.

Cata-

Catadrome.

A kind of Engine like a Crane, which Builders uſe in lifting up, and letting down any great Weights.

Catheta.

A Perpendicular, or Plumb-line, falling from the Extremity of the under-ſide of the Cimatium (of the Ionick Capital) through the Centre of the Volute.

Cavetto.

A round Concave Moulding, which has a quite contrary Effect to the *Quarter-round*; the Workmen call it *Mouth* when in its natural Situation, and *Throat*, when turn'd upſide-down.

Cavazion.

A Term of *Architecture*, ſignifying the Under digging, or Hollowing of the Earth, for the Foundation of a Building *Palladio* ſays, It ought to be the ſixth part of the Heighth of the whole Building.

Caulicoli.

The carved Scrolls (under the Abacus) in the Corinthian Capital.

Ceiling

1. *VVhat.*] In *Architecture*, is the Lathing, and Plaiſtering at the top of a Room, upon the under ſide of the Joyſts of the next Rom or upon Joyſts put up for that Purpoſe (and call'd Ceiling-joyſts) if it be in a Garret. Theſe plaiſter'd Ceilings are much uſed in *England*, beyond all other Countreys; and they have theſe Conveniencies with 'em: They make the Rooms much more lightſome; are excellent againſt raging Fire; they ſtop the Paſſage of the Duſt, and leſſen the Noiſe over head, and in Summer time the Air of the Room is ſomewhat the cooler for it.

2. *Of Meaſuring*] This Work is commonly done by the Yard, (containing 9 ſuperficial Feet:) And in taking their Dimenſions, if the Room be wainſcoted, they conſider how far the Corniſh bears into the Room, by putting a Stick perpendicular to the Ceiling, cloſe to the Edge of the uppermoſt part of the Corniſh, and meaſuring the Diſtance from the Perpendicular Stock to the Wainſcot; twice which Diſtance they always deduct from the Length, and Breadth of the Room taken upon the Floor, and the Remainder gives them the true Length and Breadth of the Ceiling; which, if it be taken in Feet (as moſt commonly it is) they multiply one into the other, and divide the Product by 9, and the Quotient gives them the Content in Yards.

3. *Price.*] In *London* the Workmanſhip (*viz.* Lathing, Plaiſtering, and Finiſhing) is commonly reckon'd about two Pence three Farthings *per* Yard. In *Rutland*, and ſome parts

parts of *Kent* (as about *Tunbridge-Wells*, &c.) I know they have 3 *d* *per* Yard. And in ſome parts of *Suſſex*, the Workmen tell me they have 4 *d.* *per* Yard. But if the Workmen find all Materials, and Lath it with Heart-oak-laths, then they commonly reckon about 1 *s. per* Yard, and with Fir-laths, about 8 *d. per* Yard.

Ceiling-Joyſts, or *Beams.*

1. *What.*] See *Ceiling.* N. 1.

2. Of *Meaſuring.*] The Work of putting up Ceiling-joyſts is meaſur'd by the Square; and therefore the length in Feet being multiply'd by the breadth in Feet, and two Places of Figures being cut off on the Right-hand, what remains to the Left hand is Squares, and what is cut off is odd Feet, of which 25 make a Quarter, 50, half, and 75 three quarters of a Square.

3. *Price.*] Putting up of Ceiling-joyſts is worth 4 or 5 *s.* (ſome Workmen tell me, they have 6 *s.*) *per* Square.

Cellars.

1. *What.*] They are the loweſt Rooms in a Houſe, the Ceilings of which lie level with the Surface of the Ground, on which the Houſe ſtands, or at leaſt but very little higher.

2. *Situation.*] Sir *Henry VVotton* ſays, They ought (unleſs the whole Houſe be Cellar'd) to be ſituated on the *North*-ſide of the Houſe, as needing a cool and freſh Air.

3. *Of Digging.*] They are commonly dug by the ſolid Yard, containing 27 ſolid Feet; and therefore the Length, Breadth, and Depth in Feet, being all multiply'd together, and the Product divided by 27, the Quotient will give the Content in ſolid Yards.

Cement.

1. *VVhat.*] In *Architecture*, is a ſtrong, ſticking, cleaving, or binding Mortar.

2. *To make.*] There are two Sorts of Cement, which ſome Bricklayers uſe in cementing of Bricks for ſome kind of Mouldings, or in cementing a Block of Bricks (as they call it) for the carving of Scrolls, or Capitals, or ſuch like, *&c.* One is called *cold Cement*, the other *hot Cement*; becauſe the former is made, and uſed without Fire, but the latter is both made and uſed with Fire. The *cold Cement* being accounted a Secret, is known but to few Bricklayers; but the *hot Cement* is common. I ſhall here ſhew how to make them both.

To make the Cold Cement.

Take half a Pound of old Cheſhire cheeſe, pare off the Rhind, and throw it away; cut or grate the Cheeſe very ſmall, and put it into a Pot; put to it about a pint of Cows-milk, let it ſtand all Night, the next Morning get the Whites of 12 or 14 Eggs, then take half a Pound of the beſt unſlack'd or Quick-lime that you can get, and beat it to Powder

Powder in a Mortar, then ſift it through a fine Hair-ſive, into a Tray or Bowl of Wood or into an Earthen-diſh, to which put the Cheeſe and Milk, and ſtir them well together with a Trowel, or ſuch like thing, breaking the Knots of Cheeſe, if there be any, then add the Whites of the Eggs, and temper all well together, and ſo uſe it. This Cement will be of a white Colour; but if you would have it of the Colour of the Brick, put into it either ſome very fine Brick-duſt, or Almegram, not too much, but only juſt to colour it.

To make the Hot Cement.

Take one Pound of Rozin, a quarter of a pound of Bees-wax, half an ounce of fine Brick-duſt, half an Ounce of Chalk-duſt, or powder of Chalk; ſift both the Brick-duſt and Chalk-duſt through a fine Hair-ſieve, (you may beat the Brick, and the Chalk in a Mortar, before you ſift it) boil all together in a Pipkin, or other Veſſel, about a quarter of an Hour, ſtirring it all the while with an Iron or a piece of Lath, or ſuch like; then take it off, and let it ſtand 4 or 5 Minutes, and it is fit for Uſe.

Note, That the Bricks that are to be cemented with this kind of Cement, muſt be made hot by the Fire before you ſpread the Cement on them, and then rub them too and fro, one upon another, as Joyners do, when they glew two Boards together.

Chambers.

1. *What.*] In a Building, all thoſe Rooms are called *Chambers*, that are ſituated between the lowermoſt (excepting Cellars) and the uppermoſt Rooms. So that in ſome Houſes there are two, in others three, or more Stories of Chambers.

2. *Situation*] Sir *Henry Wotton* tells us, That the principal Chambers of Delight (in a Houſe) ought to be ſituated towards the Eaſt.

3. *Proportions.*] The length of a well-proportionate Lodging-Chamber, ought to be the breadth, and half the breadth of the ſame, or ſomewhat leſs; but ought never to exceed that length; for the height, three quarters of the breadth will be a convenient heighth

Channel.

In the *Ionick Capital* is that part which is under the *Abacus*, and lies open upon the *Echinus*, or *Eggs*, which has its Centers, or Turnings on every ſide, to make the *Volutes*.

Chapter.

In *Architecture*, ſignifies the Top, or Head of a Pillar.

Chauncel.

Vulgarly *Chancel*, the moſt ſacred part of a Temple, or Church, ſo called from the *Cancelli*, or Lattices, which anciently us'd to ſeperate that

Part from the reſt of the Church. The *Greeks* call it *Adyton.*

Chimneys.

1. *VVhat.*] A Chimney is a particular part of a Houſe, deſign'd for the Conveniency of Fireing, with a Tube, or Tunnel to convey away the Smoak.

2. *Of Meaſuring*] Tho' Bricklayers, in making of Chimneys, do commonly agree by the Hearth ; yet they ſometimes alſo work by the Rod, as in other Brick-work ; and then their Method of taking their Dimenſions, is thus:

If the Chimney ſtand ſingly, and alone, not leaning againſt, or being in a Wall, and it be wrought upright over the Mantle-tree to the next Floor ; they gird it about the Breſt for the Length, and take the Heighth of the Story for the Breadth, and the Thickneſs of the Jambs for the Thickneſs But if the Chimney ſtand againſt (or in) a Wall, which is before meaſur'd with the reſt of the Building ; then the Breadth of the Breſt, or Front, together with the Depth of the two Jambs, is the Length, the Heighth of the Story, the Breadth, and the Thickneſs of the Jambs the Thickneſs. But if the Chimney ſtand in an Angle of a Room, and have no Jambs ; then the Breadth of the Breſt is the Breadth, the Heighth of the Story the Length, and the Thickneſs the Thickneſs Then for the Shaft, (which is that part which appears above the Tyling,) they commonly girt it about in the ſmalleſt part, for the Breadth, and take the Length of the Shaft for the Length ; and they commonly reckon the Thickneſs of both Sides for the Thickneſs, in Conſideration of the Widths, Pargeting, and Scaffolding

Note, Here is nothing to be deducted for the Vacancy betwixt the Hearth and the Mantle tree, becauſe of the Widths and the Thickning for the next Hearth above.

The Dimenſions being thus taken in Feet, the Work is thus meaſur'd : Multiply each particular Length by his Breadth, and that Product by its Thickneſs in Half-bricks, (*i.e.* by 2, for 1 Brick thick, by 3, for 1 $\frac{1}{2}$ Brick thick, and by 4, for two Bricks thick, *&c.*] Add theſe Products into one Sum which divide by 3, and the Quotient will give the Content of the whole Chimney in Feet, and the Standard-thickneſs of a Brick and half. Then divide this Content in Feet, by 272 $\frac{1}{4}$, and the Quotient will be the Content in Rods. But, becauſe 'tis difficult to divide by 272 $\frac{1}{4}$, you may do us. — Add two Cyphers to the Right hand of the Content in Feet, and then divide it by 27225, and the Quotient will be the Content in Rods, as before. And, every 100 of the Remainder, is one Foot of Work. Or 6807, of the Remainder, is $\frac{1}{4}$ of a Rod, 14613, is $\frac{1}{2}$ a Rod, and 20419 is $\frac{3}{4}$ of a Rod.

3. *Price.*]

3. *Price.*] Mr. *Leybourn* ſays, That Chimneys are ſometimes meaſur'd, and paid for by the Rod, like other Brick-work: And ſometimes, ſays he, they are paid for by the Fire-hearth, at ſo much the Fire-hearth; which ſays he is various, from 20, to 50 *s.* the Hearth And Mr. *Wing* ſays, That Building of Chimneys for ordinary Buildings, with Architrave Frieze, and Corniſh, is worth, from 15 *s.* to 20 *s. per* Hearth, according to their heighth, and ſubſtance; and without Architrave, and Frieze, from 10 *s.* to 20 *s.* But in great Buildings, ſays he, (I ſuppoſe he means in his Country of *Rutland*,) they are uſually done by the Foot, *viz.* at about 6 *d. per* Foot. I know they are commonly built in *London*, and about *Tunbridge-wells*, for about 15 *s. per* Hearth: But ſome Workmen in *Suſſex* tell me, they have 20 *s.* and ſometimes 25 *s. per* Hearth for building of 'em.

4. *Rules about Timbers near 'em.*] 1. Let no Timber be laid within 12 Inches of the fore-ſide of the Chimney-jambs. 2. Let all Joyſts on the back of any Chimney be laid with a Trimmer, at 6 Inches diſtance from the Back. 3. Let no Timber be laid within the Tunnel of any Chimney.

5. *Proportions.*] *Palladio* lays down the following Proportions, for the breadths, and depths of Chimneys, (on the in ſide) and for their heighth to the Mantle-tree.

Chimneys in	Breadth.	Heighth.	Depth.
Halls,	6, 7, or 8 Foot	4 $\frac{1}{2}$, or 5 Foot	2 $\frac{1}{2}$, or 3 Foot
Chambers,	5 $\frac{1}{2}$ 6, or 7 Foot	4, or 4 $\frac{1}{2}$ Foot	2, or 2$\frac{1}{2}$ Foot
Studies, and Wardrobes.	4, 4 $\frac{1}{2}$, or 5 Foot	4, or 4 $\frac{1}{2}$ Foot	2 or 2 $\frac{1}{2}$ Foot

Neverthelefs, in theſe Points, the Workman ought rather to be guided by the Modern Faſhions, than by the words of this ancient Architect.

6. *To prevent Smoaking.*] Mr. *Lucar* (*in his Solace*,) adviſeth to leave two Holes (one over another) on each ſide of the Chimney, one ſlopeing upwards, and the other downwards, or elſe to place two Pipes (in the ſame Poſition) on each ſide of the Chimney. Through theſe Holes, or Pipes, ſays he, the Smoak will eaſily paſs out of any Tunnel, which way ſoever the Wind blows. I cannot tell how this may take effect; but to me it ſeems but a Fancy. I think *Philippe de l'Orme*'s Advice is better, who propoſes to provide a hollow Braſs ball of a reaſonable Capacity,

pacity, with a little Hole on one ſide for the Reception of Water. (I think it were better made with a ſhort Noſe to skrew-off, when 'tis filled with Water; and then the Hole at the end of this Noſe needs not to be bigger than that at the ſmall end of a Tobaco-pipe.) This Ball being fill'd with Water, is to be placed (with the Hole upwards) upon an Iron-wire, that ſhall traverſe the Chimney a little about the Mantle-tree, at the ordinary heighth of the greateſt Heat, or Flames; and when the Water is hot, it will be rarify'd, and break out of the Hole in a windy Vapour; which will force up the Smoak, that otherwiſe might linger in the Tunnel by the way, and otentimes revert. It were good to have two of theſe Balls, one of them may ſupply the place of the other, when 'tis exhauſted; or for a need, blow the Fire in the mean time.

I have ſeen on the top of ſome Chimneys, a ſort of Fane, or Weather-cock, (ſome call it a Begger-man) whoſe back-ſide is cover'd with Plates of Tin; ſo that which way ſoever the Wind be, it can never keep down the Smoak in the Chimney, but it always comes out free, and undiſturb'd. I have known this laſt Contrivance help Chimneys, that before Smoak'd very much. But I believe the ingenious Carpenter, and Bricklayer might prevent the Smoaking of any Chimney, by a due *Situation* of the Doors of the Room, and an apt *falling back* of the Back, and convenient gathering of the Wings, and Breſt of the Chimney.

Chimney-hooks.

1. *What.*] Theſe are Hooks of Steel or Braſs, put into the Jambs of the Chimney, in each Jamb, one for the handle of the Fire-pan, and Tongs to reſt in.

2. *Price.*] The Steel-hooks are commonly about 1 *s.* the pair, and the Braſs ones, about 2 *s.* the pair in *London*; for ſo I have there bought 'em.

Chimney-jambs.

The ſides of a Chimney, commonly coming out Perpendicularly (tho' ſometimes Circularly) from the Back; on the Extremities of which the Mantle-tree reſteth. Alſo, ſee Corner-ſtone.

Chimney-pieces.

1. *What.*] Certain Mouldings of Wood, or Stone, ſtanding on the fore-ſide of the Jambs, and coming over the Mantle-tree.

2. *Price.*] *Chimney pieces of Free-ſtone*, wrought plain, are worth 10 *s.* but there may be ſuch Mouldings wrought in 'em, as with their Coves, and other Members, may be worth 20, 30, or 40 *s.* a piece. *Chimney pieces of Egyptian, or Black Fleak'd-marble*, or *of Rance*, or *Liver-colour'd marble*, are worth (of

(of an ordinary ſize) 12, or 14 *l.* a piece. *Chimney pieces of Wood*, are alſo of different Prizes, as 10, 12, or 14 *s.* to 20 *s.* a piece, more or leſs, according to their largeneſs, goodneſs of the Stuff, and Curioſity in the Workmanſhip.

3. *Painting*] They are commonly painted by the Piece, at about 2 *s.* a piece, more or leſs, according to the goodneſs of the Work, and largeneſs of the Chimney-pieces.

Chaptrels.

See Arches. N. 6.

Cima, or *Cimaiſe.*

From the *Greek Kymation*, a Wave, is what in *Engliſh* we call *Ogee*, *Ogive*, and ſometimes barely *O.-G.* Is is a *Moulding* waved on its Contour, Concave at the Top, and Convex at the Bottom, and which makes the uppermoſt Member, and as it were the *Cime* or *Top* of large Cornices. Of theſe there are two kinds; in the one, that part which has the greateſt *Projecture* in Concave, and is term'd *Doucine*, or an upright *Ogee*; in the other, the Convex part has the greateſt *Projecture*; and this is call'd the *Heel*, or inverted *Ogee*. Some write the word *Simaiſe*, and derive it from *Simus*, flat noſed; but that *Etymology* is falſe, the Beauty of this Member conſiſting in having its *Projecture* equal to its Height.

Cimbia.

See Pedeſtal. N.

Cincture.

Is a Liſt or Fillet at the Top or Bottom of a Column. That at the Top is ſometimes call'd *Colier*, and ſometimes *Annulus*.

Cilery.

A Term in Architecture, ſignifying the Drapery or Levage that is wrought upon the Heads of Pillars.

Cimeliark.

In Architecture is a Veſtry, or Room where the Plate, Veſtments, and other rich Things belonging to the Church are kept.

Ciſterns.

1. *What.*] They are Veſſels, made to ſerve as Receptacles for Rain, or other Water, for the neceſſary Uſes of a Family.

2. *To make.*] If you deſign to make your Ciſterns under your Houſe as a Cellar, which is the beſt way to preſerve your Water for culinary Uſes; then you may lay your Brick or Stone with Terrace, and it will keep Water very well. Or you may make a *Cement*, to joyn your Brick or Stone withal, with a Compoſition made of ſlacked ſifted Lime, and Linſeed Oyl, tempered together with Tow or Cotton-wool.

Or

Or you may lay a Bed of good Clay, and on that lay your Bricks for the Floor; then raiſe the Wall round about, leaving a convenient ſpace behind the Wall to ram in Clay, which may be done as faſt as you raiſe the Wall: So that when 'tis finiſh'd, 'twill be a Ciſtern of Clay, walled within with Brick, and being in a Cellar, the Brick will keep the Clay moiſt; (altho' empty of Water)that it will never crack. This (ſays Mr. *Worlidge)* I have known to hold Water perfectly well, in a ſhady place, tho' not in a Cellar. Thus in a Garden or other Place, may ſuch a Ciſtern be made in the Earth, and cover'd over; the Rain-water being convey'd thereto, by declining Channels running to it. Alſo, in, or near Houſes, may the Water that falls from them be conducted thereto.

Clamp.

1. *What*] A Clamp is a kind of Kiln built above Ground (of Bricks unburnt) for the Burning of Bricks.

2. *How made, and how Bricks are burnt in it*] An ancient and experienced Workman, that has made and burnt many Thouſands of Bricks, tells me, That they build their Clamps much after the Method that the Arches are built in Kilns, *viz.* With a Vacuity betwixt each Brick's breadth, for the Fire to aſcend by; but with this difference, that inſtead of Arching they *truſs-over*, or *over-ſpan*, as they phraſe it, *i. e.* they lay the end of one Brick about half way over the end of another, and ſo, 'till both ſides meet within half a Brick's length, and then a bounding Brick at the Top finiſhes the Arch. They make the Mouth, (where the Fire is to be put in) about 2 ½ Foot wide, and about 3 Foot high, and then they begin to truſs over, which they do for 3 Bricks in heighth; which with a bounding Brick on the Top, will cloſe up the Arch. But after they have begun, make the Place to receive to the Fuel (before it is cloſed at he top)they fill it almoſt full with Wood, and upon that lay Sea-coal; then being overſpan'd like an Arch, on all the Surface they ſtrew Sea-coal, and then they lay another Courſe of Bricks the other way, laying them at a little diſtance from one another, and ſtrewing Sea-coal upon them: And thus they continue, laying one Courſe one way, and the other another (and ſtrewing Sea-coal betwixt each Courſe) 'till they come to 8 or 10 Foot high, according as the Clamp is to be for bigneſs This being done, they fire the Wood, and that fires the Coal; which being all burnt out, the whole Clamp of Bricks is burnt.

Clamp-nails.

See Nails. N. 3.

Claſp-nails.

See Nails. N. 4.

Cleaving.

Cleaving.

Of Laths, Pales, Shingles, and Timber. See *Laths, Pales, &c.*

Cleer-ſtory-window.

See Window. N.

Clench-nails.

See Nails. N. 5.

Clinkers.

Thoſe Bricks are ſo call'd by ſome which (having naturally much Nitre, or Salt-peter in them, and lying next the Fire in the Clamp, or Kiln,) by the Violence of the Fire are run, and are glazed over.

Cloiſter.

A cloſe and ſeparate Habitation, where Friars, Monks, and Nuns live retir'd from the World. Alſo a long place cover'd with a Floor, or Platfond, ſupported by Pillars. See P. Architrave. N. 2.

Cloſet.

A general Name for any very ſmall Room. The Contrivance of Cloſets in moſt Rooms, now ſo much uſed (and ſo uſeful) is one great Improvement of Modern Architecture.

Clout-nails.

See Nails. N. 6.

Clout-brads.

See Brads. N.

Cockle-ſtairs.

As Winding-ſtairs. See Stair-caſe. N. III.

Coins.

As Quoins.

Colier.

As Cincture.

Collorino.

See Capital. N. 2.

College.

A Place ſet apart for the Society, and Cohabitation of Students.

Collar-beam.

A Beam fram'd croſs betwixt two principal Rafters.

Column.

Column.

1. *What.*] A Column is a kind of round Pillar, compos'd of a *Base*, a *Fust* or *Shaft*, and a *Capital*, and ſerves to ſupport the *Entablement*. The Column is different in the different Orders, being capable of a great number of Variations, with regard to to its Matter, its Conſtruction, Form, Diſpoſition and Uſe. The Word comes from the Latin *Columna*, which *Vitruvius* tells us was form'd from *Columen*, a Prop or Support.

2. *Parts.*] Every Column (in the largeſt Senſe) conſiſts of ſeven principal Parts, *viz.* Pedeſtal, Baſe, Body, Capital, Architrave, Frieze, and Cornice; each of which ſhall be handled in their proper Places of the Alphabet.

3. *Kinds.*] Architects reckon five Orders or Kinds of Columns, *viz.* Tuſcan, Dorick, Ionick, Corinthian, and Roman Compoſite, or Compound Order. I ſhall (here) ſay ſomething to each of theſe, in their Order.

4. *Tuſcan.*] The whole heighth of this Column, and heighth of each principal Part thereof, according to ſeveral Authors, is, as in the following Table.

Authors

Authors Names	Whole height.		Pedestal.		Base		Body.		Capital.		Architrave.		Frieze.		Cornish	
	Mo.	Mi.	Mo.	Mi	Mo	Mi.	Mo	Mi.	Mo.	Mi.	Mo.	Mi	Mo	Mi	Mo	Mi.
Vitruvius	11	5	2	20	0	30	6	0	0	30	0	30	0	03	0	30
Vignola	11	5	2	20	0	30	6	0	0	30	0	30	0	35	0	40
Palladio	10	15	1	0	0	30	6	30	0	30	0	35	0	26	0	40
Scammozzi	11	15	1	52 ½	0	30	6	30	0	30	0	31 ½	0	41	0	41

Note, (1.) That in this, and the four following Tables of the heighths of Columns, and their Parts; I have taken pains to reduce all my Author's Dimensions to *Modules* and *Minutes*; reckoning a Module the Diameter of the Body of the Column, just above the Base; and a Minute the 60th part o a Module. (2.) That the height of the Body of a Column, is reckon'd from the top of the Base, to the top of the Astragal under the Capital.

5. *Dorick.*] The whole heighth of this Column, and the heighth of each principal part thereof, according to several Authors, is as in this Table.

Auors ames	Whole height		Pedestal.		Base		Body.		Capital.		Architrave.		Frieze.		Cornish	
	Mo.	Mi.	Mo.	Mi	Mo	Mi.	Mo.	Mi.	Mo	Mi.	Mo.	Mi	Mo	Mi	Mo	Mi
Vitruviu	12	40	2	40	0	30	7	0	0	30	0	30	0	45	0	40
Vignola	12	40	2	40	0	30	7	0	0	30	0	30	0	45	0	45
Palladio	'3	0	2	20	0	30	7	45	0	30	0	30	0	45	0	35
Scammozzi	12	58	2	26	0	30	7	30	0	30	0	35	0	45	0	42

6. *Ionick.*] The whole heighth of this Column, and the heighth of each principal part thereof, according to several Authors, is as in this Table.

Authors

Authors Names	Whole height	Pedestal	Base	Body	Capital	Architrave	Frieze	Cornish
	Mo. Mi.	Mo. Mi	Mo. Mi.	Mo Mi.	Mo Mi.	Mo. Mi.	Mo. Mi.	Mo. Mi.
Vitruvius	14 15	3 0	0 30	8 10	0 20	0 37½	0 30	0 52½
Vignola	14 15	3 0	0 30	8 10	0 20	0 37½	0 45	0 52½
Palladio	13 28	2 40	0 52½	7 40	0 27½	0 34½	0 27	0 46½
Scammozzi	12 33¾	2 30	0 30	7 30	0 18¾	0 35	0 28	0 42

6. *Corint.*] The whole heighth of this Column, and the heighth of each principal part thereof, according to several Authors, is as in this Table.

Authors Names	Whole height.	Pedestal.	Base.	Body.	Capital.	Architrave.	Frieze.	Cornish
	Mo. Mi.	Mo. Mi	Mo Mi.	Mo Mi.	Mo. Mi.	Mo. Mi	Mo. Mi	Mo Mi.
Vitruvius	16 0	3 30	0 30	8 20	1 10	0 30	0 37½	1 0
Vignola	16 0	3 30	0 30	8 20	1 10	0 45	0 45	1 0
Palladio	13 54	2 30	0 30	7 55	1 5	0 36	0 28	0 50
Scammozzi	14 42½	2 30	0 30	8 5	1 10	0 39	0 31¾	0 46¾

8. *Dorick.*] *Roman Composita or Compound*]The whole heighth of this Column, and the heighth of each principal part thereof, according to several Authors, is as in this Table.

Authors Names	Whole height.	Pedestal.	Base	Body.	Capital.	Architrave.	Frieze.	Cornish
	Mo. Mi.	Mo. Mi	Mo. Mi	Mo. Mi.	Mo Mi	Mo. Mi	Mo Mi	Mo Mi.
Vitruvius	16 6½	3 30	0 30	8 20	1 10	0 52½	0 52½	0 52½
Vignola	16 0	3 30	0 30	8 20	1 10	0 45	0 45	1 0
Palladio	15 20	3 20	0 30	8 25	1 5	0 45	0 30	0 45
Scammozzi	15 20	3 20	0 30	8 25	1 5	0 40	0 32	0 48

9. Of

9. *Of Diminiſhing.*] Columns of every Order muſt be ſo formed, that the *upper* part of the Body be leſs than the *lower*; which diminiſhing, muſt be more or leſs, according to the Proportion of their Heighths; and is to begin from one third part of the whole Shaft upwards, (*i. e.* the lower third part is to be of an equal Bigneſs;) which *Philander* preſcribes, (by his own preciſe meaſuring of ancient Columns) as the moſt graceful Diminution. And for the Quantity to be diminiſhed, Architects lay down this Rule. —

That the	Tuſcan	Column, be	$\frac{1}{4}$	part
	Dorick		$\frac{1}{5}$	
	Ionick		$\frac{1}{6}$	
	Corinthian		$\frac{1}{7}$	
	Compound		$\frac{1}{8}$	

ſmaller at the Top, juſt under the Capital, than below, juſt above the Baſe, *i. e.* the Diameter of

the Top, of the	Tuſcan	Column, is	$\frac{3}{4}$	of
	Dorick		$\frac{4}{5}$	
	Ionick		$\frac{5}{6}$	
	Corinthian		$\frac{6}{7}$	
	Compound		$\frac{7}{8}$	

the Diameter of the Column below.

Colours.

The principal Colours us'd in painting of Houſes, *&c.* ſhall be treated of in their proper Places.

Compoſite Order.

See *Column* and *Order.*

Compartition.

By this Term, Architects underſtand a *graceful* and *uſeful Diſtribution* of the whole Ground-plot of an Edifice, into Rooms of Office, and of Reception, or Entertainment.

Compartment,

In *Architecture*, is a particular Square (for an Inſcription, or ſome other Device) marked out in ſome ornamental Part of a Building.

Concamerate.

To make an Arched-Roof; as in Vaults, *&c.* to Arch over.

Conclave,

In *Architecture*, is a Cloſet, or Inner-Chamber.

Conducts.

Sewers, or Gutters, to convey away the Suillage of a House. In these (says Sir *Henry Wotton*) Art should imitate Nature, in separating those ignoble Conveyances from the Sight; and (where there wants a running Water,) should place them in the most remote, and lowest part of the Foundation, with secret Vents passing up through the Walls, (like a Tunnel) to the wide Air; which all *Italian* Artists commend for the Discharge of noisome Vapours; tho' elsewhere, to my Knowledge, little practised.

Conges,

In *Architecture*, are the Rings, or Ferrils, heretofore us'd in the Extremities of Wooden-pillars, to keep them from splitting, afterwards imitated in Stone-work.

Console.

From the French *Consolider*, to close up is an Ornament cut upon the Key of an Arch, which has a Projecture or Jetting, and on Occasion serves to support little Cornices, Figures, Busts, and Vases.

Contour.

The Out-line of any Member in Architecture; as that of a *Base*, a *Cornice*, or the like; a French Word. See *Profile*.

Contignation.

A Story in Building. In Latin, *Contignatio* signifies the laying of Rafters together.

Contramure.

In *Architecture*, is an Outwall, built about the Wall of a City.

Copeing of Walls.

1. *What.*] The Copeing of a Wall is the Top, or Cover of it, made sloping to carry off the Wet.

2. *Price.*] I have known Brick-walls (of 1 ½ Brick thick) coped with Stone, for 4 *d. per* Foot, lineal, (or running) Measure; the Workman drawing the Stones into this Price.

3. *Drawing of Stones, for-*] I have known 1 *d. per* Foot given for drawing the Stones for Copeing of Walls.

Corbeilles.

From the Latin *Corbis* (a Basket) is a Piece of carv'd Work in the Form of a Basket full of Flowers or Fruits, serving in Architecture to finish some Ornament.

Corbel.

A short piece of Timber laid into a Wall, with its End sticking out some 6 or 8 Inches, more or less, according as the Occasion requires: The under-side of the End so sticking out, is sometimes cut into the Form

Form of a Boultin; ſometimes of an O-G, ſometimes of a Face, and ſometimes of other Forms, according to the Fancy of the Workman; the upper-ſide is flat and plain. *Corbels* are commonly placed (for Strength-ſake) immediately under the middle of Semi-girders of a Plat-form, and ſometimes under the Ends of the Camber-beams; but then they are commonly placed a Foot or two below the Beam, and a piece of Timber ſtands upright (cloſe by the Wall) from the Corbel to the Beam.

Corbels.

Holes left in the Walls of ancient Churches, *&c.* for Images to ſtand in.

Corinthian Order.

See *Column* and *Order.*

Corner-tiles.

See *Tiles* N. V.

Corner-ſtones.

1. *What.*] Are two Stones, (commonly of *Rigate*, or Fire-ſtone) of which there ſtands one in each Jamb of a Chimney. Their Faces are hollow in the Breadth, being a certain Sweep of a Circle. The Breadth of each Stone is equal to the Breadth of the Jamb; and their Heighth reaches from the Hearth to the Mantle-tree.

2. *Price.*] I have bought of theſe Stones in *London* for 20 *s per* Pair.

Cornice.

Comes from the Latin *Coronis*, a Crowning, and makes the third and uppermoſt Member of the *Entablement*, which is different in the ſeveral Orders. The Word *Cornice*, however, is applied to every Prominent or jetting Member that *crowns* any Body; and thus we ſay the *Cornice* of a Pedeſtal and the like. *Cornices* are alſo placed on the Top of Wainſcot, and under the Eaves of Houſes, *&c.*

2. *Kinds.*] There are as many kinds of Cornices as there are Orders of Columns, *viz.* Tuſcan, Dorick, Ionick, Corinthian, and Compoſite; to which may be added, Plain, Cantaliver, Modilion, and Coveing Cornices; Of all which I ſhall treat in their Order.

3. *Tuſcan.*] According to *Vitruvius*, the whole Heighth of the Tuſcan cornice is $\frac{1}{2}$ a Module; which Heighth being divided into four grand Diviſions, the uppermoſt of them goes to the *Boultin*, and *Fillet* under it; and this Diviſion being ſub-divided into four Parts, three of them go to the *Boultin*, and one to the *Fillet*. The two next grand Diviſions go to the *Corona*, or *Crown*, (which is flat and plain) and the lowermoſt grand Diviſion goes to the *Cimatium*; which being again divided into three parts, the uppermoſt of them goes to the *Fillet*, and the other two to the *Cima* or O-G. The Projecture of the whole Cornice, (as alſo of each Member thereof,)

I 2

of,) he makes to be equal to its Heighth; and the under-ſide of the Corona he divides into 11 Parts, whereof he gives 2 to the *Fillet*, and 1 to the *Denticle*, and ſo alternately; for 'tis fitting (ſay he) to have 3 as deep as they are large.

According to *Scammozzi*, the whole Heighth of this Cornice is 39 Minutes, and the Heighth of each particular Member thereof (beginning at the top, and deſcending orderly), is as follows; The upper *List*, or Plint of the Corniſh, 3 m. the *Supercilium*, *Liſt*, *Tinea*, or Eye-brow $1\frac{1}{2}$ m. the upper *Cima*, or O-G. 8 m. the *Liſt* under it $1\frac{1}{2}$ m. the *Corona*, or *Crown* $9\frac{3}{4}$ m. the Liſt, $1\frac{1}{2}$ m. the *Scima*, or greater O-G 6 m. (here's $1\frac{1}{2}$ m left betwixt, for the Depth of the Dentils) the *Supercilium*, or *Liſt*, $1\frac{1}{4}$ m. the *Cimatium* or little O-G 5 m. the *Liſt* 2 m

Palladio makes the whole Heighth of this Cornice 44 m. whereof the *Liſt* at the Top is $3\frac{1}{2}$ m the *Scima Recta* 10 m. the *Liſt* under him $2\frac{1}{2}$ m the Corona 10 m. the *Boultin* 9 m. the *List* $1\frac{1}{2}$ m and the *Cavetto*, or *Hollow* $7\frac{1}{2}$ m

4. *Dorick*.] *Vitruvius* makes two different Faſhions of Dorick Cornices; the whole Heighth of one of 'em is $\frac{1}{2}$ a Module, which divided into two grand Diviſions, one of them (*viz* the upper One) is again divided into 8 Parts, of which 1 part goes to the *List* at the top, and the other 7 to the O G The other grand Diviſion is ſubdivided into 4 parts, of which the uppermoſt, and lowermoſt parts go to the 2 *Cimatiums*, and the 2 middle Parts go to the *Corona*, the *Liſt* of each of thoſe *Cimatium*'s is $\frac{1}{3}$ of the whole *Cimatium*. The whole Heighth of the other faſhion'd Cornice is 40 m. which divided into 9 Parts, 2 ſhall go to the 2 *Facia*'s, 1 to the *Thorus*, or *Boultin* above 'em, 2 to the *Modilions* above that, 2 to the *Crown*, and 2 to the *Cima*, or O-G at the top. The *Modilions*, as alſo the *Crown* being divided, each into 3 Parts, one of 'em ſhall go to their reſpective *Cimatium*'s, of which their *Lists* are each $\frac{1}{3}$ of of the whole.

According to *Scamozzi*, the whole Heighth of this Cornice is 42 m. whereof the *Liſt* at the Top is 2 m. the great O-G 7 m. the *List* 1 m. the little O-G 3 m. the *Corona* 8 m. the *Liſt* 1 m. the *Caſement* 2 m. the *Boultin* 5 m. the *Liſt* 1 m. the *Square* 7 m the *List* 1 m. and the *Boultin* 4 m.

Palladio, in his verbal Deſcription of this *Cornice*, makes the whole Heighth of it to be 35 m but in his Figure 'tis but $33\frac{1}{4}$ m of which the *List* at the Top is $2\frac{1}{4}$ m. the *Scima Recta*, or O-G $6\frac{3}{4}$ m. the *Liſt* 1 m. the *Scima Reverſa* $3\frac{1}{4}$ m. the *Corona* 8 m. the *Ovolo*, or *Boultin* 6 m. the *Liſt* 1 m. and the *Caſement* at the bottom 5 m.

5 *Ionick*.] The whole heighth of this Cornice, according to *Vitruvius*, is about $52\frac{1}{2}$ m. He deſcribes two faſhion'd Corniees in this Order; in one of them he divides the whole Heighth into 11 parts, the two uppermoſt of which go

go to the *Cimatium*, and the *Boultin* under it, and this Space being ſub-divided into 6 parts 2 of them go to the *Fillet* of the *Cimatium*, 3 to the *O.G*, and 1 to the *Boultin*. The next two grand Diviſions go to the *Corona*. The next 3 grand Diviſions go to the *Cartouſes*, and the *Cimatium* over 'em ; and this Space being divided into 5 Parts, 1 of 'em makes the *Cimatium*, of which the *Fillet* is $\frac{1}{3}$ of the whole. Then 1 $\frac{1}{2}$ of the next grand Diviſion goes to the *Boultin*, and *Fillet* over it, of which the *Fillet* is $\frac{1}{7}$ part of the whole. Again, 1 $\frac{1}{2}$ of the next grand Diviſion goes to the *Caſement*, and *Fillet* over it, of which the *Fillet* is $\frac{1}{4}$ of the whole. And the laſt grand Diviſion goes to the *Cimatium*, of which the *Fillet* is $\frac{1}{3}$ part of the whole In the other faſhion'd Cornice, he divides the whole Heighth into 6 Parts, the uppermoſt of which goes to the O G, whereof its *Fillet* is $\frac{1}{6}$ part, the next grand Diviſion being ſub-divided into 3 parts, the uppermoſt of 'em goes to the *Cimatium*, (of which its *Fillet* is $\frac{1}{3}$ part,) and the other two to the *Corona* The next two grand Diviſions are ſub-divided into 5 Parts, the uppermoſt of which goes to the *Cimatum* (of which its *Fillet* is $\frac{1}{3}$ part) and the other 4 to the Cartouſes. The next grand Diviſion being ſub-divided into 4 Parts, 3 of them go the *Boultin*, and 1 to the *Fillet* under it. And the laſt grand Diviſion being ſub-divided into 4 parts, 3 of 'em go to the *Caſement*, and 1 to the *Cimatium*, of which its *Fillet* is $\frac{1}{7}$ part.

Scammozzi make the whole Heighth of this Cornice 42 m. whereof the *Liſt* at the Top is 2 m the *Scima Recta* 5 $\frac{1}{2}$ m the *Liſt* 1 m. the *Scima Reverſa* 2 $\frac{1}{2}$ m the *Corona* 6 $\frac{1}{2}$ m. the *Scima Reverſa* 2 $\frac{1}{2}$ m. the *Cartouſes* 7 m the *Boultin* 4 m the *Liſt* 1 m. the *Square* 5 m. the *Liſt* 1 m and the *Boultin* 4 m.

Palladio makes the whole heighth of this Cornice 46 $\frac{1}{2}$ m. whereof the *Liſt* at the top is 2 $\frac{1}{2}$ m. the *Scima Recta* 7 m. the *Liſt* 1 $\frac{1}{4}$ m. the *Scima Reverſa* 3 $\frac{1}{2}$ m the *Corona* 8 m. the *Scima Recta* over the Modilions 3 $\frac{1}{4}$ m. the *Modilions* 7 $\frac{1}{2}$ m the *Liſt* 1 m the *Ovolo*, or *Boultin* 6 m. the *Liſt* 1 $\frac{1}{2}$ m. and the *Cavetto*, or *Hollow* 5 m.

6. *Corinthian*.] The whole Heighth of this Cornice, according to *Vitruvius*, is about 1 Module He deſcribes two different faſhion'd Cornices in this Order ; in one of which he divides the whole Heighth into 5 Parts, the uppermoſt of which goes to the O-G, of which its *Fillet* is $\frac{1}{6}$ part. Then 1 $\frac{1}{4}$ of the next grand Diviſions goes to the *Corona* and *Cimatium* over it, of which ſpace the *Cimatium* is $\frac{1}{3}$ part, and its *Fillet* $\frac{1}{3}$ of that. Then 1 $\frac{3}{4}$ of the next grand Diviſions goes to the *Modilions*, and *Cimatium* over them, of which ſpace the *Cimatium* is $\frac{1}{7}$ part. And the laſt grand Diviſion goes to the *Boultin*, and *Fillets* over and under it ; and this being divided into 3 parts, the lowermoſt goes to the Fillet, and the other two being again divided into 6 parts, 5 of them go to the

the *Boultin*, and the other to the *Fillet* over it; in the other fashion'd Cornice, he divides the whole Heighth into 9 parts, of which the two uppermost being divided into 4 parts, 3 of 'em go to the *O.G*, (whose Fillet is $\frac{1}{6}$ of the whole) and the other to the *Cimatium* over the Corona, (whose *Fillet* is $\frac{1}{5}$ of the whole.) The next two grand Divisions go to the *Corona*. The next two grand Divisions go to the *Modilions*, and the *Cimatium* over 'em, $\frac{1}{4}$ of this space goes to the *Cimatium*, (whose *Fillet* is $\frac{1}{3}$ of the whole Cimatium) and the rest to the *Modilions*. The next two grand Divisions go to the *Boultin*, and *Fillet* over and under it, which Fillets are each $\frac{1}{7}$ of the whole. And the last grand Division goes to the *Cima* at the foot of the Cornice.

According to *Scammozzi*, the whole Heighth of this Cornice is 46 $\frac{3}{4}$ m. whereof the *List* of the *Scima Recta* is 2 m. the *Scima Recta* 6 $\frac{1}{2}$ m. the *List* of the *Scima Reversa* 1 m. the *Scima Reversa* 3 $\frac{1}{4}$ m. the *Half-round* 1 $\frac{1}{2}$ m. the *Corona* 7 $\frac{1}{2}$ m. the *Cimatium* 3 $\frac{1}{2}$ m. the Modilions 8 $\frac{1}{2}$ m. the *List* 1 m. the Boultin 5 m. the *List* 1 m and the *Scima* 5 m.

The whole Heighth of this Cornice, according to *Palladio*, is 50 m. whereof 2 $\frac{1}{3}$ m. goes to the *List* of the *Scima Recta*; the *Scima Recta* is 8 $\frac{1}{3}$ m the *List* $\frac{2}{3}$ m. the *Scima Reversa* 3 m. the *Corona* 7 $\frac{1}{3}$ m the List of the O.G over the Modilions $\frac{2}{3}$ m. the O G 2 $\frac{2}{3}$ m the Modilions 8 $\frac{1}{2}$ m. the Boultin 4 $\frac{1}{2}$ m. the List 1 m. the Boultin 5 $\frac{1}{2}$ m. the List 1 m. and O.G 4 $\frac{1}{2}$ m.

7. *Roman Composita*, or *Compound*.] The whole Heighth of this Cornice, according to *Vitruvius*, is equal to the Diameter of the Column above, which is about 52 $\frac{1}{2}$ m. He describes two different fashion'd Cornices in this Order; one of which he divides into 2 Parts, the uppermost of which goes to the O- G, (whose *Fillet* is $\frac{1}{7}$ of the whole,) and the undermost to the Corona and Cimatium over it; and this Space being divided into 4 parts, 3 of them go to the Corona, and one to the Cimatium, whose Fillet is $\frac{1}{3}$ of the whole *Cimatum*.

Scammozzi makes the whole Heighth of this Cornice 48 m. and *Palladio* 45 m. but for the Heighth of each particular Member, they leave us very much in the dark; for according to either of them the Sum of the Particulars will never make the whole Heighth; and besides, *Palladio* sets down no Dimensions to several of the Members of this Cornice. So that I think, a Man is but little the wiser for what any of these Authors say of this Cornice.

8. *Cantaliver*.] Workmen tell me, that those are call'd Cantaliver-cornices, that have Cantalivers under them. See *Cantalivers* N. 1.

9. *Modilion*.] Workmen tell me, That Modilion cornices are such as have Modilions under them. See *Modilions*.

10. *Coveing*.] Workmen tell me, That they call that a Coveing-cornish, which has a great

great *Casement*, or *Hollow* in it, which is commonly lathed and plaister'd upon Compass, Sprockets, or Brackets.

11. *Price*] Some Cornices (says Mr. *Leybourn*) are valued by the piece, dearer, or cheaper, according to their Largeness, Goodness of the Stuff, and Curiosity of Workmanship: Others are measur'd, and rated by the Foot *Running measure*, i. e. by the Number of Feet in length only. Experienced Carpenters tell me, That for making of plain Cornices (without any Carving) under the Eves of a House, they commonly have 1 *s per* Foot, Running-measure. Mr. *Wing* tells us, That Cornices are valu'd according to their Nature, and Bigness; a Modilion-cornice (of Free-stone) of 18, or 20 Inches thick, is worth (says he,) 5 or 6 *s. per* Foot, Running-measure He also tells us, (in Joyners Work,) That a Modilion-cornice, with its carved Work, is worth 7 *s. per* Foot. And a plain Modilion-cornice of 12, or 14 Inches, (says he) will be worth 3 *s.* 6 *d.* or 4 *s. per* Yard, running-measure. A Brick-cornice, (as some Workmen tell me) 2 *s.* 6 *d. per* Foot.

Corona, *Coronis*, or *Crowning*.

These Words are indifferently applied to any Thing that finishes an Ornament in Architecture; as for Instance to a *Cornice* or *Pediment*, &c.

Corsa.

This Word, as well as *Fascia* and *Tænia*, in *Vitruvius*, signifies what we call Plat-band. See *Plat-band*.

Coveing.

1. *What*. See *Cornice*, N. 10. Also, Workmen tell me, That Coveing is also used in this Sense; *viz*. When Houses are built projecting forth over the Ground-plot, and that is (as commonly 'tis) turn'd with a Quadrant of a Circle, (or Semi-arch) of Timber, which is lathed and plaister'd; (under which People may walk dry; as 'tis much us'd at *Tunbridge-Wells*, on the upper Walks;) I say, such Work is commonly call'd *Coveing*.

2. *Price*.] Mr. *Wing* (in his *Geodetes Practicus Rederius*) says, That the Carpenter's Work of Coveing, is worth 4 *s. per* Square.

Coussnet.

The first Stone whence a Vault or Arch commences, is so call'd. The little Cornice or Plinth that crowns the Peer and supports the *Coussinet*, is call'd *Impost*.

Crown,

As *Corona*.

Crown-post,

Is that Post, which (in some Buildings) stands upright in the middle, between two principal

cipal Rafters, from which there goes Struts, or Braces to the middle of each Rafter. It is alſo call'd a King-piece, or Joggle piece.

Croſs-garnets.

See *Hinges*. N. 2.

Croſs-grain'd.

Timber is ſaid to be croſs-grain'd, where a Bough, or ſome Branch ſhoots out on that part of the Trunk of the Tree; for the Bough, or Branch ſhooting forwards the Grain of that Branch ſhoots forward alſo, and ſo runs a-croſs the Grain of the Trunk; and if it be well grown together, it will ſcarce be perceiv'd in ſome Stuff, but only in Working.

Croſs-Multiplication.

1. *What.*] Croſs-Multiplication is the multiplying of Feet and Inches by Feet and Inches, or Feet, Inches, and 12th parts of Inches, by Feet, Inches, and (12th) parts of Inches. 'Tis ſo call'd, becauſe they multiply a-croſs, as I ſhall ſhew how in the following Number. This way of Multiplication is much us'd by Workmen, in meaſuring their Work; but, I think, none of'em are ſo nice, as to take their Dimenſions to parts of Inches, except Glaziers.

2. *How perform'd*] Set the Multiplicand over the Multiplier, as is done in the following Examples, and then multiply as the Lines direct; obſerving to ſet down the particular Products under Feet, Inches, or Parts reſpectively, according to theſe Rules.

1. Feet multiply'd by Feet, produce Feet.

2. Feet by Inches, produce Inches.

3. Feet (by 12th) parts, produce parts.

4. Inches by Feet, produce Inches.

5. Inches by Inches, produce Primes, (or 12th) Parts (of an Inch.)

6. Inches by (12th) parts, produce ſeconds, or 12th parts of the 12th. part of an Inch.

7. Parts by Feet, produce (12th) parts.

8. Parts by Inches, produce Seconds.

9. Parts by Parts, produce Thirds, (or 12th. Parts of a Second.)

But *Note*, That in ſetting down the Products of each Denomination, (except the Feet) you muſt ſet down only the odd ones above 12, or 12's, carrying all the 12's as ſo many Unites to the next greater Denomination.

Exam-

Example 1.

Let it be required to Multiply 5—3 (F. In.) by 2 F. 4 In. set down the Numbers thus

	F. In. oo
	5—3—oo
	1+1
	2—4—o

Say 2 times 5 is 10 Feet	10—0—0
Then 2 times 3 is 6 Inches	6—0
Then 4 times 5 is 20 Inches, or 1 F. 8 In.	1—8—0
And lastly, 4 times 3 is 12 parts, or 1 Inch.	0—1—0
The whole Sum is	12—3 0

or 12 Feet and a quarter.

Example.

Let it be requir'd to Multiply 5 Foot 3 Inches, and 6 Parts, (or a half) by 2 Foot, 4 Inches, and 6 Parts.

Set down the Numbers thus
Then I say,

	F. In. P. S. T
	5—3—6
	1—1—1
	2—4—6

2 times 5 Feet is	10—0—0—0—0
2 times 3 Inches is	6—0—0—0
2 times 6 Parts is	1—0—0—0
4 times 5 Feet is	1—8—0—0—0
4 times 3 Inches is	1—0—0—0
4 times 6 Parts is	2—0—0
6 times 5 Feet is	2—6—0—0
6 times 3 Inches is	1—6—0
6 times 6 Parts is	3—0
The whole Sum is	12—6—9—9—0

that is 12 Feet 6 Inches and $\frac{3}{4}$ of an Inch, and $\frac{3}{4}$ of a Twelfth part of an Inch.

Cubicle.

A Bed-chamber.

Culinary,

Of, or belonging to the Kitchin.

Culvertail,

As Dovetail.

Cupola,

In Architecture, is a small Room (either Circular, or Polygonal)

lygonal) ſtanding on the very top of a Building; ſome call it a Lanthorn.

Cy-mace-macium,

As *Cimatium*.

Dado.

SEE Capital. N. 2.

Deals.

Of Dreſſing.] Dreſſing of Deals' [*i. e.* rough-plaining them over with a Fore-plain, that they may dry,] is worth (ſays Mr. *Wing*,) 1 *s. per* ſcore: and ſo I know ſome Workmen have; tho' others tell me, they have known them done for 9 *d. per* ſcore.

Deal-floors.

Of Laying.] The Laying of ordinary Deal-floors, [*i e.* plaining, and joining 'em, *&c*] is worth 5 *s per* Square. But if they are laid with Dove tail, or Key joynts, without Pins or Nails, ſome Workmen tell me, They have 10 *s. per* Square. And if the Workman find Deals, and lay them the ordinary way, 'tis worth from 24 to 30 *s per* Square, according to the goodneſs of the Deals. But if the Deals are very good, and laid either with Dovetail, or Key-joints, (without Nails, or Pins) 'tis worth 35 *s*, or 40 *s*. the Square. See Pl. Floors.

Deck-nails.

See Nails. N. 6.

Decor.

This Word is perfect Latin, and ſignifies the keeping of a due Reſpect between the *Inhabitant* and *Habitation*. Whence *Palladio* concludes, That the *Principal Entrance* muſt never be regulated by any certain Dimenſions, but by the Dignity of the Perſon that is to live in it; yet to exceed, rather in the more, than in the leſs, is a Mark of Generoſity, and may be excus'd with ſome noble Emblem, or Inſcription, as that of the *Conti di Bavilacqua*, over his large Gate at *Verona*, (where perchance had been committed a little Diſproportion:) *Patet Janua, Cor magis*.

Denticles.

Ornaments in a Cornice, cut after the manner of Teeth. (from *Dens*, a Tooth.) Theſe are particularly affected in the *Doric* Order, and the Square Member wherein they are cut is called the *Denticule*, in Latin *Denticulus*.

Diaſtyle.

An Edifice, where the Columns are placed at the diſtance of three of the Diameters from one another.

Diamond-glaſs.

See Glaſs-quarry.

Diamond.

Diamond-pavement.

See Paving. N. 10.

Die.

This Term is applied to any ſquare Body, as the *Trunk* or *Naked* of a *Pedeſtal*, which is that Part included between the *Baſe* and *Cornice* thereof

Digging.

Of *Meaſuring*.] The digging of the Ground for Cellars, and for the Foundations of Buildings, is commonly done by the Yard ſolid, containing 27 ſolid Feet; and that is uſually counted a Load. Therefore the Dimenſions being given in Feet, multiply the length by the breadth, and the Product by the depth, dividing this laſt Product by 27, and the Quotient will give the Content in ſolid Yards.

Diminiſhing.

Of *Columns*.] See Columns N. 9.

Diptere.

Among the Ancients, a kind of Temple or other Edifice encompaſs'd round with a double Row of Columns. It ſignifies in *Greek* two-wing'd The Pſeudo (or falſe) Diptere was the ſame; excepting, that inſtead of the double Row of Columns, this was only encompaſs'd with a ſingle one.

Dogg nails.

See Nails. N. 7.

Dome.

Alſo a Cupola, a round piece of Architecture, (reſembling the Bell of a great Clock) ſet upon the top of a Building, particularly upon Cathedral Churches, where it ſerves for the Bell-tower.

Dorick-order.

See Column, and Order.

Doors.

1. *What*.] Doors are thoſe parts of a Building, that are ſerviceable for the Paſſage *in* and *out* of Perſons.

2. *Situation of*.] Firſt, ſee that the Doors of a Houſe be as few in number, and as moderate in Dimenſions, as may poſſibly conſiſt with other due Reſpects: For in a word, *all Openings are Weaknings*. Secondly, That they do not approach too near the Angles of the Walls; for 'twere a moſt eſſential Soleciſm to weaken that part, which muſt ſtrengthen all the reſt: A Precept well recorded, but ill practiſed by the *Italians* themſelves, particularly at *Venice*. Thirdly, Let the Doors, if poſſible, be right over one another, that the void may be upon the void, and the full upon the full; which will be a great ſtrengthning to the whole Fabrick. Fourthly, Let them (if poſſible)

ble) be placed oppoſite to one another, in ſuch manner, that one may ſee from one end of the Houſe to the other; which will not only be very graceful, but alſo moſt convenient, in reſpect 'twill cool the Houſe in Summer, by letting the Air through the Houſe, and in Winter to keep out the Wind, which way ſoever it ſit. Fifthly, 'Tis not only Ornamental, but very ſecure to turn Arches over the Doors, which will diſcharge them in a great meaſure, from the Super-incumbent weight, which might otherwiſe preſs upon them too much.

3. *Dimenſions of.*] Inner-doors in large Buildings ought to be three Foot broad and upwards and their heighth twice their breadth. And Inner-doors in leſſer Buildings, ought never to be leſs than 2 ½ Foot broad, and 5 ½ Foot high.

4. *Price of.*] *Doors made of plain whole Deal, and Rabited* are for Stuff Nails, and Workmanſhip, valu'd at 3 *d.* or 4 *d.* the Superficial Foot; the Workmanſhip only, about 2 *s.* or 2 *s.* 6 *d. per* piece; as ſome Workmen tell me. But *double-doors*, Batton'd, and made Wainſcot Faſhion, may be worth (for Workmanſhip and Materials) 7 *d.* the Foot, and for the Workmanſhip alone, about 4 *s.* or 5 *s. per* piece. *Folding-doors* and *Caſes*, (as ſome Workmen tell me) are worth about 20 or 30 *s. per* Pair; and *Balcony-doors* and *Caſes*, the ſame. *Ordinary doors* without Plaining, are worth making and hanging up, about 1 *s. per* piece. In Stone and Brick-buildings, *Architrave-door Caſes* are worth, according to the breadth of the Mouldings, 1 *d.* an Inch, *i. e.* if the breadth of the Moulding, (from the out-ſide to the inſide of the Frame) by 9 Inches, 'tis worth 9 *d per* Foot running-meaſure; if 10 Inches, 10 *d. per* Foot; and ſo proportionable, more or leſs. And *Frontiſh-doors* in great Buildings, with their Ornaments, as Pilaſters, *&c.* are worth, (according to their Magnitude, and Variety of Workmanſhip included) ſome 3 *l* ſome 5 *l.* ſome more, to 10 or 20 *l per* piece; and perhaps more. See *Batten-door*. N. 2.

Dormant-tree.

In Architecture is a great Beam lying croſs a Houſe, otherwiſe call'd a Summer. See Summer.

Dor-man mer,

In Architecture is a Window made in the Roof of a Houſe, it ſtanding upon the Rafters. Dormers are commonly rated at ſo much *per* piece, according to their bigneſs, *&c.*

Dorman-tiles.

See Tiles. N. VIII.

Dor-tor-mitory.

A Sleeping-place.

Doucine

Doucine.

See *Cima.* It is an upright O-G.

Dove-tails.

A ſort of Joynts, or Hinges, ſo call'd, becauſe they reſemble the Tail of a Dove, or Pigeon.

Dove-tailing.

In Architecture, is a manner of faſtning Boards, (or any Timber) together, by letting one piece into another, in the Form of a Dove's Tail.

Drag.

In Architecture, a Door is ſaid to drag, when in opening and ſhutting it drags upon the Floor.

Dragon-beam.

Dragon-beams are 2 ſtrong Braces, or Struts, that ſtand under a Breſſummer, meeting in an Angle upon the Shoulder of the King-piece.

Drapery.

A Term in Architecture, and Painting it being a Work wherein Cloaths are repreſented. Alſo as Cilery.

Drip.

This is call'd *Larmier* in *French* (in *Engliſh Weeper*) from *Larme* (a tear) and that from the *Latin Lachrymæ.* It is the bottom of the *Corona*, (the Corona is that large ſquare Moulding immediately under the Cymaiſe.) It is call'd *Larmier*, becauſe the Rain-water is by means thereof forced to fall drop by drop on the Ground; dripping like Tears.

Draught, or Draft.

1. *What.*] A Draught, or Draft, is the Picture of an intended Building deſcribed on Paper; wherein is laid down (by Scale, and Compaſs) the deviſed Diviſions, and Partitions of every Room, in its due proportion to the whole Building.

2. *Its Uſefulneſs.*] As 'tis uſual, ſo 'tis (alſo) very convenient for any Perſon before he begins to erect a Building, to have Deſigns, or Draughts drawn upon Paper, or Vellum; in which Draughts the Ground-plot, or Ichnography of each Floor, or Story, is delineated, and repreſented: As alſo the Form and Faſhion of each Front, together with the Windows, Doors, and Ornaments, (if they deſign any) are to be ſhewn in the Orthographies, or Draughts of the Uprights.

Sometimes more Fronts than one are ſhewn perſpectively in a Draught, and then 'tis called Scenography; but this is not eaſily underſtood, except by thoſe

thoſe that underſtand the Rules of Perſpective. And therefore 'twill be more intelligible to the ſeveral Workmen to have a Draught of each Front, in a particular Paper by it ſelf; and alſo to have a Draught of the Ground-plot, or Ichnography of every Floor, or Story, in a Paper by it ſelf; because many times the Conveniencies, or Contrivances in one Story differ from thoſe in another, either in the bigneſs of the Chimneys, or Diviſions of the Rooms, ſome being larger in one Story than in another, and ſometimes having more Chimneys in one Story than in another, &c.

All which things being well conſider'd, and drawn on Paper, before the Building is begun; theſe Draughts will be a great Guide to the Workmen, and ſave them a great deal of time in contriving their Work; and beſides there will be no need of Alterations, or tearing, or pulling the Building to pieces after 'tis begun; which, beſides the Hindrance of the Procedure, makes the Building lame and deficient; nothing being ſo well done, when 'tis put up, and pull'd down, and ſet up again, as if it were well done at firſt. Beſides, it makes the Workmen uneaſie to ſee their Work, (in which they have taken a great deal of Pains, and uſed a great deal of Art) to be pull'd down again. See Building. N. II. 2.

The drawing of Draughts is moſt commonly the Work of a Surveyor, tho' there be many Maſter-workmen that will contrive a Building, and draw a Draught, or Deſign thereof, as well almoſt (and better than ſome) Surveyors. But whoever makes a Draught of a Building, ought to be very well skill'd in the Theorical Part of Architecture.

Drips.

In Architecture, are certain kind of Steps (made on a flat Roof) to walk upon, a way of Building much us'd in *Italy*. The Roof is not quite flat, but a little rais'd in the middle; and thoſe Steps, or Drips, lie each a little inclining to the Horizon.

Drops.

In Architecture are an Ornament in the Pillars of the Dorick Order, underneath the Trigliphs, repreſenting Drops, or little Bells.

Dutch Bricks.

See Bricks. N. 5.

Eaves.

IN Architecture, is the Margin of the Roof of a Houſe; that part of the Roof that hangs over without the Walls.

Eaves-

Eaves-lath.

That thick Feather-edg'd-board, generally nail'd round the Eves of a House, for the lowermost Tiles, Slate, or Shingles to rest upon. Eaves-laths are commonly sold for 1 *d.* ½ or 2 *d.* a Foot, (running-measure) according as they are of goodness.

Echinus.

Is sometimes use to signifie the *Quarter Round*, but more commonly that part of it which includes the *Ovum* or *Egg*. It comes from the *Greek Echinos*, the Shell of a Chesnut.

Eggs.

See Anchors, and Quarter-Round.

Elaboratory.

A Place to work in; properly a Chymist's Work house, or Shop.

Embossing.

In Architecture, is a kind of Sculpture, or Engraving, wherein the Figure sticks out from the Plain whereon it is Engraven, and according as it is more or less protuberant, is call'd by the *Italians Basso Mezzo* or *Alto-relievo*, and by the *English* Bas-relief, Mean-relief, or High relief.

Embrasure.

In Architecture, is the Enlargement that is made in a Wall, on the inside of a Window, or Gate, to give the more Light, or for the more Convenience of the Gate, or Window.

Entablement, or Entablature.

By *Vitruvius* and *Vignola* is called *Ornament*, and signifies the *Architrave*, the *Freeze*, and the *Cornice* together. It is likewise call'd *Trabeation*, and is different in the different Orders. The word seems borrowed from the *Latin*, *Tabulatum*, a Ceiling, because we suppose the *Freeze* to be form'd by the Ends of the *Joysts* which bear upon the *Architrave*.

Entresole.

Sometimes call'd *Mezanine*, is a kind of a little Story, contrived occasionally at the top of the first Story, for the conveniency of a Ward-robe, *&c.*

Entry.

In Architecture, is a Room design'd only (or chiefly) for *Passage* to and fro betwixt other Rooms, or from the outer Door into the House.

Epis-tyle-tylium.

As Frieze.

Eurithymia.

A Term of Architecture, us'd by *Vitruvius*, by which he intends only, That agreeable Harmony, that ought to be between

tween the length, breadth, and heighth of each Room in a Fabrick.

Euſtyle.

Is the beſt manner of placing Columns with regard to their Diſtance, which *Vitruvius* maintains to be of two Diameters and a Quarter. The word is compounded of the *Greek Eus*, well, and *Stylos* Column.

Eye-brow.

As Liſt, or Fillet. See Capital.

Eye,

In Architecture, is the middle of an Ionick Volute, or Scrol, cut in the Form of a Roſe.

Fabrick.

1. WHat] A Church, a Houſe, or any other Building.

2. Of *Cenſuring*] I am deſirous, (ſays Sir *Henry Wotton*) to ſhut up theſe Elements of Architecture with ſome methodical Directions for cenſuring of Fabricks already rais'd; for indeed, without ſome way to contract our Judgment, which among ſo many Particulars, would be loſt by Diffuſion; I ſhould think it harder to be a good Cenſurer, than a good Architect; becauſe the *working part* may be helped by *Deliberation*, but the *Judging* muſt flow from an *Extemporary Habit*. Therefore, (not to leave this laſt piece of Architecture without ſome Light,) I could wiſh him that comes to examine any noble *Work*, firſt of all to examine himſelf, whether the ſight of many brave Things before (which remain like impreſſed Forms in his Mind,) have not made him think nothing *good*, but that which is the beſt; for this Humour were too ſowre. Next, before he ſettle any Opinion upon the Work, let him by all Means ſeek to inform himſelf of the Age thereof. And if he find the apparent Decays to exceed the Proportion of Time; then let him conclude, That either the Materials were too ſlight, or that the Seat is naught.

Now, after theſe Premiſſes, if the Building be found to bear his Years well, then let him ſuddenly run backwards, (for the Method of *Cenſuring* is contrary to the Method of *Compoſing*) from the *Ornaments*, (which firſt allure the Eye) to the more *eſſential Members*, till at laſt he be able to form this Concluſion; That the Work is Commodious, Firm, and Delightful; which are the three capital Conditions requir'd in good Building, by all Authors, both ancient and modern. And this is, (as I may term it,) the moſt ſcientifical way of Cenſuring. There are two other ways which I muſt not forget. The firſt, which you may find in *Georgio Vaſſari*, before his laborious Work of the Lives of of Architects,) is to pass a running Examination over the whole Edifice, according to the Properties of a well-shapen Man. As whether the *Walls* stand upright upon a clean Footing and *Foundation*; whether the *Fabrick* be of a beautiful Stature; whether for the *Breadth* it appear well burnished; whether the *principal Entrance* be in the middle of the Front, or Face; whether the *VVindows*, (as our Eyes) be set in equal Number and Distance on both sides, and whether the Offices (like the Veins in our Bodies) be usefully distributed, *&c* The second way you may find in *Vitruvius* himself, *Lib.* 1. *Cap.* 2. Where he summarily determines 6 Considerations, that accomplish this whole Art, *viz. Ordinatio, Dispositio, Eurythmia, Symmetria, Decor & Distributio*, each of which see in their proper Places.

Face,

In *Architecture*, is any Member that has a great Breadth, and but small Projecture, as the Architrave in the Front of a Building.

Face of a Stone.

By the *Face of a Stone*, Workmen mean that Superficies or Plain of the Stone that is to lie in the Front of the Work; which is very easily known when the Stones are scapt'd; for the *Face* is always opposite to the *Back*, and the *Back* goes rough as it comes from the Quarry. But in rough Stones, Workmen generally choose to make one of those sides the *Face*, which in the Quarry lay perpendicular to the Horizon, and consequently the breaking (and not the cleaving) way of the Stone. For a clearer understanding of this, see *Stone*, N. 4.

Faceing.

Of Timber Buildings with Brick.] Some Workmen tell me, That they have sometimes faced Timber-buildings with Brick; which, say they, is thus done. — All betwixt the Timber, the Wall is a Brick's-length thick; (or a 9 Inch Wall,) and against the Timber but half a Brick, or 4 ½ Inch Wall. But Workmen do not approve of this way of facing of Timber-buildings, by reason the Mortar doth so extreamly burn the Timber.

Facia, or Fascia.

M *Perrault* says It signifies any flat Member; as the Band of an Architrave, *&c.* There are some who write it *Fasce*, grounded upon the Latin Word *Fascia*, a large Turban, which *Vitruvius* makes use of on the like Occasion. In short, it is no more but a broad List, or Fillet. See *Fillet*. They are commonly made in Architraves, (see *Architrave*;) and in the Cornice of Pedestals, see *Pedestal*. In Brick-buildings, *Fascia's* are certain Juttings out of the Bricks, over the Windows of each Story, except the upper one. And these are sometimes plain, like those

thoſe of Columns; but ſometimes they are moulded; which ſhew very handſome: And this Moulding is commonly a *Scima-reverſa* at the bottom, above which are two plain Courſes of Bricks, then an Aſtragal, and laſtly a Boultin, or as Workmen (by Corruption) call it a Boultel, or Boltel. In Stone-buildings 'tis the ſame as in Brick, and they are alſo ſometimes plain and ſometimes moulded with a *Scima-reverſa* or O-G. The *Price* of *Faſcia*'s, if the Workmen find Materials, is commonly about 10 *d. per* Foot Running-meaſure, and the Workmanſhip only about 6 *d.* or 8 *d. per* Foot By the Word *Faſcia*, (as alſo *Tænia* and *Corſa*) in *Vitruvius*, is ſignified what we call Plat-band. See *Plat-band.*

Feather-edg'd.

Boards, or Planks, that are thicker on one Edge, than on the other, are call'd Feather-edg'd Boards, *&c.*

Felling of Timber.

See *Timber.*

Fencing

1. *With Pale.*] Some Workmen tell me, That for Paleing with 3 Rails, Cleft pails, Rails, and Poſts, cleaving and ſetting up; they have 3 *s.* 6 *d. per* Rod, felling the Timber and all. But then their Materials are laid down to their hand.

2. *With ſingle Rail and Poſts.*] Some Workmen tell me, That Fencing with ſingle Rail and Poſts, Felling, Cleaving and ſetting up, is commonly done for 8 *d.* or 10 *d. per* Rod; but then their Materials muſt be laid down to their hand, that they may have no carrying. Others tell me, That they have known it done for 4 *d.* 5 *d.* or 6 *d. per* Rod, Felling, Cleaving, and ſetting up; but then the Fence muſt be croſs a Field, or the like, where it is eaſie digging the Poſt-holes, (and where there is a pretty many Rods together,) and the Materials muſt alſo be laid down to their hand,) and not in Gaps, in Hedges, and the like, where 'tis difficult digging, and but a little at a place; for there 'tis worth 8 *d.* 10 *d.* or 1 *s. per Rod.*

Fence-walls.

Walls of Brick, or Stone, made round Gards, *&c.* See *Walls*, N. V.

Feſtoon.

An Ornament of carv'd Work, in the Manner of a Wreath, or Garland of Flowers or Leaves twiſted together, thickeſt at the middle, and ſuſpended by the two Extremes, whence it hangs down perpendicularly. The Word *Feſtoon* may probably be derived from *Feſtus*, as being uſually employ'd on Feſtival Occaſions.

Fillet.

Fillet.

Is any little ſquare Moulding, which accompanies or crowns a larger.

Fire-ſtone.

1. *VVhat.*] *Rigate-ſtone*, commonly call'd Fire-ſtone, is a ſort of Stone very good, (and much us'd) for Chimney-fire Hearths, Ovens, Stoves, *&c*

2. *Price.*] Mr. *Miller*, Stone-cutter in *Cold harbour*, *London*, tells me, That they uſually ſell Fire-ſtone Hearths, at 1 *s. per* Foot. And Chimney-corner-ſtones of Fire-ſtones at 20 *s. per* Pair. And Blocks to ſet up Coppers, each being about 3 Foot long, 1 ½ Foot broad, and 8 or 9 Inches thick, at 6 *s* 8 *d. per* Piece.

Flat-head-nails.

See *Nails.* N.

Flat-point-nails.

See *Nails.* N. 9.

Flemiſh-bricks.

1. *VVhat*] They are a ſort of Bricks brought out of *Flanders*, and uſed for Paving; being much neater and ſtronger than common, or Clay-bricks. They are of a yellowiſh Colour, and each Brick is 6 ¼ Inches long 2 ½ Inches broad, and 1 ¼ Inch thick. Now, allowing ¼ of an Inch for the Joynt, 72 of 'em will pave a Yard Square; but if they be ſet edge-ways, then to pave a Yard Square will require 100 Bricks.

2. *Price*] They are commonly ſold for 2 *s.* the Hundred.

Flint-Walls.

See *VValls.* N. VII.

Floors.

1. *VVhat.*] A Floor in Architecture is the under-ſide of a Room on which we walk. Floors are of ſeveral ſorts; ſome are of Earth, ſome of Brick, ſome of Stone, and ſome of Wood. Carpenters, by the Word *Floor*, underſtand as well the fram'd Work of Timber, as the Boarding over it.

2. *Earthen.*] Earthen-floors are commonly made of Lome, and ſometimes (for Floors to make Malt on) of Lime, and Brook-ſand, and Gun-duſt, or Anvil-duſt from the Forge; the particular Method of both which I muſt at preſent omit; but I cannot paſs by that Receipt (given us by the Ingenious Sir *Hugh Plat*, *To make an Artificial Compoſition, wherewith to make ſmooth, gliſtering and hard Floors, and which may alſo ſerve to plaiſter VValls with.* Take, (ſays he) Ox-blood, and fine Clay, and tempering them well together, lay the ſame in any Floor, (or Wall) and it will become a very ſtrong and binding Subſtance; as I have been told by a Gentleman and Stranger, who affirm'd to me, that the ſame is of great Uſe in *Italy*.

In

In

3. *Brick and Stone.*] These I shall refer to *Paving.* See *Paving* N. 1. to 9.

4. *Boarded.*] Concerning Boarded-floors, 'tis to be observ'd, that tho' Carpenters never Floor their Rooms with Boards till the Carcass is set up, and also enclos'd with Walls, least the Weather should wrong the Flooring; yet they generally rough plane their Boards for Flooring, before they begin any thing else about the Building, that they may set them by to *season*; which they do thus. They lean them one by one on End a-slant, with the Edge of the Board against a *Balk*, (or, as 'tis call'd in some parts of *Sussex* a *Perch*;) somewhat above the Heighth of half the length of the Board, and set another Board in the same Posture on the other side of the *Balk*, so that above the *Balk* they cross one another; then on the first side they set another Board in that Posture, and on the second side another, and so alternately, till the whole Number of Boards are set on end: Being set in this Posture, there is left the Thickness of a Board between every board, all the length, but just where they cross one another, for the Air to pass through to dry and shrink them, against they have Occasion to use them: But they set them under some cover'd Shed, that the Rain or Sun comes not at them: For if the Rain wet them, instead of shrinking them, it will swell them; or if the Sun shine fiercely upon them, it will dry them so fast, that they will *tear* or *shake* 'em, as they Phrase it, that is, in plain *English*, split or crack They have another way to dry and season their Boards for Floors, *viz.* By laying them flat upon 3 or 4 Balks, each Board about the Breadth of a Board asunder, the whole Length of the *Balks*. Then they lay another Lay of Boards athwart upon them, each Board also the Breadth of a Board asunder; then another Lay athwart the last, and so till all are thus laid. So that in this Position also they lie hollow for the Air to play between them.

5 *Of Measuring*] Floors boarded are commonly measur'd by the Square (of 100 superficial Feet) by multiplying the Length of the Room in Feet, by the Breadth in Feet, and the Product is the Content in Feet; then measure the Chimney-ways, and Well-holes for Stairs by themselves, and deduct their Content in Feet from the whole Content in Feet, and from the Remainder cut off two Figures on the Right-hand and what remains on the Left-hand is Squares, and what is cut off are odd Feet of the Content of Flooring in that Room.

6. *Price*] The *Framing of Floors* in ordinary Buildings, (says Mr. *Wing*) is worth 7 or 8 *s. per* Square, in great Buildings, 10 or 11 *s.* But several Workmen in *Sussex* tell me, That

That they commonly have but 4 *s. per* Square, for framing of Floors in ordinary Buildings. And ſome Workmen (in *Suſſex*) tell me, That if they Frame the Joyſts the whole Depth of the Girder, and pay for ſawing the Timber, they have 9 or 10 *s. per* Square.

The Price of *Laying*, [*i e* Boarding] of *Floors* (ſays Mr. *Leybourn,*) is various, according to the Goodneſs of the Stuff, from 12 *s.* to 20 *s.* the Square; but if the Boards be found by the Builder, then they commonly allow for Plaining, Joynting, and laying of Boards, 4 or 5 *s. per* Square, beſides Nails, of which 200 is a competent Allowance for one Square of Flooring But ſome Workmen in *Suſſex* tell me they will lay Deal floors braded, and plain Joynts broken at every 4 or 5 Boards, for 3 *s. per* Square; and if they break Joynt at every Board, then 6 *s.* others ſay 6 *s.* 8 *d.* or 7 *s per* Square.

Plaiſter-floors running, the Workman finding all, is worth, (ſays Mr. *Wing*) 1 *s* 4 *d. per* Yard but the working part only is worth 4 *d* 5 *d* or 6 *d. per* Yard. See Pl. Deal-floors.

Flooring brads.

See *Brads*: N. 4.

Flutes, or *Fluteings.*

In *Architecture*, are the Hollows made in the Body of a Column The Dorick, Ionick, Corinthian, and Compoſite Columns are commonly fluted, or made with Flutes or Hollows, running along the Body of the Column, from the Baſe to the Capital. Each Column has 24 Flutes, and each Flute is hollow'd in, exactly a quarter of a Circle In the Dorick-Column, the Flutes join together, without any Interſpace; but in the Ionick; Corinthian, and Compoſite Columns, there runs a Liſt betwixt every two Flutes The Account given by M. *Perrault* of *Fluteings*, is as follows. They are certain, perpendicular Cavities, cut length wiſe around the Shaft of the Column, and rounded at the two Extremes. The French call them *Cannelures*, Channellings. Their Number was at firſt limited to 24 in the Ionic, and 20 in the Doric Order; but that Limitation, ſome of our modern Architects have taken the Liberty to diſpenſe with. Theſe Cavities are frequently fill'd up with ſomething not unlike a Flute, whence 'tis probable they take their Names: as the French Term *Cannelures*, ſeem borrowed from the Cavities themſelves, which reſemble Canals. *Vitruvius* tells us, they were at firſt intended to repreſent the Folds of a Garment.

Flyers,

Are Stairs made of an Oblong-ſquare, whoſe fore and back ſides ſtand parallel to each other; and ſo are their Ends; the ſecond of theſe Flyers ſtands parallel behind the firſt, the third behind the ſecond

cond. and ſo of the reſt. If one Flight carry them not to the intended heighth, then there is a broad Half-pace, from whence they begin to fly again, as at the firſt.

Foliage,

In Architecture and Sculpture is Work wrought in Branches and Leaves.

Foot-pace,

Or as ſome call it, *Half pace*, is a part of a pair of Stairs, whereon, after 4, or 6 Steps, you arrive to a broad Place, where you may take two or three paces before you aſcend another ſtep, thereby to eaſe the Legs in aſcending the reſt of the Steps.

Fornication.

In Architecture, is an Arching, or Vaulting.

Foundation.

1. *What*.] The loweſt part of a Building, (generally laid under Ground) upon which the Walls of the Superſtructure are rais'd. This word is alſo ſometimes taken for a publick Building, erected for pious Uſes.

2. *Digging for, and laying of*.] Concerning Digging for, and laying of the Foundation of a Building, there are ſeveral things to be well conſider'd and taken notice of; the moſt material of which I ſhall extract from the beſt Architects, Ancient and Modern.

(1.) *This*, (ſays that great Architect, Sir *Henry Wotton*,) requires the exacteſt Care; for if the Foundation happen to dance, 'twill marr all the Mirth in the Houſe: Therefore, that we may found our Habitation firmly, we muſt firſt examine the *Bed of Earth*, upon which we build; and then the *Underfillings* or *Subſtruction*, as the Ancients call'd it. For the former we have a general Precept in *Vitruvius*, twice repeated by him, as a Point indeed of main Conſequence. Firſt, *Lib*. 1. *Cap*. 5. And again, *lib*. 3. *cap*. 3. in theſe words, (as *Philander* does well correct the vulgar Copies) *Subſtructionis Fundationes fodiantur*, (ſays he) *ſi queant invenire ad ſolidum, & in ſolido*. By which words I underſtand him to commend to us, not only a diligent, but even a jealous Examination what the Soil will bear; adviſing us, not to reſt upon any Appearance of Solidity, unleſs the whole Mould, thro' which we we cut, hath likewiſe been ſolid: In Order to which, Architects ought to uſe their utmoſt Diligence; for of all Errors that may happen in Building, thoſe are the moſt pernicious which are committed in the Foundation; becauſe they bring with 'em the Ruin of the whole Fabrick; nor can they without great Difficulty be amended.

(2) If the *Foundation* happen to be on a Rock, or hard Gravel; theſe (without digging, or other artificial Helps,) are of

of themselves excellent Foundations, and most fit to uphold the greatest Buildings.

(3) If the Place where you Build, be firm solid Earth, you may dig for the Foundation, so far as a discreet Architect shall think requisite for the Quality of the Building, and Soundness of the Earth; but how deep we should dig, *Vitruvius* has no where to my Remembrance determin'd, as perhaps depending more upon *Discretion* than *Regularity*, according to the weight of the Building; yet *Palladio* has fairly ventur'd to reduce it to Rule, allowing for the *Cavazione*, (as he calls it, *i. e.* the hollowing of the Earth for the Foundation) a sixth part of the heighth of the Fabrick; and if the Building be Cellar'd, he would have us, (as it should seem) to dig somewhat lower. *Palladio* lays down several Rules, to know if the Earth be firm enough for the Foundation (without Artificial Helps) by Observations from the digging of Wells, Cisterns, and such like, (which he would have to be done in the first place) and from Herbs growing there, if there be such as usually spring up only in firm Ground; also, if a great weight be thrown on the Ground, it neither sounds nor shakes, or if a Drum being set on the Ground, and lightly touched, it does not resound again, nor shake the Water in a Vessel set near it; these (says he) are signs of firm Ground. But the best way to discover the Nature of the Soil, is to try it with an Iron Croe, or else with a Borer, such as Well diggers use.

(4) If you build upon mossie, and loose Earth, then you must dig till you find sound Ground. This sound Ground (fit to uphold a Building) is of divers kinds; for (as *Alberti* well observes) in some places 'tis so hard, as hardly to be cut with Iron, in other places very stiff, in others blackish, in others whitish, (which is accounted the weakest,) in others like Chalk, and in others Sandy; but of all these that is the best which is cut with most Labour, and when wet does not dissolve into Dirt.

(5.) If the Earth you build on be very soft, as in moorish Grounds; then you must get good pieces of Oak, whose length must be the breadth of the Trench, or about 2 Foot longer than the breadth of the Wall; these must be laid cross the Foundation about 2 Foot asunder; and being well ram'd down, lay long Planks upon them; which Plank need not lie so broad as the peices are long, but only about 4 Inches of a-side wider than the Basis or Foot of the Wall is to be, and pinn'd or spiked down to the pieces of Oak, on which they lie. But if the Ground be so very bad, that this will not do, then you must provide good Piles of Oak, of such a length as will reach the good Ground, and whose Diameter must be about $\frac{1}{12}$ part of their length; these Piles must be drove or forced down with a Commander, or an Engine for that purpose, and must be placed

placed as close as one can stand by another; then lay long Planks upon them and Spike, or Pin them down fast.

(6.) If the Earth be faulty but in here and there a place, and the rest be good Ground, you may turn Arches over those loose Places, which will discharge them of the Weight. You must observe to place your Piles, not only under the outer Walls, but also under the inner Walls that divide the Building; for if these should sink, 'twould be a means to make the outer Walls crack, and so ruin the whole Fabrick.

(7.) Thus much for the *Bed of Earth* on which we Build. We are next to consider the *Substruction*, as the Ancients call'd it; but modern Artists generally call it the *Foundation*. This is the Ground-work of the whole Edifice, which must sustain the Walls, and is a kind of *Artificial Foundation*, as the other was *Natural*: About which these are the chief things to be remember'd. First, That the bottom be precisely level, where the *Italians* therefore commonly lay a Platform of good Board. Secondly, That the lowest Ledge or Row be meerly of Stone (the broader the better) closely laid *without Mortar*; which is a general Caution for all parts of a Building that are contiguous to Board or Timber; because *Lime* and *Wood* are utter Enemies; and if any where unfit Confiners, then most especially in the Foundation. Thirdly, That the breadth of the *Substruction*, be at least double to the breadth of the Wall to be raised thereon. Yet here *Discretion* is freer than *Art*, and you may make it broader or narrower, according as the goodness of the Ground, and the weight of the Fabrick shall require. Fourthly, That the Foundation be made to diminish as it rises; yet so, as that there may be as much left on one side, as on the other; so as the middle of that above may be Perpendicularly over the middle of that below: Which ought to be also observed in diminishing the Walls above Ground; for so the Building becomes much stronger than it would be, by making the Diminution any other way. Fifthly, That you never build upon the Ruins of an old Foundation; unless you are very well assur'd of its Depth and that its Strength is sufficient to bear the Building. Lastly, I find (in some ancient Architects) a curious Precept. *That the Stones in the Foundation should be laid as they lay naturally in the Quarry*: They supposing them to have most Strength in their natural Posture. But this Precept is generally observ'd by all good modern Artists, not only in the Foundation, but also in all parts of the Superstructue; and that for a better Reason than bare Conjecture, *viz.* Because they find the Stones to have a cleaving Grain, (or be subject to cleave) that way of the Stone that lay Horizontal in the Quarry: And there-

therefore, if the Horizontal Position of the Stone in the Quarry should be placed Vertical in the Building, the Superincumbent weight should be apt to cleave them, and so render the Fabrick Ruinous; for as *Philip de l'Orme* observeth, the breaking or yielding of a Stone in the Foundation, but the breadth of the back of a Knife will make a Cleft of more than half a Foot in the Fabrick a-loft. See Stone, and Bed, and Face of a Stone.

3. *How to value.*] There are several ways, (says Mr *Philips*) by which Men value the Foundations (or Ground plots) of Houses. (I. Suppose he means, in Cities and great Towns.) As——

First, Some value them by their length or breadth toward the Street, reckoning every Foot in Front to be worth 4, 5, 6, 8, or 10 *s* Yearly, according to the Street, or Place they stand in; and this Yearly Value they reckon at 20 Years Purchase and so every Foot in Front is worth 4, 5, 6, 8, or 10 Pounds. But this is a very uncertain way, by reason of the great Difference in the depth of Houses, *&c.*

Secondly, Others value Foundations by their length and breadth, measur'd by the Foot; reckoning every Foot to be worth 3 or 4 *s*. But this way will deceive you as much, or more than the other, if you don't set a good Rate upon each Foot of Ground. For Ground being scant in a City, each Foot of it *there* may be worth 8 or 10 *s*. which in the Country is not worth ½ a Farthing; tho' you reckon Land at 10 *s*. an Acre, and 20 Years Purchase; for so 'tis worth but one penny a Yard, and every Yard has 9 Feet.

Thirdly, But the way that I shall prescribe (says my Author) as more general and certain, to value these Foundations, is to get a true and indifferent Estimate of the Yearly Rent these Houses formerly went at, at a moderate rack Rent, without any Abatement or Diminution thereof by Fines, or any other Considerations: Which being known, you may reckon the true value of these Foundations to be 4, 5, or 6 Years Purchase, according to the said Yearly Rent, that is about the third part of the full worth or purchase of the Fee simple of the House. But if you will more exactly judge of, and determine the true worth of these Foundations; it will be best to range them into three sorts, reckoning the first and lowest sort of Houses which yield least Rent, at 4 Years Purchase; the second sort which yield a moderate Rent, at 5 Years Purchase, and the third sort which yield the biggest Rent, at 6 Years Purchase. My Author, (the afore mentioned Mr *Wing*) demonstrates the *Meliority* of this way of Valuing Foundations, above any other; but I have been already too long upon this Theam, and therefore I shall defer the rest of this Ingenious Discourse on this Subject to another Opportunity.

Foun-

Fountain.

An artificial Spring of, (or Well to contain) Water in a Garden; whither the Water is brought in Pipes of Lead, *&c.* and commonly made to ſpout out of the Mouths, or other parts, of Images.

Framing.

1. *Of Houſes*] I know ſome Workmen in *Suſſex* that do all the Framing in a Houſe, *viz.* The Carcaſe, Flooring, Partitioning, Roofing, Ceiling-beams, Aſhtoring, *&c* all together, and make the Windows, and Lantherns, and hew and ſaw the Timber for 12 *s. per* Square.

2. *Carcaſe of a Houſe.*]Mr. *Leybourn* ſays, That Carpenters commonly work by the Square of 10 Foot, in erecting the Carcaſe, that is, (ſays he)framing and ſetting up with the Partitions, Floors, Rafters, and ſuch like; for which (ſays he) they have (in running Buildings) from 15 to 20 *s.* the Square, and ſome may deſerve 30 *s.* or more, (and to a ſquare of good Carcaſe, (ſays he) 20 Foot of Ground rough Timber may be allow'd.) But I know not whether he means that the Carpenter fells, and hews and ſaws the Timber into that Price; for ſome Workmen in *Suſſex* tell me, That for framing the Carcaſe of a Houſe, and ſawing the Timber, they have but 8 *s. per* Square, and without ſawing timber but 4 *s.* 6 *d.* others ſay but 4 *s per* Square.

3. *Carcaſe of a Barn.*] Some Workman tell me, That they have for Framing of Barns 3 *s.* 6 *d. per* Square. They alſo tell me, That the Charge of the Carcaſe of a Barn may be thus computed, *viz.* 4 *s. per* Square for ſawing the Boards, conſidering the Slabbing, and the Boards lying one over another, 2 *s. per* Square for ſawing the Timber, 3 *s.* 6 *d per* Square for Framing, and 4 *s. per* Square for the Timber, reckoning at 12 *s. per* Tun, and 1 Tun to make three Square of Framing. So that the whole Charge of the Carcaſe will be at leaſt 13 *s.* 6 *d per* Square; for if the Timber be more than 12 *s. per* Tun, then the whole Charge will be more than we have computed.

4. *Partitions.*] Tho' ſome Workmen reckon Partitions into the Carcaſe, as was ſaid, Number 2. yet others reckon them by themſelves for which, and ſawing the Timber, they tell me, They have 6 *s* or 7 *s. per* Square; and for the Workmanſhip only, 2 *s.* 6 *d. per* Square.

5. *Roofs*] Mr. *Leybourn* ſays, That Carpenters commonly reckon 4 or 5 *s.* in the Square more for framing of Roofs, than for the reſt of the Building. I know not how he means; for I am ſure ſome ingenious Workmen in *Suſſex* tell me, That for framing of Roofs, and ſawing the Timber, they have but 8 or 9 *s. per* Square, and for the Workmanſhip only but 4 *s* 6 *d. per* Square.

6. *Floors*] See Floors. N 6.

7. *Through.*

7. *Through*] Some Workmen tell me, That for Thorough-framing, (as they call it, that is framing all, and making Doors and Windows,) they have 5 *s. per* Square, for the Workmanship only.

8. *By the great Square.*] Some Carpenters tell me, That in Brick Buildings they sometimes work by the *great Square*; and then besides framing the Floors, Partitions, Roof *&c* they also make Doors, Windows, Cornishes, Stair-cases, and (in general)all that is Carpenters Work, and sawing of Timber. Yet I think they told me, They were particularly paid for making the Modilions, or Cantalivers. And for this work they have 6 *l. per* Square. But 'tis to be noted, That in this way of working, they measure only the Ichnography, or Ground-plot, only to the Dimensions they add one of the Projectures in Front, and not in Flank, and so cast it up.

9. *Of Measuring*] This kind of work is measur'd by the Square, as Floors. See Floors N. 5.

Fret work.

See Plastick-Art.

Free-Masons Work.

See the Particulars in their proper Places of the Alphabet

Freeze, or *Friese*.

1. *What.*] A large Flat Member, which separates the Architrave from the Cornice. See its Etymology under the word *Zophoros*

2. *Kinds.*] There are as many kinds of Freezes, as there are Orders of Columns, *viz.* Tuscan, Dorick, Ionick, Corinthian, and Composite; of all which in their order.

3. *Tuscan*] *Vitruvius* makes this Freeze flat and plain, and in heighth 30 Minutes. *Vignola* also makes it flat and plain, but in heighth 35 Minutes. *Palladio* makes it convex or swelling, and in heighth but 26 Minutes. *Scammozzi* makes it plain, and in heighth 42 Minutes.

4. *Dorick.*] *Vitruvius*, (and so also *Vignola)* makes this Freeze flat, only Carv'd with Trigliphs and Metops, and its heighth 30, or 45 Minutes. *Palladio* and *Scammozzi* also make it like *Vitruvius*, and in heighth 45 Minutes.

5. *Ionick*] *Vitruvius* makes this Freeze flat, but commonly Carv'd with Acanthus Leaves, Lions, and Men, *&c.* And in heighth 30 Minutes. *Vignola* makes him flat also and in heighth 45 Minutes. *Palladio* makes him Convex or Swelling, and in heighth but 27 Minutes. *Scammozzi* makes him flat, and in heighth 28 Minutes.

6. *Corinthian.*] *Vitruvius* makes this Freeze flat, but Carv'd with Acanthus Leaves, and Men, *&c.* and in heighth 37 $\frac{2}{3}$ Minutes. *Vignola* makes it like *Vitruvius*, but in heighth 45 Minutes. *Palladio*, and *Scammozzi* also make it like *Vitruvius*, but *Palladio* makes it in heighth

heighth 28 Minutes, and *Scammozzi*, 31 $\frac{1}{4}$ Minutes.

7. *Composite.*] *Vitruvius* makes this Freeze flat, but beset with Cartouses, and Carv'd betwixt each Cartouse, and in heighth 52 $\frac{1}{2}$ Minutes. *Vignola* makes it like *Vitruvius*, but in heighth but 45 Minutes. *Palladio* makes it convex, or swelling, and in heighth but 30 Minutes *Scammozzi* makes it like *Vitruvius*, and in heighth but 32 Minutes.

Fresco.

1. *What*] A way of Painting or Plaistering (or rather both) upon Walls, to endure the Weather, and representing Birds, Beasts, Herbs, Fruits, *&c.* in relief.

2. *Of Painting in-*] Painting in *Fresco* is thus perform'd. Grind your Colours with Lime-water, or Milk or Whey, and so temper and mix them together in Pots, as in Size-colouring. Also, take the Powder, of old rubbish Stones mix it with well burnt Flint, (or Lime) and Water; but wash out the saltness of the Lime, by often pouring out the Water, and putting fresh, the oftener the better; and this makes the Plaister or Compost. Avoid moist Weather, for that has great influence on the Walls: And to make the Work the more durable, strike into the Joints of the Brick or Stone Wall, stumps of Horse-nails, about 6 Inches asunder; for this will keep the Plaister from pealing off.

Then, with this Compost Plaister the Wall a good thickness, letting it dry; then (your Colours being ready prepar'd and mingled,) Plaister again over the former, the thickness of a Half-crown, so much as you intend presently to work upon; and whilst it is wet, work your Colours therein, which will mix and incorporate with the Plaister, so as never to wash out.

Work your Painting quick with a free Hand; for there can be no alteration after the first Painting; and therefore make your Colour high enough at the first; you may deepen, but not easily heighten.

Avoid Mineral Colours, Earthy Colours are the best, as all *Okers Spanish brown, Terra-vert Spanish white*, &c.

Your Brushes and Pencils must be long and soft; otherwise they will rake and rase the Painting; your Colours must be full, and flowing from the Brush; your Design perfect in the Image, or Paper Copy; for in this Work you cannot alter or add upon any Colour.

3. *History.*] This kind of Painting was the ancient Grecian way of Painting, and since much us'd by the Romans. *Plutarch* tells us, That *Aratas* the great Commander under *Ptolemy* King of *Egypt*, (in a Complement to the Emperor's Affections that ways) spared the sacking of a wealthy City, meerly for the Excellency of the *Fresco Painting* upon the Walls of the Houses.

There

There have been several whole Towns of this work in *Germany* rarely done, but now ruin'd by war.

At *Rome* there are three Chambers (in the Pope's Pallace) of *Fresco*, done by *Raphael Urbin*, and *Julio Romano* (his Disciple) who finished his Master's VVork, which is yet call'd *Raphael*'s Designs Other Places there are done by *Andrea del Sexto*, and *Michael Angelo*, and some other Artists.

At *Fountain bleau* in *France* is most excellent *Fresco-work*. It is the continu'd Travels of *Ulysses* in 60 pieces, done by *Bollameo*, *Martin Rouse* a Florentine, and others.

French-glass.

See Glass. N. III.

Friese.

As Freeze.

Frigeratory.

A Place to make or keep things cool in.

Free-stone.

See Stone. N. 1.

Front, or *Frontispice.*

1. *What.*] The Face, or Foreside of a House. See *Portial.*

2. *Of Setting.*] The *Setting*, (that is making) of the Fronts of great Buildings, *viz.* Ashlar, (or Stones) Architrave VVindows and Doors, with the Ground-table, Fascia's, and other Members, is worth from 3 *l.* 10 *s* to 5 *l* *per* Rod, (says Mr. *Wing*) according to the goodness of the Work.

Frontish doors.

See Doors. N. 4.

Frowey.

Timber is by some Workmen said to be *frowey* when it is evenly temper'd all the way, and works freely without tearing.

Funnels of Chimneys.

1. *VVhat.*] The Funnel of a Chimney is the Shaft, or smallest part of it, from the Wast (where 'tis gather'd into its least Dimensions) upwards.

2. *Of making*] *Palladio* tells us, That the Funnels of Chimnies must be carry'd through the Roof, 3, 4, or 5 Foot at least, that they may carry the Smoke into the Air. And here you must take care, (says he,) That they be made neither too wide, nor too narrow; for if they be too wide, the Wind will drive back the Smoke into the Room; and if they be too narrow, the Smoke (not having free Passage) returns back also. Therefore 'tis that Chamber-chimneys are not made narrower than 10 or 11 Inches, nor broader than 15, which is the ordinary depth of the Funnels of great Kitchin-chimneys, whose Breadth is 4 or 5 Foot within the Work, from the Place where the Brest ends, to the Top of the Funnel Now, the said Brest reaches from the Mantle

Mantle-tree, to the Ceiling, or pitch of the Arch, always diminiſhing within the Work, till you come to the Meaſures of Depth and Breadth, before-mentioned; and from thence to the End of the Funnel, it muſt be carry'd up as even as you can poſſibly; for failing in this, it often happens the Smoke is offenſive.

Furrs.

Furrings.

In *Architecture*, *Furring* is the making good of the Rafter-Feet in the Cornice. That is, when Rafters are cut with a Knee, theſe Furrings are pieces that go ſtraight along with the Rafter from the Top of the Knee to the Cornice. Alſo when Rafters are rotten, or ſunk hollow in the middle, and pieces (cut thickeſt in the middle, and to point at each End) are nail'd upon them to make them ſtraight again; the putting on of thoſe Pieces, is call'd *Furring the Rafters*; and thoſe Pieces ſo put on, are call'd *Furrs.*

Fuſt.

From the Latin *Fuſtis*, a Club, ſignifies the Trunk or Shaft of a Column, being that Part comprehended between the Baſe and the Capital.

Gable-end.

1. WHat.] In *Architecture*, the Gable-end of a Houſe is the Upright Triangular End of the Roof.

2. *To Meaſure.*] To meaſure a Gable-end, multiply the Breadth at the bottom, by half the Perpendicular, or Line from the Angle at the top to the middle of the bottom; or multiply half the former by the whole of the latter, and the Product will give the Content in ſuch Meaſures as the Dimenſions were taken in.

Gain.

The bevelling Shoulder of the Joyſt, or other Stuff. 'Tis alſo us'd for the lapping of the End of the Joyſt, &c. upon a Trimmer or Girder, and then the Thickneſs of the Shoulder is cut into the Trimmer alſo bevelling upwards, that it may juſt receive the Gain, and ſo the Joyſt and Trimmer lie even and level upon their Superficies. This way of working is us'd in a Floor or Hearth.

Galleries.

Are long narrow Rooms, made on the Sides or Fronts of Houſes; they ſerve for Walking, Eating, and other Divertiſements. Their Length (ſays *Palladio*) ought to be at leaſt 5 times their Breadth; they may be 6, 7, or 8 times their Breadth, but muſt not exceed.

Gard-manger.

A Store-houſe, or Room to ſet Meat in.

Gates.

Gate.

1. *What.*] 'Tis a thing ſo well known, that it needs no Deſcription; for all know it to be a Place for Paſſage of Perſons, or Horſes, Coaches or Waggons, *&c.*

2. *Of their Proportion.*] The principal Gates for Entrance, through which Coaches and Waggons are to paſs, ought never to be leſs than 7 Foot in Breadth, nor more than 12 Foot, which laſt Dimenſion is fit for large Buildings.

As to the Heighth of a Gate, it ought to be 1 $\frac{1}{2}$ the Breadth, or ſomething more.

But for common Gates in Inns, where Waggons loaded with Hay and Straw go under, their Heighth may be twice the Breadth.

3. *Of the Price of ſome ſorts.*] As to the Price of *Gates*, it is various, according to the ſorts of *Gates*, which again will differ according to the Dimenſions and Workmanſhip. Thoſe which we ſhall mention at preſent, will be only *Palliſado*, and *Pold Gates*.

And firſt of *Palliſado Gates*; Mr. *VVing* ſaith, in *Rutland*, That if the *Gates* be 6 or 7 Foot high, and the Workman find Timber and Workmanſhip, they are worth about 9 or 10 *s per* lineal Yard; but if he find only Workmanſhip, then 'tis worth 6 or 7 *s. per* Yard.

I have obſerved, That if they are *Semi-palliſado*, with Kneeling-rails at the Top, handſomely moulded on both Sides, and ſquare Palliſades, Raiſed Pannels, and Biſection Mouldings on both ſides, the Gates about 8 Foot high, and the Poſts a Foot ſquare, open'd in the Front, or revailed with a Moulding ſtruck in it on both ſides the Revail, a Baſe and Capital laid on the Poſts, and the Heads cut into one of the Platonick Bodies; as ſuppoſe an Icolaedron, and the Poſts were about 10 or 11 Foot above Ground, the Workmanſhip is worth 12 or 13 *s. per* Yard lineal; but if the Workmen find Timber, it will be worth more than 20 *s per* Yard lineal; in ſuch Gates, to find all Iron-work, Painting, *&c.* it would be worth above 30 *s. per* Yard lineal.

Secondly, Of *Pold Gates*, (which are ſuch as are ſet in Fences for to ſhut up the Paſſages into Fields, and other Incloſures.) Theſe are of two ſorts, either of ſawed, or cleft Timber; for to make a ſawed one, and ſet him up, and his Poſts, the Price in different Places is from 3 *s.* 6 *d.* to 5 *s.* but if the Carpenter pay for the ſawing, then the Price is from 5 *s* to 6 *s.* 6 *d.* Such a *Gate*, Timber and Work, is worth from 7 to 10 *s.* according to their Goodneſs; but with Poſts from 12 to 15 *s.* But *Gate* and Iron-work from 10 to 13 *s.* But Gate, Iron-work, and Poſts, from 15 *s* to 18 *s.* but Cleft pold *Gates*, cleaving, and making, and hanging, from 4 to 5 *s* and ſo proportionably for all Timber, Iron, and Poſts, *&c.* The Reaſon why the Prices are thus diffe-

different, is, becauſe 'tis according to the Cuſtoms of different Places where I had my Information.

Gavel.

A Word uſed by ſome, by which they mean the ſame as *Gable*, which ſee.

German-glaſs.

See *Glaſs.* Num. V.

Girding-beams.

'Tis uſed by ſome Architects, to ſignifie the ſame as

Girders.

1. *VVhat*] Are ſome of the largeſt Pieces of Timber in a Floor, the Ends of them are for the moſt part framed into the *Summers* or *Breſt-ſummers*, and the *Joyſts* are framed in at one End to the *Girders*.

2. *Of their Size or Proportion*] The *Scantlings*, or Size of *Girders* and *Summers*, upon the Rebuilding of *London*, after a Conſultation of experienced Workmen, were reduced to an Act by the Parliament, and are thus ſet down, as fit for all Fabricks, great and ſmall, *viz.*

Girders and Summers, in length.	From	to	muſt be in	
	F. In.	F. In	Breadth, Inches	Depth, Inches.
	10. 0	15. 0	11	8
	15. 0	18. 0	13	9
	18. 0	21. 0	14	10
	21. 0	24. 0	16	12
	24. 0	26. 0	17	14

3. *How to be laid in the Brick-work.*] No *Girder*, or *Summer* ought to lie leſs than 10 Inches into the Wall, and their Ends muſt be laid in Loam.

That *Girders* and *Summers* be of good hearty Oak, as free from Knots as may be; becauſe that will be leaſt ſubject to break and may with more Safety be relied on in this croſs and tranverſe Work.

Girt.

See *Fillet.*

Glaſs.

1. *VVhat.*] All know it to be a diaphanous, or tranſparent Body made by Art, of Sand and Nitre, ſaith *Pliny*: 'Tis alſo made of white gliſtring Flints, mixt with Sal-Alkali,

Alkali, or the Salt of the Herb *Glass work*, or Salt of Fern-ashes for common *Glass* some say. *Monsieur Blancourt* saith, That the *Venetians* also use white Flints, and also a rich Sand, and likewise a sort of white Marble; he also saith, That all white transparent Stones that will not burn to Lime, are fit to make *Glass*; and that all Stones that are fit to strike Fire, are capable to be made into Glass.

II. *The Sorts of Glass.* There are various sorts of Glass which are made use of in the World, but at present I shall confine my self to speak only of those Sorts which *Glaziers* commonly work upon here in *England*, which are these following, *viz* *Crown-Glass*, which is of two sorts, *Lambeth* and *Ratcliff* 2 *French* or *Normandy Glass*. 3. *German Glass* of two sorts white and green. 4. *Dutch Glass* 5. *Newcastle Glass*. 6 *Staffordshire Glass* 7 *Bristol Glass* 8. *Looking glass*. 9 *Jealous Glass*; of which sorts I shall succinctly treat in their Order.

III *Of Crown-glass.*] Is of two sorts, *Ratcliff* and *Lambeth Crown glass*, of both which Sorts I will briefly speak. And

1. *Of Ratcliff Crown-glass.*] That sort of Glass which goes by this Name, is the best and clearest sort of *Crown-glass*; which Sort was at first made at the *Bear-garden* on the *Bank side*. In the Year 1691, I had it publish'd in the *Gazette* for *June* 15, *&c.* where it is commended in this Manner, and called *Crown-window glass* much exceeding *French Glass*, in all its Qualifications. But now at the *Bear garden Looking glass-plates* are made; the Maker of this best sort of *Crown-glass*, being now removed to *Ratcliffe*, and upon that Account it now bears the Name of *Ratcliff Crown glass* as it did at first of *Bear-garden Crown Glass*.

This sort of *Crown-glass* is of a light Sky-blew Colour, which may be very distinctly seen, if it be laid on a piece of white Paper.

I have been informed, That an *English* Glass-maker went over into *France* on Purpose to learn the *French* Way of making *Glass*, which when he had attained to, he came over again into *England*, and set up making of *Crown-glass*, and therein much out-did the *French* his Teachers, as *English-men* usually do. I have been told by some *London Glaziers*, that there is 24 Tables of this *Crown glass* to the *Case*, the *Tables* being of a Circular Form, of about 3 Foot 6 or 8 Inches Diameter, and by Consequence, each *Table* will be in Area about 9 or 10 Foot, and the *Case* betwixt 220, and 240 This *Glass* is brought from *Ratcliff* in such kind of Frames as *Newcastle glass* is brought up into the Town, only the *Newcastle glass* is brought on Shipboard, and this *Ratcliff glass* upon a Staff betwixt two Men, as they carry a small Vessel of Beer, or the like.

I

I have known this *Ratcliff-crown-glaſs* ſold about 9 *d per* Foot in *London* ready cut into Squares. And when wrought in Lead, and ſet up, for about 18 *d. per* Foot.

2. *Of Lambeth Crown-glaſs.*] This ſecond ſort of *Crown-glaſs* takes its Name alſo from the Place where 'tis made; 'tis of a darker Colour than *Ratcliff Crown-glaſs*, inclining ſomething to a Green.

This ſort is ſold in *London* for about 8 *d per* Foot cut into Squares, and being wrought and ſet up in Windows with Lead, its Price is about 16 *d. per* Foot.

IV *Of French glaſs*] By ſome call'd *Normandy glaſs*, becauſe it was formerly made at *Cherbourg* in *Normandy.* But I am informed that the Work-houſes have within theſe few Years been removed, for certain Reaſons of State to *Auxerre* in *Burgundy* They alſo make Glaſs at *Nevers* in *Orleans*, and likewiſe at St. *Gobin* near *La Fere* in *Picardy*; but from which of theſe Places our *French-glaſs* comes which we uſe in *England*, I cannot certainly tell. But it is a thinner and more tranſparent ſort of Glaſs than our *Newcaſtle-glaſs*, and when 'tis laid on a piece of white Paper, it appears of a dirtyiſh green Colour. It uſed to be of a middle Price, betwixt *Crown* and *Newcaſtle-glaſs*, and I have known it ſold for 12 *d. per* Foot, wrought in Lead, and ſet up; but ſome ſay 'tis now (we have War with *France*,) near as dear as *Crown glaſs*.

Of this Glaſs there is but 25 Tables to the Caſe.

V. *Of German glaſs*] Of this ſort of Glaſs there are two Kinds, *viz. White* and *Green.*

1. *White German glaſs.*] This Glaſs is of a whitiſh Colour, and free from thoſe Spots and Blemiſhes which our *Newcaſtle Glaſs* is ſubject to, but it hath commonly ſome fine or ſmall carved Streaks, or Lines, as the *Newcaſtle-glaſs* hath.

2. *Green German-glaſs*] This you may well ſuppoſe to be of a greeniſh Colour by its Name; it is ſubject to have thoſe fine Lines, or Streaks as the White is; but both this and the white *German* is ſtraighter, and not ſo crooked and warped as *Newcaſtle* Glaſs is: Both theſe ſorts of Glaſs are brought over from *Germany*, and yet it is generally as cheap as *Newcaſtle* Glaſs.

VI. *Dutch-glaſs*] It differs not much from *Newcaſtle* Glaſs in its Colour, and I have obſerved ſome of it that hath been very crooked: I am informed that the Tables of this ſort of Glaſs are but ſmall; 'tis not much uſed now in *England.* This Glaſs hath its Name alſo from thoſe that make it, *viz.* The *Dutch*, for I am told 'tis brought out of *Holland*; 'tis alſo about the ſame Price with *Newcaſtle-glaſs.*

VII. *Newcaſtle-glaſs.*] This ſort of Glaſs is of a kind of an Aſh-colour; it is the Glaſs that

that is most in use here in *England*, but 'tis subject to have Specks and Blemishes, and Streaks in it, and it is very often warped and crooked. Of this Glass, Mr. *Leybourn* says there are 45 Tables to the Case; but if I did not mistake, a *London* Glazier told me, That they had but 35 Tables to the Case, and Mr. *Leybourn* also saith, That each Table contained 5 superficial Feet, and by Consequence a Case of 45 Tables to the Case will contain 225 Foot. The Glazier before-mention'd, said there was 6 Foot in a Table, and if but 35 Tables to the *Case*, that would amount to but 210 Foot. But I was informed by one that told me he had taken the Dimensions of some Tables of *Newcastle* Glass, and he found them to contain 7 Foot at least; for, saith he, they are of this Form; The upper Edge, as they stand in the Cases or Frames, is Circular, about the 4th or 5th part of a Circle, the Cord of which, saith he, was about 3 $\frac{1}{2}$ Foot, the lower Side was straight, about 18 or 19 Inches, the Perpendicular from the bottom to the top, about 3 Foot: From this Observation, a Case of 35 Tables would amount to 245 Foot These Tables of Glass are brought in Cases, or slight Frames of Sticks fixed at some Distance one from another, into 4 Corner-pieces which are stouter. The Ends of these Frames are made tapering nearer one another at the bottom than they are at top, according to the Form of the Glass; but the sides are parallel; the Glass is set on some Straw, which is laid on the bottom of the Frame, and there is some Straw also put on the sides, and top of each Case, but none betwixt the Tables These Cases are brought to *London* in the Coal-ships, they being set on End in the Coles more than half its depth, by which Means they are kept steady from falling and being broke by the Motion, and Rowling of the Ship.

Mr. *Leybourn* saith, That a Case of 45 Table, 5 Foot to a Table, equal to 225 Foot, doth weigh about 200 ℔. and by Consequence 9 Foot will weigh about 8 ℔.

He also saith, the Price of *Newcastle* Glass is uncertain; for when Coals are plenty, then Glass is cheap, and when the Coals are dear at *London*, then *Newcastle* Glass is so likewise, not that they want Coals at *Newcastle*; but because they have no other Conveyance for it to *London*. So that sometimes it is at 30 *s. per* Case, and other Times 40 *s*

But I was informed by a *London* Glazier, that the most constant Price was 34 *s. per* Case.

To cut a Case of this Glass into *Quarries* Diamond fashion (with halfs, and quarters, and 3 quarters of *Quarries*, as the Glass falls out) some say it is worth 6 or 7 *s.* but I did hear a Glass-cutter say he would do it for 3 *s.* or 3 *s. 6 d.*

Newcastle-glass cut into *large Squares* are sold for 22 to 25 *s. per* 100 Foot, according to their Size.

And

And *ſmall Squares*, from 19 to 22 *s. per* 100 Foot.

And *Quarries*, of *Newcaſtle* Glaſs, for about 16 *s. per* 100 Foot.

Glazing done with this *New-caſtle* Glaſs with *Quarries* banding, ſoddering, pinning the Caſements being included, the uſual Price is 5 *d*. or 6 *d. per* Foot in *London*, and there abouts: but in ſeveral Parts of the Country they have 6 *d. per* Foot, and will be paid for pinning of the Caſements beſides.

VII Glazing, in ſome Places of *England*, as in *Rutland*, and other *Northern* Parts is done, with Quarries of *New-caſtle* Glaſs for 4 *d.* ½ or 5 *d. per* Foot. And Squares wrought into Lead, and ſet up for 6 *d. per* Foot.

But then again in *Suſſex* and *Kent*, the *South* Parts of it they will not work ſo cheap, becauſe their Glaſs is ſomething dearer to them; in theſe *Southern* Parts, they commonly reckon 7 *d. per* Foot for Glazing with Squares of *Newcaſtle-glaſs*, beſides they will be paid for pinning of the Caſements.

VIII. *Staffordſhire glaſs*] This ſort of Glaſs which is made in *Staffordſhire*, I could never yet learn any certain Account of; for 'tis a ſort of Glaſs but ſeldom uſed in theſe parts of the Kingdom.

IX. *Briſtol glaſs*] This is made at the City of *Briſtol*; but by reaſon they have not the Conveniency to ſend it by Sea, (as they have from *New-caſtle* by Coal-Ships,) it is very rare to have any of it in *London*, tho' it be as cheap, and better than *Newcaſtle* Glaſs.

X *Looking glaſs*.] As to *Looking glaſs Plates*, they are made at the *Bear garden* on the *Bank-ſide*, *London*, (where Crown-glaſs was firſt made.) I do not certainly know whether this ſort of Glaſs be not made with the ſort of Sand, mentioned by Dr *Grew* in his *Muſæum Regalis Societatis*, 346 P. Fine Sand, ſaith he, from a Sand-pit near *Bromley* in *Kent*, of this is made the cleareſt and beſt *Engliſh-glaſs*; it conſiſteth of ſome Grains as clear as Cryſtal; with which others obſcure being mixed, give a whitiſh Aſh-colour to the whole Maſs.

Theſe *Looking glaſs* Plates are ground ſmooth and flat, and poliſhed; they are ſometimes uſed in *Saſhes* or *Saſh-windows*; but 'tis a dear ſort of Glaſs; for they ask 4 *s. per* Foot for ſuch Squares, and if they are large, 'tis much more.

There is a Way that ſome uſe to try which is the whiteſt and cleareſt Glaſs, which is thus, *viz.* they take it up cloſe by one Edge, betwixt the Edges of the middle and Fore-finger; and then looking againſt the Cut, or Broken Edge, the Eyes being thus skreen'd by the Edges of the two Fingers, they ſay 'tis eaſy by this Method to diſcern which is the whiteſt and cleareſt Glaſs.

Looking-glaſſes foil'd being in Vogue for Ornaments over Chimneys in Parlours, *&c.* I ſhall briefly ſay ſomething concerning them. Sir *William Petty* ſaith, That the Value of Looking-

Looking-glass plates consists in a duplicate proportion of their sides to their Squares. Because you shall not be left quite in the dark as to this matter, I will exhibit the Price which I have known set down upon two Sizes of *Looking glasses*, *viz.* One of 5 Inches long, and 12 bread in a Frame to place over a Chimney 40 *s.* some of 10 and 8 Inches, in Walnut-tree frames for about 4 *s. per* piece, if they have Dimond-cut; but if not, this size is about 6 *d per* piece cheaper. I could here have shew'd the Method of Grinding, Polishing, and Foiling of these *Glasses*; but it being foreign to my purpose, shall omit it.

XI. *Jealous Glass.*] This is a sort of wrinkled Glass of such a Quality, that one cannot distinctly see what is done on the other side of it; but yet it admits the Light to pass thro' it. 'Tis made of the same sort of Materials, as *Looking glass-plates* are. This *Jealous-glass* is cast on a Mould, and is composed all over its Surface with a multitude of Oblong Circular Figures, (which are Concave,) somewhat resembling Weavers Shuttles; this is on one side of it, but the other consists of Figures a little Convex, and this last side is the side they cut it on, when the Squares are too big for the present use, it being found to be very difficult to cut it on the Concave side. Some sorts of this *Jealous glass* have a Convexity rising in the middle of the Concavity; so that one side, or Surface of it doth much resemble the Boats which Boys used to make by folding of Paper; only in this Glass the Concavities, and Convexities are both more obtuse and blunt. But there are various sorts of this Glass, in respect either to the Form or Size of the Figures, of which this *Glass* is composed; for I have seen some of this *Glass* have the Shuttle-like Figures, much larger than other some, and some of it with the Points (as it were) of the Shuttles very curved, and sometimes these Figures are in a Position perpendicular to one Edge of the Square, and other some are oblique to it.

I am informed that this *Glass* is about 18 *d. per* Square, each Square being about 12 or 14 Inches broad, and 15 or 16 Inches long. The reason why they are so dear, is, (as I am informed) because the *Looking-glass-plate makers* do not care to make these sort of Squares, but only when their Pots of Metal are almost out, and they are a little at leisure; for they say it wastes *Glass* too much for their Profit. I heard a *London* Glazier say that he hath sometimes stayed a Month for some of it, before he could have it to use. This sort of *Glass* is commonly used, in and about *London*, to put into the lower Lights of Sash-windows, *&c.* where the Windows are low against the Street, to prevent People's seeing what is done in the Room as they pass by: It is sometimes set in Lead in such Places, where they would not have their Actions overlook'd.

Now it is very plain (to any reasonable Man, that knows

knows any thing of Refractions,) that this sort of Glass must needs prevent People's seeing through it, into a Room as they pass by; because the Rays, or Species of a visible Object, are by reason of such a variety of Refractions, (caused by the inequality of the Glasses Surface) broken and confused when they arrive at the Retina, or Fund of the Eye.

XII. *Woolwich*, or *Woolledge*] This also was one kind of our *English Glass*, which did receive its Name from the Place of its make; but by reason they met with some Discouragement in their Proceedings there, they have laid it down for some time, and do not now make there.

Glazier's-work, or *Glazing*.

I. *What*] *Glazier's-work*, or *Glazing*, all know to be a manual Art, whereby pieces of Glass (by the Assistance of Lead) are so fitted and compacted together by *Straight* or *curved* Lines, that it serves as well for the intended use, (almost) as if it were one entire piece; nay in some respects far better and cheaper, *viz.* in case of breaking, *&c*

These two Heads of *Straight* or *Curved*, will admit of several Sub-divisions, and first of *Straight*, which contains a square Work, whose Angles are right ones, as almost all Window-Lights are in Timber Window-frames, and so likewise are the Squares, (if it is Glazed with such of which the Lights are composed.

2. *Miter*, or making an Angle of 45 Deg. this but seldom happens in this Profession, unless it be in some piece of *Fret-work*.

3. *Bevel*, this is the most common, especially in the Country, and ordinary Houses, (which all know to be most numerous,) for most such are Glazed with *Quarries*, which is Bevel Work, so likewise is a great deal of *Fret*, and all *Snip-work*.

Curved Work, consists either of *Circles*, *Ovals*, or some distorted Arches; *Circles* and *Ovals* are commonly used for Lights at some particular place in a Building, as in a *Pediment*, over a Door, or the like, in the middle of a Front, *&c.* I have also observed a Light over a Door in the Front of a Building that did consist of two Arches of a pretty large Circle like a Weavers Shuttle, lying along, and the whole Light was Glazed with one piece of Glass both Parts, Circles, and Ovals, and sometimes some distorted Arches are made use of in crocket Windows, *&c.* And also both whole and parts of Circles, and Ovals in their Fret, or Ornamental Works.

II. *Of Glazier's Draughts.*] the most ingenious sort of *Glaziers*, both in the City and Country work by Design, (and not by Guess, like the common Blunderers in most Professions relating to Building;) for they make a *Draught* of all their Windows on Paper, in which they

they set down the Demensions of each Light, both of heighth and breadth, and the number of Squares, both in breadth and heighth, in each Light; and also the number of Lights in each Window, after the following manner, *viz.*

	1			2	
3. 6. 0	3. 6. 0	3. 6. 0	4. 0. 0	4. 0. 0	4. 0. 0
$\frac{6}{4}$	C	$\frac{6}{4}$	$\frac{7}{4}$	C	$\frac{7}{4}$
2. 1. 0	2. 1. 0	2. 1. 0	2. 1. 0	2. .1. 0	2. 1. 0

	3		4	5		6
4. 50	4. 50	4. 50	4. 50	4. 25	3. 75	3. 75
$\frac{8}{3}$	C	$\frac{8}{3}$	C $\frac{8}{3}$	$\frac{7}{3}$	C	$\frac{6}{3}$
1. 50	1. 50	1. 50	1. 50	12. 5	1. 75	1. 25

N. B. Note, That here are six distinct Windows, *viz.* the two upper ones are three light Windows, and of the four lower ones there is one of, three Lights, two single Lights, and one double one.

N. B That a number standing at the top (of the Oblong Figure in the Scheme above) is the heighth of the Light; that at the bottom the breadth, and that number in the middle, the upper one for the number of Squares in heighth, and the lower one for the number in breadth.

N. B. Also that the first and second Windows, (which are three light Windows,) have their Dimensions set down in Feet, and Duodecimal Parts of Feet, E. G in the first Window you have this Number, 3. 6. 0 at the top, which signifies the heighth of the Light to be 3 Foot, and 6 Duodecimal Parts of a Foot, in the middle there is $\frac{6}{4}$ which signifies 6 Squares in heighth, and 4 in breadth, (equal to 24 in the whole Light,) and below their stands 2. 1. 0 which signifies 2 Foot, and 1 Duodecimal Part of a Foot: In the second, or middle Light there is a C set to shew that there must be a *Casement* in that Light, and by Consequence the upper Squares, and lower ones must be cut something shorter, because of the Frame of the Casement,) and the side Squares must be cut something narrower, and the four corners ones both shorter and narrower.

Now by such a Draught, the *London Glazier* when his Country Customers send to him for such a certain parcel of *Glass*, he knows immeditely how to cut it to fit his Work, and the Country *Glazier* knows how to work

work up his Glafs by it, fo that it fhall fit each Window, though he be forty Miles remote from it, as well as if he were by it.

The *London-glafs cutters* commonly mark (with a Letter or Figure over them,) all the Windows that are of one Size, and write the fame Mark on a piece of Paper, the which is put in among that parcel of Squares which belong to thofe Lights which are all of one fize; this piece of Paper is fo put in, that the Character is vifible above the edges of the Squares: By which diftinguifhing Character the Country Glazier readily knows which Squares to take for any Window.

I fhall only add one thing more to this of *Draughts* and fo conclude with this Head: And that is, that fuch Glaziers as underftand it, fet down their Dimenfions in Decimals; which fits as well or better for the *London Glafs-cutters*, becaufe they have their Rules Centefimally divided for the purpofe. I have here therefore, (for fatisfaction to the curious) fet down the Dimenfions of the four lower Windows, in Feet and Centefimal Parts. As for Example, in the third Window, at the top, you have thefe numbers 4. 50, which fignifies the height of the Light is 4 Foot and 50 Centefimal Parts, and at the bottom there is this number 1 50. which is 1 Foot 50 Centefimal Parts, and fo of the reft.

III. *Of Meafuring Glaziers Work.*] In difcourfing of this, I fhall do thefe two Things; firft, Confider the *Cuftoms* ufed amongft them, (for Cuftom is to be the greateft Guide in all manner of Meafures) And, fecondly, of taking the *Dimenfions*, and computing the Quantity.

1. Before we proceed to taking of *Dimenfions*, which one would think fhould be the firft thing, in order to meafure any Superficies or Solid; Yet before we can pretend to take thefe *Dimenfions*, we muft be informed of all Cuftoms that are claimed, and have been tolerated by long ftanding, *&c.* in any Profeffion. And therefore

Note, 1. That in Glazing when Windows have a *Semicircular* top, (or any other *Curved* Form) the Cuftom is to take the full heighth as if it were fquare.

2. That all Windows confifting of intire *Circles*, *Ovals*, or any other *Curved* Form, the Dimenfions are taken the two longeft ways at Right Angles one to another, (which we may call the Diameters, and from thefe Dimenfions the Areas are found the fame as if they were Square.

3. That all *Crocket-windows* in Stone-work are all meafured by their full Dimenfions in heighth and breadth, as if they were Square and not Curved.

4. That there is very good Reafon for all thefe Cuftoms, if we confider the trouble in taking Dimenfions to make them by; 2*dly*, the wafte of Glafs in working it to thefe Forms, and 3*dly*, the extraordinary time expended

expended in ſetting it up, more than in ſquare Lights. I ſay if theſe things be duly weigh'd and conſider'd, they will be found of more Value than the Glaſs which would cover a ſquare Superficies of that Dimenſion. Having thus diſpatched the firſt thing, *viz* *Of Cuſtoms*, I ſhall now proceed to the

2. *Of taking Dimenſions, &c.*] In doing of which, in this Porfeſſion of *Glazing*, it is generally taken to parts of Inches, and ſo computed to the Nicety of a Fraction of an Inch, which may be done ſeveral ways. 4 of which being practiſed by ſome Surveyors and Workmen, I ſhall juſt mention here, *viz* 1 By *Vulgar Fractions*, 2. By *Croſs Multiplication* of Feet, Inches, and Parts, 3 By *Duodecimals*, and 4. By *Decimals*. There is another way by Logarithms, which is more expeditious than either of the former; but I cannot here ſtay to treat of theſe which will require too much Time and Room for this Undertaking, or Deſign.

But becauſe in Glaziers work, they uſually take Dimenſions to the parts of an Inch, the beſt and readieſt way to compute the Area's, is to take the Dimenſions with a Sliding rule, ſuch as the *Glaziers* generally uſe; which Rule is divided Centeſimally, the Dimenſions being thus taken, and ſet down, are Multiplied one into the other as eaſily as Vulgar Arithmetick in whole Numbers is.

As the the Method of taking Dimenſions, and ſetting them down in a Pocket-book, or the like. See *Building*, Numb V. §. 2, 3, 4, 5. where alſo you will find a Bill of Meaſurement of Glazier's Work.

And for the manner of Computing the Quantity. See *Croſs-Multiplcation*

IV. *Of the Price of divers ſorts of Glazier's work.*] The different ſorts of Glazier's Work which we ſhall here mention, are theſe following, *viz Glazing* with Squares, and with Quarries.

And, 1. Of *Glazing* with Squares for the Price of *Crown*, *French*, *German*, *Dutch*, and *Engliſh Glaſs* wrought in Lead, and ſet up. See Numb. III, *&c.* of Glaſs.

As to the Price of Square-work, the Maſter finding Glaſs, and the Glazier Lead, Sodder, and Workmanſhip, 'tis worth about 2 ½ *d. per* Foot; but they will be (in the Country) paid 3 *d per Caſement* for pinning of them, (which is putting of Lead-pins thro' the Iron-frame, and ſoddering them, thereby to fix the Glaſs to the Frame) *viz. Caſements* of 4 ½ Foot long, and ſo proportionably, if they find Lead or Sodder for it.

But to work up Squares and ſet up, finding nothing but Workmanſhip, its worth about 1 *d.* or 1 ½ *d. per* Foot.

2. *Of Glazing* with Quarries, which is for the moſt part done with *Newcaſtle-glaſs* which ſee for the Price of new Work and Materials. Numb. VII.

But if the Glazier find only Lead, Sodder, and Work, 'tis worth about 3 *d per* Foot. But if if they find nothing but work, then 1 ½ *d.* or 2 *d.* is a sufficient Price.

For taking down *Quarry-glass*, scouring it, and soddering it anew, and banding, and setting up again, the usual Price is 1 ½ *d. per* Foot.

But if it be in Churches where they say they have usually more Banding, *&c.* their Price is 2 *d per* Foot. They have also 2 *d. per* Foot for taking down, scouring, soddering, banding, and setting up again of the old fashion'd Work, which is composed of pieces of Glass of different Sizes and Figures.

As to the Quantity of Lead used in any Number of Feet of Glass. See Lead. Num. 10.

I find by Mr. *Leybourn*, that in *London* they generally use that size of Quarries, call'd 12*s.* the which he thus describes. Quarries, (saith he) for the most part are 6 Inches in length from one Acute Angle to the other, and in breadth from Obtuse Angle to Obtuse Angle 4 Inches; so that each Quarry, saith he, contains 12 Superficial Inches; which sort is that which they call long Quarries. See P. *Quarries.*

N. B. That there are several Appellations given to the various Dimensions, *&c* of Quarries, *viz.* 1. The *Range* which is a Perpendicular let fall from one of the Obtuse Angles to the opposite side. 2. And the Length is the longest Diagonal from one acute Angle to the other. 3 The Breadth is the shortest Diagonal, which is drawn between the two Obtuse Angles; as for the Sides and Area of a Quarry, I think all know that.

You will find in the word *Quarries*, that there have been, or still are 12 sorts of *Quarries*; from whence there arises divers Propositions, of great use to Glazier's. As 1. To find any of the five fore-cited Dimensions, as Range, Side, Length, Breadth, and Area, of any of the sort of Quarries. 2 To find the Area of any sort of Quarries. 3. Having any of the Dimensions given, *viz.* Range, Side, Breadth, or Length, to find the Name, or Denomination of the Size, *viz.* Whether 8*s.* 10*s.* 12*s* *&c.* 4. Having the Area of a Quarry given to find of what sort or size it is. 5. To find whether a Window be Glazed with those they call square Quarries or long ones; for you must note there are six sorts of sizes or Quarries, and six sizes of long Quarries, which makes 12 sorts in all, as I told you before. Altho' I am sensible that these five Propositions (but just now mentioned) would be acceptable to Glazier's, and some other ingenious Persons, yet I must here evade them as not being necessary to the Design of this Book. *Glazier's* Bill. See Building. See 7.

Gothick.

Gothick, or *Modern Architecture*, is that which is far remov'd from the Manner and Proportions of the *Antique*, having its *Ornaments* wild and chyme-

chimerical, and its Profiles incorrect. However, it is oftentimes found very ſtrong, and appears very rich and pompous ; as, particularly in ſeveral of our *Engliſh* Cathedrals. This Manner of Building came originally from the North, whence it was brought by the *Goths* into *Germany*, and has ſince been introduced into other Countries.

Gradatory.

'Tis derived from the Latin, and is uſed to ſignifie a Place to which we go up by Steps, particularly an Aſcent from the Cloiſter to the Quire in ſome Churches.

Granary.

A Place to lay up Corn in. Sir *Henry Wotton* adviſes to make it toward the *North*, as much as may be ; becauſe that Quarter is moſt Cool and Temperate.

Mr. *Worlidge* ſaith, That the beſt *Granaries* are built of Brick with Quarters of Timber wrought in the in-ſide, whereto to nail the Boards, with which the in-ſide of the *Granary* muſt be lined ſo cloſe to the Bricks, that there be no room for Vermine to ſhelter themſelves. There may be many Stories one above another, and let them be near the one to the other ; for the ſhallower the Corn lieth, the better, and it is the eaſier turned, which it muſt be ſometimes

Some have had two *Granaries*, one above the other, and filled the upper with Wheat, or other Corn, this upper one had a ſmall Hole in the Floor, by which the Corn deſcended into the lower one, like the Sand in an Hour-glaſs, and when it was all come down into the lower *Granary*, it was then carried up again into the upper one ; and ſo it was kept continually in Motion, which is a great Preſervation to the Corn.

A large Granary, full of ſquare Wooden-pipes, may keep Corn long from heating.

Grange,

From the Latin Word *Grana*, a *Building* which hath Barns, Stables, Stalls, and other neceſſary Places for Husbandry.

Grates.

See *Iron* Numb. 4.

Great Bricks.

See *Bricks*. Numb. IV. §. 9.

Groteſque, or *Groteſca.*

See *Antick*.

Ground.

To *build* on] See *Foundation*, Numb. II. §. 1, 2, 3, 4, 5, and 6.

Ground-guts.

See *Alder*. Numb. III.

Ground-

Ground-Plat, or *Plot.*

1. *What.*] A piece of Ground which a Building is to be erected upon

2. *Of Valuing*] See *Foundations* Numb. 3.

Ground-plate.

See *Sell.*

Grove.

A Term used by *Joyners*, to signifie the Channel that is made by their Plow in the Edge of Molding, or Stile, or Rail &c. to put their Pannels in in Wainscotting.

Cuttæ

See *Architrave.* Numb. 6.

Gutters.

What] Those which we shall here treat of, are *Vallies* in the Roofs of *Buildings*, and these are of two Kinds in respect of their Position; for they are either something near a Paralleism with the Horizon, or inclining towards a Vertical Position to the Horizon.

Of the first kind of *Gutters*, (which, for Distinction sake, I will call Parallel) there are three sorts which are cover'd with Lead; for 1st Either it is a *Gutter* betwixt two Roofs, which stand Parallel to each other, being made upon the Feet of the Rafters of two Roofs which meet together. Or, secondly, A *Gutter* where a *Building* hath a Cantaliver, or Modilion Cornice, which projects 1 ½ Foot, or two Foot, (according to the Design of the *Building*) beyond the Walls, then the Roof is set with the Feet of the Rafters, no farther out than the Walls, but rather within it; so that the Joysts of the upper Floor lie out beyond the Walls, and also beyond the Feet of the Rafters which is yet cover'd with Lead. The third sort of these parallel *Gutters*, are in these Roofs that are flat, commonly called *Platforms*, where are also *Gutters* for the Water that runs from the Platform to descend to, which is from thence conveyed off from the *Building*, either by Spouts or Pipes.

Having described the *Parallel Gutters*, we are next to treat of the *Vertical Gutters*. By *Vertical Gutter*, I mean such a one as is made by two Roofs meeting at Right Angles, one to the other, or which is the same Thing (but in other Words) it is made by the end of one Roof joyning to the side of another: As for Example, If a Building be erected in the Form of a Roman L, 'tis then common to have 1 *Gutter* on the inside of the L, But if it resembles a T, it hath two *Gutters*, but if like an H, then 4. These *Gutters* also are of two sorts, *viz.* Either *Lead* or *Tile*. Of all which we shall speak in their Orders. Having given the Definition of *Gutters*, I shall now proceed to treat

2. *Of laying Parallel Lead Gutters.*] In speaking to this Head, before I proceed to treat

treat of laying the *Lead*, I muſt here give a neceſſary Caution, which is this, *viz.* To take care that the *Gutter* Boards, *&c* lie not too near a Paralleiſm with the Horiſon, but in ſuch a Poſition, that it may have a good *Current* (as Workmen phraſe it;) for if it be laid too near a Level the Water is very ſubject to ſtand in Splaſhes, if the Gutter chance to ſink a little in the middle, which ſome *Gutters* are moſt apt to do; but this is as they are poſited on the Building.

I have obſerved ſome Gutters to have a Layer of Sand for the *Lead* to lie upon; but this way I do not approve of for two Reaſons, which are theſe, *viz* 1. Becauſe ſome ſorts of Sand I have obſerved corrode and decay the Timber that lies near it very much. 2. When 'tis laid on Sand, a very little Squatting *viz* by jumping upon it with the Heels of ones Shoes will dent it, and there the Water immediately ſtands, which is a Means to decay the Lead the ſooner.

In laying of Lead for *Gutters* upon *Boards*, if they are ſo long that one Sheet of Lead will not reach them, then 'tis common (for ſome Plumbers) to ſodder them; for which Purpoſe they cut a Channel croſs the Gutter boards at the End of the Sheet, where the Soddering muſt be; and into this Channel they beat down the Ends of both the Sheets (that are to meet there) into the Channel; which, when they have done, there will remain a little Cavity, which the Sodder fills up level with the reſt, when it is ſoddered.

The Lead which they commonly lay in Gutters, is that which weighs about 8 or 9 ℔. to the Foot. Of theſe *Gutters*. V. P. *Lead* Num. 6.

III. *Of Vertical Gutters.*] Theſe *Gutters*, as I told you before, are of two ſorts, *viz. Lead*, and *Tile*: As to the *Lead* ones I ſhall ſay nothing here, being almoſt the ſame in Effect as the Parallel ones Only unleſs the Builder will be at the Charge, the Lead need not be altogether ſo thick for theſe *Vertical* ones; for if they are laid with Lead of but 6 or 7 ℔. to the Foot, theſe *Vertical Gutters* will laſt as long as the Parallel ones with Lead of 8, 9, or 10 ℔. to the Foot.

Gutters laid with *Tiles*, are alſo made of two kinds of *Tiles*, viz *Concave*, (or *Gutter*) and plain *Tiles*; I ſhall here ſay little of the Concave, or *Gutters*, but refer you to the Deſcription of *Gutter-tiles*.

Plain Tile Gutters, are alſo diſtinguiſhed into two Sorts, *viz.* 1. *Plain-tile gutters*, (properly ſo call'd) and ſecondly, 3 Point Gutters, of both which I will treat in their Order. And,

1. *Of Plain-tile-gutters, (properly ſo call'd)* in theſe *Plain-tile-gutters*, there is a Gutter-board laid which raiſes them from pointing to an Angle: And in laying on the Tiles, the Workman begins at one ſide of the *Gutter*, and ſo works croſs it as if it were plain Work,

Work, and then brings the next row of Tiles back again; ſo that he works forth and back, or to and fro from right to left. So that *Gutters* laid after this Method are not Angular, but of a kind of diſtorted *Curvelinear* Form; by which Means they are not ſo ſubject to be furred up with the Mortar which waſhes out of the adjacent Tiles.

2. *Of Three Point gutters.*] This is the ſecond ſort of *Gutters* which are laid with plain Tiles: In laying of which, they begin and lay one Tile on one part of the Roof, (it matters not which part firſt,) and lay one corner of the Tile juſt in the middle of the *Gutter*, and then they lay another on the other part of the Roof with his Corner juſt in the middle of the *Gutter* alſo; ſo that the Corner of the ſecond Tile is contingent with the firſt; and then lay another Tile in the Gutter, with his Corner, (as it were) betwixt the other two, and to them: Having ſo done, they perſiſt in their Work, and lay a Tile on each part of the Roof, as before, and another betwixt them in the Gutter: After this manner they go on with their Work, till they have finiſhed the *Gutter*: And this is what they call a *Three Point Gutter*; for there always comes three Points, (or Angles) of Tiles together, (*viz.* 1 Angle of three diſtinct Rules) which makes it very uniform and handſome. You are here to note, That only three Inches ſquare of the middle Tile is viſible, (if the Gage be 7 Inches,) the reſt of that *Tile* being cover'd with the next row of Tiles above it.

Tho' theſe *Gutters* are very handſome, and if well done very ſecure alſo; yet if they let the Water into the Houſe, (by reaſon of ſome Stoppage, or broken Tile in the Gutter,) they are very troubleſome to mend.

IV. *Of Meaſuring Gutters,*] or *Vallies.* In different Parts of the Kingdom there are commonly diverſe Cuſtoms, as to this matter of meaſuring *Vallies*, or *Gutters* in *Tiling*: For, firſt, At ſome Places, they but ſeldom or never allow any thing for the *Gutters*, but include it with the reſt of the *Roof*, at flat and half: I am informed, That at *London* they but ſeldom meaſure the *Gutters*; but only as they are part of the Roof, ſo they are included in the Flat and $\frac{1}{2}$ Meaſure. And I know ſome Workmen at *Tunbridge-Wells* never demand any other, but only as it is included in the Plain Meaſure; which is an Area found by Multiplication of twice the length of the Rafters, by the length of the Building. Or which is the ſame Thing (when it is $\frac{3}{4}$ Pitch,) the Flat and $\frac{1}{2}$ Flat.

2. In laying of *Gutters* with Concave-tiles, the Workmen in ſome parts of *Suſſex* and *Kent*, have gotten a Cuſtom to be allowed ſo many Foot more than the plain Meaſure, as there are *Gutter tiles*, (and alſo including *Corner-tiles*, *Ridge-tiles*, *Dorman tiles*) in the whole Roof.

3. At

3 At ſome other Places, I know they claim ſo many Feet more to be added to the Plain-meaſure, as the *Gutters*, (and alſo Corners) are in Length, including *Gutters* at the Sides of *Dormans* and *Lutherns*, if there be any *Dorman-tiles* uſed.

4. In ſome Places the Workmen claim a Cuſtom of having double Meaſure for *Plain-tile*, (eſpecially three Point) Gutters, e. g. If there were but 1 Gutter in a Roof, and this Gutter 15 Foot long, then their Cuſtom is to have 30 Foot more than the Area of the Roof amounts to; and this Method ſome Workmen claim as a Cuſtom in both ſorts of Gutters with *Plain-tiles* Either of theſe *Plain tile-gutters* are cheaper to the Maſter Builder, than *Concave* ones; becauſe *Plain tiles* are cheaper than *Gutter tiles*, being not above $\frac{1}{4}$ part of the Price in many Places. And then again, in caſe the Workman be allowed ſo many Foot more (than the Area of the Roof,) as there are *Gutter-tiles*, that will be $\frac{1}{2}$ as much more as the double Meaſure; for if it be gaged ſo ſlight as 8 Inches, than in a Gutter of 15 Foot long, there would be 45 Tiles, which will be reckon'd 45 Foot; whereas at double Meaſure it did amount to but 30 Foot.

5. I find there is yet another way of computing double Meaſure; for the Account of which I will refer you to *Slating*. Numb. 5.

I am informed, that at *London* Plain-tiles are uſed much for Gutters'

Guttering.

In Carpentry, is commonly done by the Lineal Foot, which ſome *Londoners* value at 1 *s. viz.* Materials and Workmanſhip.

Gutter-tiles.

1. *What.*] Are whilſt they are flat and plain, (before they are bent fit for the Work for which they are deſign'd) ſeemingly at a diſtance a kind of a Triangle with one Convex ſide. But tho' they ſeem ſo at a diſtance, they are not really ſo; for they are Quadrangular Figures, conſiſting of two ſtraight ſides of about 10 or 10 $\frac{1}{2}$ Inches long (they ought to be) and of two circular Sides, one Convex, the other Concave, the Convex ſide is about 14 Inches, and the Concave one about two Inches, this is their Form as to their Edges or Sides, and I will next deſcribe the Form of them in reſpect of the Plain; at the little End they are bent Circular, and ſo likewiſe at the Convex great End, at firſt like a *Corner tile*, but then they bend the Corners of the broad End back again; ſo that if one look againſt the broad Ends Edge, it conſiſts of a Circular Line betwixt two ſtraight ones, like the upper part of the Character for the Sign of *Libra*, thus ♎; this you muſt note is when you hold the Concave ſide of the Tile downwards These Tiles are laid with their broad Ends, and hollow ſides upwards.

2. Of

2. *Of their Weight.*] An Experiment was made to find the Gravity of some of these Tiles, and I found by Computation, That 100 of these Tiles, whose Dimensions were as followeth, *viz.* 10 Inches on the straight Edges, 14 Inches on the greater Convex Edge, when pressed down flat, as when they were in the Mould, and two Inches at the Concave Edge, and about $\frac{5}{8}$ Inch thick: I say 100 of these weigh'd about 321, or 322 ℔ and by Consequence 1000 will weigh about 3210, or 3220 ℔. which is near 29 C weight. So by Consequence about 682 would be a Tun weight.

3. *Of their Price.*] Mr. *Leybourn* saith, That at *London* they are sold at 1 *d.* $\frac{1}{2}$, or 2 *d.* *per Tile*, or between 10 and 15 *s. per* 100. in some Places their constant Price is 1 *d* $\frac{1}{2}$ *per* Piece, or 12 *s. per* Hundred.

Hack.

WHat. See *Bricks* Numb. 12.

Hair.

1. *What*] The Hair here mentioned is Bullocks, *&c* to put in *white Mortar*, of the Quantity of a certain Quantity of Lime. See *Mortar*, Numb 4

2 *The Price*] As to the Price, 'tis various, according to the Plenty or Scarcity of it. At some Places in *Kent* I have known it sold for 7 *d. per* Bushel, and at other Places, *viz.* in *Sussex*, I have known it sold for 10 *d.* and 12 *d per* Bushel; so that its Price by the Load (for a Horse), which is 60 Bushels, is from 30 to 60 *s.*

Half-round.

What.] See *Capital.* Numb. 2.

Halls.

Of their Dimensions] According to a noted *French* Architect, ought to be in Length twice and $\frac{1}{4}$ its Breadth, at least, and in great Buildings he saith, You may well allow the Length to be three Times the Breadth: He further adds, That this last Length will be the most beautiful and convenient.

As to the Heighth of *Halls*, it may be $\frac{2}{3}$ of the Breadth, or about 16 or 18 Foot in noble *Buildings*

In great *Buildings*, the *Hall*, and other Rooms of the first Story may be arched, by which Means they will become much handsomer, and less subject to Fire: The Heighth is found by dividing the Breadth of the *Hall* into six Parts, and five of those shall limit the Height of the Room from the Floor to the under side of the Key of the Arch.

Hangs over.

What] See *Batter.*

Hanse.

What.] See *Arches.* Numb 6.

Heads,

Heads.

What.] A Term used by Bricklayers, by which they mean ½ a *Tile* in Length, but to the full Breadth of a *Tile*; these they use to lay at the Eaves of a Roof.

Heading Architrave.

See *Architrave* Numb. 2.

Healing.

What.] By this Word is understood the covering of the Roof of any Building, which is of various Kinds, *viz.* 1 *Lead*, 2. *Tiles*. 3. *Slate*. 4. *Horsham Stone*. 5. *Shingles*. 6. *Reed*. 7. *Straw*. An Account of all these sorts of *Healing*, you may find under these Heads, *viz. Lead*, under that Head. *Tiles*, under *Tiling*; *Slate* under *Slating*; *Horsham stone*, see the same Word; *Shingles*, and *Shinglings*, *Reed* and *Straw*; see *Thatching*.

Hearth-stones.

See *Fire-stones*.

Heel.

An inverted Ogee. See *Cima*.

Helix.

Helix, or *Urilla*, is a little *Volute*, *Caulicole*, or Stalk under the Flower of the Corinthian Capital. The Word is mere Greek, and signifies a kind of Ivy, whose Stalk is twisted like the Vine.

Hinges.

1. *What.*] Are those necessary Irons, by Means of which, all Doors, whether of Houses, (or other Buildings) or of Pews, Cupboards, *&c.* all Lids of Boxes, Chests, Trunks, *&c* Folds of *Tables*, Beds, Bureaus Scrutores, *&c.* make their Motion, whether of opening or shutting, or of Folding, *&c.*

2. *The Sorts or Kinds.*] The Species of Hinges are many, *viz Bed*, *Box*, *Butts*, *Casement*, *Lancashire*, and smooth filed; *Casting*, *Chest* black *Lancashire*, smooth filed, *Coach*, *Desk*, *Dovetails*, *Esses*, *Folding*, *Garnets*, *Dozen-ware-long*, *Dozen-ware-short*, *Weighty long* *Weighty-short*, *Lamb-heads*, *Port*, *Side-Lancashire*, *Side-smooth-filed*, *Side* with rising Joints, *Lancashire* and *Smooth-filed*, *Side* with Squares, *Screw*, *Scuttle*, *shutter*, *Lancashire*, and Smoth-filed, *Stall*, *Trunk* of sundry sorts. *Joints*, *Lancashire* Dozen-ware, with *Hooks*, Dozen-ware long, Dozen-ware short, weighty long, weighty short.

3. *The Price of some Sorts of Hinges.*] As to the Value of *Hinges*, I shall not now stay to insist upon the usual Price of all sorts, but shall give the Reader a brief Account of some sorts that are pretty commonly used: And first, of *Butts*, of which there are different Sizes, and by Consequence of various Prizes, *viz.* from 2 *s.* or 2 *s.* 6 *d.* to 5 *s. per* Doz.

2. *Bed-hinges*, from 5 to 7 *s. per* Doz.

3. Box-hinges, from 12 *d.* to 4 *s. per* Doz.

4.

4. — Small Brass ones, from 2 *s.* to 2 *s.* 6 *d. per* Doz.

5. Dovetails, from 12 *d.* to 4 *s. per* Doz.

6. Hooks and Hinges, *&c. per* ℔. from 3 ¼ *d.* to 4 *d.*

7. Side-hinges, from 3 *s.* to 16 *s. per* Doz.

8. — With a Square, from 20 *s.* to 36 *s. per* Doz.

9. Screw-hinges, from 30 *s* to 48 *s per* Doz.

See *Iron.* Numb. 4.

Hip.

1. *What*] Are those Pieces of Timber which are at the Corners of a Roof, they are a great deal longer than the Rafters, by reason of their oblique Position, and they are placed not with a Right or Square-Angle, but a very oblique one; and by Consequence they are not, (or at least ought not to be) square at any Angle, (as Rafters are not at all) but bevel at every one of them; and which is yet more, as Rafters have but 4 Plains, these commonly have 5. They are commonly by Countrey-workmen call'd *Corners*, and some call them *Principal Rafters*, and others *Sleepers*. The Truth is, *Hips* and *Sleepers*, are almost the same; only the *Sleepers* lie in the *Vallies*, (and join at the Top with the *Hips*): But those Surfaces, or Plains which make the back of the *Hip*, are the under-sides of the *Sleeper*.

2 *Back, or Back of a Hip, what*] By the *Back of* a *Hip*; you are to understand, is meant those two Superficies, or Plains on the outside of the *Hip*, which lie Parallel, (both in respect of their Length and Breadth,) with the Superficies of the adjoining Side and End of the *Roof*.

3. *Mould, what.*] By the word *Hip mould*, some mean the same as *the Back of the Hip*. But some others understand it to mean, the Prototype, or Pattern, (which is commonly made of a piece of thin Wainscot), by which the Back and Sides of the *Hip* are set out.

4 *Of finding the Lengths and Backs of Hips*, &c. *in square Frames.*] I shall here, not only give you the Method of finding the Length of *Hips*, or *Sleepers*; but also of the *Rafters*, *Diagonals*, ½ *Diagonal* and *Perpendicular*. And that in a brief Analogical Table, say,

	Foot		
As 20	15, 00	:: Breadth of the House:	Len. of the Rafter.
	18, 00		Len. of the Hip.
	11, 18		Perpendicular.
	28, 28		Diagonal.
	16, 63		Nearest Dist.

Hip Angles	at foot—38—22	Rafter Angles at	top 41—50
	at top—51—28		foot 48—10
	at back-116—12		

The

The Angles are always the ſame in all ſquare Frames that are true Pitch.

5. *Roof, what.*] By a *Hip Roof*, you are to underſtand ſuch a *Roof*, as hath neither Gable-heads, nor *Shread-head*, or *Jirk-in-head*, (by which we mean ſuch Heads as are both *Gable* and *Hip* at the ſame End; for 'tis a Gable, or upright as high as the *Collar-beam*, and then there is two ſhort Hips which ſhut up with their Tops to the Tops of a pair of Rafters, which Country Carpenters call *Singlars*) For a *Hip Roof* hath Rafters as long, and with the Angles at the Foot, *&c.* at the ends of Building, as it hath at the Sides, and the Feet of the Rafters on the Ends of ſuch Buildings as have *Hip-roofs*, ſtand on the ſame Plain, (*viz* Parallel with the Horizon) and at the ſame Heighth from the Foundation with the Rafters on the Sides of the Roof. Theſe *Hip-roofs*, ſome call *Italian Roofs*.

6 *Of meaſuring Hip-roofs*.] As to meaſuring of *Hip-roofs*, if they are ¾, or true Pitch, as it is commonly called; then 'tis only to multiply the Length of the Building by the Breadth, and to the Area thus found, add half as much more, or elſe multiply the Length by the Breadth and ½; or the Breadth by the Length and ½, either of theſe three ways will produce the flat and ½, which is equal to the Content of the Roof in plain Meaſure, if there be nothing allowed for *Hips* and *Vallies*: But if the Roof have no Cornice, but the Rafters have Feet, then they muſt be added, and alſo the Eaves Board in a Bill of Meaſurement.

Or to meaſure ſuch a Roof, you may multiply the Length of it by the Length of the Rafter, and it ſhall give you half its Content; or elſe multiply the length of the Building by twice the length of the Rafter; and then you will have the whole Content.

Hip tiles.

See *Corner tiles*, or *Tiles*. N. V.

Hollow.

A Term in Architecture, by which is meant a Concave Molding, being about a Quadrant of a Circle; by ſome it is call'd a *Caſement*, by others an *Abacus*.

Hooks.

Are a neceſſary Utenſils which are made uſe of for ſeveral Purpoſes in Buildings, *&c.* They are of various ſorts; ſome of Iron, and others of Braſs: I ſhall here mention the Names of ſome, which take as followeth, *viz.* 1. *Armour-hooks*, (theſe are generally of Braſs, and are to lay up Arms upon; as Guns, Muskets, Halberts, Half-pikes, Pikes, Javelins, *&c.* 2. *Caſement hooks*. 3. *Chimney-hooks*, which are made both of Braſs and Iron, and of different Faſhions: Braſs ones I have known ſold from 2 *s.* to 2 *s.* 6*d. per* pair, the Iron ones from 12 *d.* to 1 *s.* 6 *d. per* pair, their uſe is to ſet the Tongs and Fire-

Fire-shovel against. 4. *Curtain-hooks.* 5. *Hooks* for Doors, Gates, *&c.* These are from 3 *d.* ¼ to 4 *d. per* ℔. 6. *Double-line-hooks*, large and small. 7. *Single-line-hooks*, large and small. 8. *Tenter-hooks*, of various sorts, *viz.* 2 *d.* 3 *d.* 4 *d.* 6 *d.* 10 *d.* 20 *d.* and 40 *d.*

Horsham-stone.

1. *What.*] Is a kind of thin broad Slate of a greyish Colour, much used in some parts of *Sussex* formerly, not only to heal, or cover Churches and Chancels, but some great Houses also; it is call'd *Horsham-stone*, in that County, because it is for the most part brought from a Town there called *Horsham*; this sort of Stone, or Slate rather, is laid of different Sizes, *viz.* From 8 or 9 Inches, to 24 Inches, or more in length, or breadth, *&c.* It is commonly from ½ Inch, to 1 Inch thick.

2. *Of the Price of Horsham-stone.*] The value of them is according to the distance from the Quarry, *viz.* From 10 to 20 *s. per* Load, I have known a Load of them laid in for 17 or 18 *s* at 18 or 20 Miles distance from the Quarry. A Load of these, (as I have been inform'd by those that have made some Observations on this Matter,) will cover about ¾ of a Square.

3. *Of Laying.*] The Price of laying a Square and Pointing, (which is striking Mortar under the lower ends) in new Work, 5 or 6 *s.* But to rip it from old, and new lay and point it, not less than 6 or 7 *s. per* Square, which is the lowest I ever knew it done for.

4. *Of the Weight of this sort of Healing.*] I have been informed by an observing Mechanick, that a Square of this kind of covering will weigh about 33 or 34 C. weight: Whereas, saith he, a Square of *Tiling* doth not weigh above 16 or 17 C. weight. Nay, he is confident not above 18 C. weight, if it be gaged at 6 Inches, and the Tiles not exceeding 10 Inches long. (Nay, I know that in many Places they fall short of that.)

5. *Of the Properties of this sort of Covering.*] From what hath been said before, you may easily gather that this sort of Healing is dearer than Tiles; for the Charge of a Square of Tiling, is from about 23 *s.* to 30 *s.* or some will have it from 24 to 28 *s. per* Square; whereas I find by Calculations from some Observations, a Square of Healing with *Horsham-stone*, will be worth from 32 *s* to 38 *s.* And besides for this sort of Covering, the Timber for the Roof, ought to be considerably stouter and stronger, because a Square of this sort of Stone is almost as heavy again as a Square of Tyling. Now that which I suppose to be the cause why these Stones have been so much in use for Churches, *&c.* must be, because they are far more durable than Tiles; which makes some Recompence for the Charge; for some say these Stones are very durable, being for the most part very hard, so that no Weather

ther will do them any hurt, as it will Tiles.

House.

1. *What.*] 'Tis a thing so well known, that every one that knows any thing, is sensible of the Contrivance whereby Men preserve themselves and their Goods from the Injuries of the Weather, and other Inconveniencies.

In treating of this word, *House*, I shall here do these four things. *First*, Discourse concerning the Situation of a Country-house. *Secondly*, of the Ground-work of houses *Thirdly*, Concerning Building in *London*. *Fourthly*, and *lastly*, Discourse of Party-walls.

The Reason why I shall add no more under this Title, House, is this; because I sufficiently treat of these six things, *viz. Situation Contrivance*, *Receipt*, *Strength*, *Beauty*, and *Form*, or *Figure*, in the Word *Building*.

2 *Of the Site of a Country house*] To what I have said, concerning the Situation of a Country-house, in the word *Building*, I shall here add, That Woods, as well as Water, ought to be near your Country Habitation; they being the principal Things that adorn a Rural Seat: But if you cannot conveniently Seat your *House* among the Trees, yet there are but few Places, but you may speedily raise Trees about your House; according to Mr *Evelyn*'s, or Mr. *Worlige*'s Directions.

It is far better to have a House, defended by Trees than Hills, for Trees yield a Cooling, Refreshing, Sweet, and Healthy Air, and Shade, during the Heat of the Summer, and very much break the cool Winds and Tempests from every Coast in the Winter. The Hills, according as they are situated, defend only from some certain Winds; and if they are on the *North* side of your House, as they defend you from the cold Air in the Winter, so they also deprive you of the oool refreshing Breezes which are commonly blown from thence in the Summer. And if Hills are situated on the *South* side, it then proves also very inconvenient. Besides, they yield not the Pleasures and Contentments, nor the Varieties of Oblectations to the ingenious Rustick, as the tall Plumps of Trees, and pleasant Groves do. Yet Hills which are cloathed with Coppices, or otherwise improved are pleasant Objects, if they stand not too near a House.

Let not your House be too low seated, lest you lose the Conveniency of Cellars; but if you cannot avoid Building on low Grounds, set the first Floor above the Ground in your House the higher, to supply what you want to sink in your Cellar in the Ground; for in such low and moist Grounds, it conduceth much to the dryness, and healthiness of the Air to have Cellars under the House, so that the Floors be good and cieled underneath.

Not

Not to ſpeak of the Building of Places, or Seats, for the Nobility, or Gentry, but only of plain and ordinary *Farm-houſes*. I have thus much obſerved, (ſaith Mr. *Worlige)* that Houſes built too high in places obvious to the Winds, and not well defended by Hills, or Trees, require more Materials to build them, and more alſo of Reparations to maintain them, and are not ſo commodious to the Inhabitants as the lower built *Houſes*, which may be made at a much eaſier rate, and alſo as compleat and beautiful as the other.

3. *Of the Ground-work of Houſes.*] Buildings, or Houſes, that are not above two Stories with the Ground-room, and not exceeding 20 Foot to the Raiſon-place, and upon a good Foundation, the length of two Bricks, or 18 Inches for the Heading courſe, will be ſufficient for the *Ground-work* of any common Structure, and 6 or 7 Courſes above the Earth to a Water-table, where the thickneſs of the Walls are abated; (or taken in) on either ſide the thickneſs of a Brick, namely, 2 ¼ Inches.

But for large and high Houſes, or Buildings, of 3, 4 or 5 Stories with the Garrets: The Walls of ſuch Edifices ought to be from the Foundation to the firſt Water-table. 3 Heading courſe of Bricks, or 28 Inches at the leaſt, and at every Story a Water-table, or taking in on the inſide for the Summers, Girders, and Joyſts to reſt upon, laid into the middle, or ¼ of the Wall, at leaſt for the better Bond. But as for the innermoſt, or Partition-wall, one ½ Brick will be of a ſufficient thickneſs: And for the upper Stories a 9 Inch (or Brick length) Wall will a very well ſuffice.

4. *An Act concerning Building of Houſes in London*] What here follows is ſo much of the Act only as relates to the *Bricklayers'-work, viz.* The heighths and numbers of Stories, and thickneſs of Walls of the four ſeveral Rates of Houſes, which is as follows.

And be it further Enacted, That the Houſes of the firſt and leaſt ſort of Building, fronting by Streets or Lanes, as aforeſaid, ſhall be of two Stories high, beſides Cellars and Garrets, that the Cellars thereof be 6 ½ Foot high, if the Springs of Water hinder not; and the firſt Story be 9 Foot high from the Floor to the Ceiling, and the ſecond Story be 9 Foot high from the Floor to the Ceiling; that all the Walls in Front and Rear, as high as the firſt Story, be of the full thickneſs of the length of two Bricks, and thence upwards to the Garrets of the thickneſs of one Brick and half; and that the thickneſs of the Garret Walls on the back part be left to the Diſcretion of the Builder, ſo that the ſame be not leſs than one Brick a length; and alſo that the thickneſs of the *Party walls* between theſe Houſes of the firſt and leſſer ſort of Buildings be 1 Brick and half as high as the ſaid Garrets, and that the thickneſs of the Party-walls in the

Garret

Garret be of the thickneſs of the length of 1 Brick, at leaſt.

And be it further Enacted, That the Houſes of the ſecond ſort of Building fronting Streets, and Lanes of Note, and the River of *Thames* ſhall conſiſt of three Stories high, beſides Cellars and Garrets, as aforeſaid; that the Cellars therof be ſix Foot and an half high, (if the Springs hinder not) that the firſt Story contain full 10 Foot in heighth from the Floor to the Ceiling: The ſecond full 10 Foot: The third 9 Foot: That all the ſaid Walls in Front and Rear, as high as the firſt Story, be two Bricks and an half thick; and from thence upward to the Garret-floor, of one Brick and an half thick; and the thickneſs of the Garret Walls on the back part be left to the Diſcretion of the Builder; ſo that the ſame be not leſs than one Brick thick: And alſo that the thickneſs of the Party-walls between every Houſe of this ſecond, and larger ſort of Building be two Bricks thick as high as the firſt Story, and thence upwards to the Garrets, of the thickneſs of one Brick and an half.

Alſo, that the Houſes of the third ſort of Buildings, fronting the high and principal Streets, ſhall conſiſt of four Stories high, beſides Cellars and Garrets, as aforeſaid: That the firſt Story contain full ten Foot in heighth from the Floor to the Ceiling, the ſecond ten Foot and an half, and the third nine Foot, the fourth 8 Foot and an half: That all the ſaid Walls in Front and Rear, as high as the firſt Story be two Bricks and an half in thickneſs, and from thence upwards to the Garret-floor, of the thickneſs of one Brick and an half: That the thickneſs of the Garret-walls on the back part be left to the Diſcretion of the Builder, ſo as the ſame be not leſs than one Brick: And alſo that the Party-walls between every Houſe of this third and larger ſort of Building, be two Bricks thick as high as the firſt Floor, and thence upwards to the Garret floor, the thickneſs of one Brick and an half.

And be it further Enacted, That all Houſes of the fourth ſort of Building, being *Manſion-houſes*, and of the greateſt Bigneſs, not fronting upon any of the Streets, or Lanes, as aforeſaid, the number of Stories, and the Heighth thereof, ſhall be left to the Diſcretion of the Builder, ſo as he exceeds not five Stories.

Alſo the ſame Act enjoyns, That no *Timber* be laid within twelve Inches of the fore-ſide of the *Chimney jambs*, and that all *Joyſts* on the back of any *Chimney* be laid with a *Trimmer*, at ſix Inches diſtance from the Back: Alſo that no Timber be laid within the Tunnel of any Chimney, upon Penalty to the Workman, for every Default 10*s*. and 10 *s*. every Week it continues unreformed.

Thus far the Act.

Note, further, When you lay any *Timber* on *Brick-work*, as Taſſels, (or Torſels) for Mantle-Trees to lie on, or *Lintels* over

over Windows, or *Templets* under Girders, or any other Timbers; lay them in *Loam*, which is a great Preſerver of Timber; for Mortar eats and corrodes the Timber: Likewiſe the Joyſt ends, and Girders which lie in the Walls, muſt be loamed all over to preſerve them from the Corroding of the Mortar. Some Workmen pitch the ends of the Timber that lie in the Walls to preſerve them from the Mortar.

5. *Concerning Party-walls.*] In diſcourſing of this Matter, I will preſent the Reader with two different Methods of valuing ſuch Walls according to two different Surveyors, *viz.* Mr. *Leybourn*, and Mr. *Phillips.*

And firſt according to Mr. *Leybourn*.

Now, (ſaith he) foraſmuch as the *Buildings* in *London* joyn one upon another, and almoſt every ſeveral houſe hath a diſtinct Proprietor, the Parliament hath decreed, That the Wall dividing the Proprietor's Ground, ſhalt be built at the equal charge of both the Owners; it will not therefore be unneceſſary to ſhow how theſe Party-walls are to be valued.

All Brick-work, whether it be 1, 2, 3, 4, or any other number of Bricks lengths in thickneſs, are all to be reduced to the thickneſs of one Brick and an half.

It hath been obſerved (ſaith he) that about 4500 of Bricks, 100 and a quarter of Lime, 2 ½ Loads of Sand will compleatly raiſe one Rod of Brick-work, of a Brick and an half thickneſs.

	l.	*s.*	*d.*
Now 4500 of Bricks at 16 *s. per* 1000, is	3	12	0
A Hundred and quarter of Lime at 10 *s per* C.	0	12	6
Two Load and ½ of Sand at 3 *s. per* Load.	0	7	6
In all	4	12	0

And thus much will a Rod of *Party-wall*, (the Materials only) reduced to Brick and ½ thick, amount unto at the former ſuppoſed Rates, to which may be added for Workmanſhip.

	1	8	0
The whole Sum is	6	0	0

So that for every Rod that is in a Party-wall, between Proprietor and Proprietor, they are to allow 3 *l.* a piece for every Rod of Party-wall. So that if a Party wall meaſured, and the meaſure reduced to a Brick and an half, ſhould be found to contain 16 Rods, that 16 being multiplied by 3 *l.* giveth 48 *l* and ſo much is the one Proprietor to allow the other.

But

But note by the way, That altho' this Rule here deliver'd be general, yet the Price of the Party wall shall be more or less, according as Materials rise or fall. For sometimes a Rod of Wall of a Brick and an half thick, will cost but 5 *l.* 10 *s* and then each Proprietor must pay but 2 *l.* 15 *s* *per* Rod. Thus far Mr. *Leybourn* I will now add Mr. *Phillips*'s Way.

Now, (saith he) having the Dimensions, both in length and heighth of the Cellar, and all other Stories in the House, then the following Tables will shew (according to the thickness of the Wall,) how many Bricks your Neighbour is to pay for towards his Party-wall.

For which purpose the insuing Tables will serve very well; for these *Walls*, according to the Act of Parliament, for this purpose are to be made part of them two Bricks thick, part of them one Brick and half thick, and part of them one Brick thick.

Now knowing the number of Bricks which go to the making of the Wall, you may easily compute the Charge of the Mortar and Workmanship thereof, and from thence find the whole Charge, which you will find, (saith he) to be about 30 *s.* for every 1000 of Bricks, (I think Mr. *Phillips*'s Price is too great; for I think 25 or 26 *s.* *per* 1000 is very well, but he acknowledges that Bricks then were something dear, *viz.* about 18 or 20 *s.* *per* 1000)

He then proceeds to an Example; suppose a House of the third Rate, the Party-wall thereof being thirty Foot long, and you would know how many Bricks are to be paid for towards this Party wall.

First, measure the Cellar where the Party-wall is to be two Bricks thick, the length whereof is thirty Foot, and the depth seven Foot; find this length in the side, and the depth in the top of the Table, and in the Square of meeting in the Table for one Brick thick, you will find 2314 Bricks are to be paid for.

Then proceed to the first Story, which will be likewise 30 Foot long, and 10 Foot high, and also 2 Bricks thick, the same Table shews the allowance for this 3306

The second Story also is 30 Foot long, and 10 $\frac{1}{2}$ high, but the Party-wall is to be but a Brick and $\frac{1}{2}$ thick, the $\frac{1}{2}$ whereof is $\frac{3}{4}$ of a Brick, and this in the Table of $\frac{3}{4}$ of a Brick, yields for 30 Foot long, and 10 Foot high. 2479

And for the $\frac{1}{2}$ Foot more in heighth. 124

The third Story is 9 Foot high, and 30 Foot long, being likewise a Brick and half thick; and for this the Table shews the half to pay for it. 2231

The fourth Story is 8 Foot and a half high, and 30 Foot in length, for the 8 Foot the Table shews. 1983

and for the half Foot 124

All which added together, make } 12559

which

which are to be paid for the half of the Party wall, which at :6 *s*. *per* Thousand, comes to *86 l.—6 s.— 6 d.*

Thus you may ſee what any Party-wall comes to, tho' your Neighbouers Houſe joyns never ſo little, or much to yours, as readily as you can by meaſuring by the Rod.

And whereas the Floors of the ſeveral Stories add ſomewhat to the heighth, you may add ſomewhat for them according as you find them in thickneſs.

Laſtly, For the Garrets, the Walls thereof being but one Brick thick , you may take half the number in the Table of one Brick's thickneſs, and add to the reſt of the Account.

All the difference that can be between Neighbours herein, will be about the Price of Bricks, and the Lime, and Workmanſhip ; but if Neighbours build together, they will eaſily determine it ; but if they do not, yet the firſt Builder is ſufficiently provided by his Workmen to rectifie his Charge, and by ACT of Parliament is allowed full Satisfaction, with Intereſt from the Time of his Building.

A TABLE

A TABLE *for* 1 *Brick in Thickneſs, or the half of* 2 *Bricks.*

The Walls heighth in Feet.

Foot long.	½ Bricks	I. Bricks	II. Bricks	III. Bricks	IV. Bricks	V. Bricks
1	5	11	22	33	44	55
2	11	22	44	66	88	110
3	16	33	66	99	132	165
4	22	44	88	132	176	220
5	27	55	110	165	220	275
6	33	66	132	199	264	331
7	39	77	154	231	309	386
8	44	88	176	264	353	441
9	50	99	198	298	397	496
10	55	110	220	331	441	551
11	61	121	244	364	485	606
12	66	132	264	397	529	661
13	72	143	286	431	573	716
14	77	154	309	462	617	771
15	83	165	331	496	661	826
16	88	176	355	529	705	882
17	94	187	375	562	749	937
18	99	198	397	595	793	992
19	105	209	419	628	837	1047
20	110	220	441	661	882	1102
21	116	231	463	694	926	1157
22	121	242	485	726	970	1212
23	127	253	507	760	1014	1267
24	132	264	529	793	1058	1322
25	138	275	551	826	1102	1377
26	143	286	573	860	1146	1432
28	154	309	617	926	1234	1543
30	165	331	661	992	1322	1653
40	220	441	881	1322	1763	2204
50	275	551	1102	1652	2204	2755

A

A TABLE *for* 1 *Brick in Thickneſs, or the half of* 2 *Bricks.*

The Walls heighth in Feet.

Foot long.	VI. Bricks	VII. Bricks	VIII Bricks	IX. Bricks	X. Bricks
1	60	77	85	99	110
2	132	154	176	198	220
3	198	231	264	298	331
4	264	309	353	397	441
5	331	386	441	496	551
6	397	463	529	595	661
7	463	540	617	694	771
8	529	617	705	793	882
9	595	694	793	893	992
10	661	771	882	992	1102
11	727	848	970	1091	1212
12	793	926	1058	1190	1322
13	859	1003	1146	1289	1433
14	926	1080	1234	1388	1543
15	992	1157	1322	1488	1653
16	1085	1234	1410	1587	1763
17	1124	1311	1499	1686	1873
18	1190	1388	1587	1787	1983
19	1256	1466	1675	1884	2094
20	1322	1543	1763	1983	2204
21	1388	1620	1851	2083	2314
22	1455	1697	1939	2182	2424
23	1520	1774	2028	2281	2534
24	1587	1851	2116	2380	2645
25	1653	1928	2204	2479	2755
26	1719	2006	2292	2578	2865
28	1857	2160	2458	2777	3085
30	1983	2314	2645	2975	3306
40	2645	3085	3526	3967	4408
50	3306	3857	4408	4959	5510

A TABLE *for 3 Quarters of a Brick Thick, being the half of a Brick and half.*

The Walls heighth in Feet.

Foot long.	½ Bricks	I. Bricks	II. Bricks	III. Bricks	IV. Bricks	V. Bricks
1	4	8	17	25	32	41
2	8	17	33	50	66	83
3	12	25	50	74	99	124
4	17	33	66	99	132	165
5	21	41	83	124	165	207
6	25	50	99	149	198	248
7	29	58	116	174	231	289
8	33	66	132	198	264	331
9	37	74	149	223	298	372
10	41	83	165	238	331	413
11	45	91	182	273	364	455
12	50	99	198	298	397	496
13	54	107	215	322	430	537
14	58	116	231	347	463	578
15	62	124	248	372	496	620
16	66	132	264	397	529	661
17	70	140	281	421	562	702
18	74	149	298	446	595	744
19	79	157	314	471	628	785
20	83	165	331	496	66[illegible]	826
21	87	174	347	521	694	868
22	91	182	369	545	727	909
23	95	190	380	570	760	950
24	99	198	397	595	793	992
25	103	206	413	620	826	1033
26	107	215	430	645	860	1074
28	116	231	463	694	926	1157
30	124	248	496	744	992	1240
40	165	331	661	992	1322	1653
50	207	413	826	1240	1653	2066

A TABLE *for 3 Quarters of a Brick thick, being the half of a Brick and half.*

The heighth of the Walls in Feet.

Foot long.	VI. Bricks	VII. Bricks	VIII. Bricks	IX. Bricks	X. Bricks
1	50	58	66	74	83
2	99	116	132	149	165
3	149	174	198	223	248
4	198	231	264	298	331
5	248	289	331	372	413
6	298	347	397	446	496
7	347	405	463	521	579
8	394	463	529	595	661
9	446	521	595	660	744
10	496	579	661	744	826
11	545	636	727	818	900
12	595	691	793	893	992
13	645	752	860	976	1074
14	694	810	926	1041	1157
15	743	868	992	1117	1240
16	793	926	1058	1189	1322
17	843	983	1124	1264	1405
18	893	1041	1190	1339	1488
19	942	1099	1256	1413	1570
20	992	1157	1322	1488	1653
21	1041	1215	1388	1562	1736
22	1091	1273	1455	1636	1818
23	1140	1331	1521	1711	1901
24	1190	1388	1587	1785	1983
25	1240	1446	1652	1860	2066
26	1290	1504	1709	1934	2149
28	1338	1620	1851	2083	2314
30	1488	1736	1983	2231	2479
40	1983	2324	2625	2975	3306
50	2479	2893	3306	3719	4132

A

6. *I shall here add an Abstract of the Act.*] Being a Table of Proportions for Brick walls, Number and Heighth of Stories, *&c.* In the Building of Three sorts of *Houses*, (*viz.* The Three First Rates) in the City of *London*, according to the ACT of Parliament.

The Rates, or sorts of Houses.	Several Stories.	Feet in heighth. (The depth of the Cellars, and heighth of the several Stories)	Brick length in thickness. (The thickness of the Walls of each House in Fr. and Rear.)	Brick length in thickness. (The thickness of the Party-walls between House and Hou.)
The first or least sort of Houses, fronting by Streets and Lanes, are to have	Cellars,	$6\frac{1}{2}$	2	$1\frac{1}{2}$
	1 Story,	9	2	$1\frac{1}{2}$
	2 Stories,	9	$1\frac{1}{2}$	$1\frac{1}{2}$
	Garrets.		1	1
The second sort of Houses fronting lesser Streets, and Lanes of note, and the *Thames* Side are to have	Cellars,	$6\frac{1}{2}$	$2\frac{1}{2}$	2
	1 Story,	10	$2\frac{1}{2}$	2
	2 Stories,	10	$1\frac{1}{2}$	$1\frac{1}{2}$
	3 Stories,	9	$1\frac{1}{2}$	$1\frac{1}{2}$
	Garrets.		1	1
The third sort of Houses fronting high Streets are to have	Cellars,	$6\frac{1}{2}$	$2\frac{1}{2}$	2
	1 Story,	$10\frac{1}{2}$	$2\frac{1}{2}$	2
	2 Stories,	10	$1\frac{1}{2}$	$1\frac{1}{2}$
	3 Stories,	9	$1\frac{1}{2}$	$1\frac{1}{2}$
	4 Stories,	$8\frac{1}{2}$	$1\frac{1}{2}$	$1\frac{1}{2}$
	Garrets.		1	1

More concerning Building of *Houses* you may find under the Head *Architect*; also a very ample Account under the Head *Building*, where are these following Paragraphs and Sections, *viz.* 1. Considerations about *Building*: 2. *Aphorisms*, which are subdivided into the following seven Sections.

1. *Situation*, in respect of the whole.

2. *Contrivance*, with some Precautions.

3. *Receipt.*

4. *Strength*, with Directions about it.

5. *Beauty*, in the whole and parts.

6. *Form, Figure*, or *Fashion*, and what Figure is strongest and most convenient.

III.

III. The third Paragraph, contains a Comparison betwixt the modern and ancient Way of Building in *England*.

IV. Contains some general Rules which ought to be observed in Building all Houses, both in the City, and the Country.

V. Contains a Method of Surveying of Buildings, or Houses, and also the Method of taking of Dimensions, and setting them down in a Pocket-book, or the like; and also the Form of a *Bill of Measurement*.

VI. This sixth Paragraph is of the Method of Measuring all Artificer's Works, relating to Building of Houses, *&c.*

VII. In the seventh Paragraph, there is shewed a Method how you may nearly value all sorts (almost) of Buildings, whether they are great or small.

VIII. You have in the eighth Paragraph a Method of *Censuring* Buildings, or Houses, *viz.* Directions how to pass one's Judment on a Fabrick (that is already built) whether it be well and firmly compacted, and whether well contrived, as to the whole, and all its parts separately, for Use, or Conveniency; and as to its Beauty, whether its Parts be placed in an uniform Order, and whether the Designer, or Master-Builder, observed a due Symmetry, or Proportion of the Parts, in respect of one another, *&c.*

IX. And lastly, under the Head Building, you have some Directions concerning advising with Workmen about the Charge of Building any House: And how much a Builder, (or Gentleman that is going to build) is the wiser for such Men's Advice commonly, I mean if he advise with such as are to do the Work: Tho' otherwise perhaps he may be well inform'd by some ingenious Workmen that understand the Speculative Part of Architecture, or Building: But of these knowing sort of Artificers there are but few, because but few Workmen look any further than the Mechanical, Practick, or working Part of *Architecture*; not regarding the Mathematical, or Speculative Part of Building, thinking it to be of little or no Use; this I know to be true, because I have heard some Workmen, who thought themselves no Ignoramusses, tho' at the same time I had the Liberty to think as I pleas'd, which was in truth almost quite contrary; for as they thought themselves Men of Science [or Skill] I thought so too almost, only I thought that the Particle *Ne* was wanting before the word Science; for if that had been prefix'd, then it might have been very well applied to have signified the acquired Intellectual Wits of such Men as affirm, That the Theory, or Speculative Part of Architecture was of no Use, because, say they, It is false: But 'tis the Humour of Men to slight and contemn such Things as they are not Masters of, and do not know which way to attain them, as being beyond their reach

reach. Like the Fox in the Fable, who despised Grapes, because they grew too high for him to come at.

Thus I have shewed you, That what is wanting under this Head, you may find supplied in the word *Building*.

Housing.

A Term used by some Bricklayers, for when a *Tile*, or *Brick* is warped, or cast crooked or hollow in burning, they then say such a *Brick* or *Tile* is *Housing*; they are apt to be *Housing*, or hollow on the struck side, (or that which was upmost in the Mould) and *Bricks* on the contrary side.

Also some observe that *Tiles* are always smoothest, when burnt on the struck-side, by reason the Sand sticks to the under-side, which they strew on the Stock of the Mould, to prevent the Earth's adhering to it.

Hyperthyron.

A *Greek* word, signifying, the Lintel, or Cap-piece of a Door-case. 'Tis also used to signifie a large Table in manner of a Frieze above Dorick Gates.

Hypotrachelium, or *Hypotrachelion.*

What.] See Capital. N. 2. and 3.

Jack.

SEE Architrave, N. 2.

Jambs, or *Jaums.*

Door-posts, also the upright Posts at the ends of Window-frames, are so call'd. Also, Bricklayers call the upright sides of Chimneys, from the Hearth to the Mantle-tree, by this Name. *Jambe* is a *French* word, and signifies a Leg.

Ichnography,

A Description, or Draught of the Plat-form, or Ground-work of a House, or other Building.

Impost.

Is a Term in Architecture, which the Writers of this Science leave very much in the dark; especially as to any account of its use: I understand *Imposts* to be the Capitals of Pilasters that support Arches. Conformable to this Notion is M. *Perrault*'s Definition of it, *viz.* a *Plinth* or little *Cornice*, that crowns a *Peer*, and supports the first Stone, whence a Vault or Arch commences. It comes from the *Italian Imposto*, *surcharged* or *burthen'd with*, or *laid upon*. Take it either way, the Name expresses the thing.

Insulate Columns.

See it explain'd in *Parastatæ*.

Inter-

Intercolumnation.

Is the Space between two Columns, which in the *Dorick* Order is regulated according to the Diſtribution of Ornaments in the Freeze; but in the other Orders, according to *Vitruvius*, is of five different kinds, *viz.* Picnoſtyle, Siſtyle, Euſtyle, Diaſtyle, and Aræoſtyle. (All which ſee in their Places.) This the Latins expreſs by their word *Intercolumnium.*

In the { Tuſcan / Dorick / Ionick / Corinthian / Compoſite } Order.

The Intercolumnation muſt be { 4 / 3 / 2 / $2\frac{1}{4}$ / $1\frac{1}{2}$ } Diameters of the Body of the Column below.

Inter-ties,-duces.

In Architecture, are thoſe ſmaller pieces of Timber that lie Horizontally betwixt the Summers, or betwixt them and the Sell, or Reſon.

Jobents.

See Nails. N. 10.

Joggle-piece.

As Crown-poſt.

Joyners { *Brads,* / *Rivets.* }

See { Brads. N. / Rivets. N. }

Joyners-work.

See the Particulars in their proper Place of the Alphabet.

Joyſts.

1. *What.*] *Joyſts,* in Architecture, are thoſe pieces of Timber (fram'd into the Girders and Summers,) on which the Boards of the Floors are laid.

2. *Scantlings.*] *Joyſts* at full length (to bear in the Wall,) I ſay, their full length

being { 12 Foot, / 11 Foot, 6 Inches, / 10 Foot, 6 Inches, } they ought to be in their

Squares { 8 Inches and 3 Inches. / 7 Inches and 3 Inches. / 6 Inches and 3 Inches. }

And

And Binding, or Trimming-Joysts being

in length { 7 Foot, / 9 Foot, / 11, or 12 Foot, } ought to be in their

Squares { 6 Inches, and 5 Inches. / 7 Inches, and 5 Inches. / 8 Inches, and 5 Inches. }

3. *Distance* and *Position.*] (1.) No Joysts ought to lie at a greater Distance from each other than 10, (or at most than 12) Inches. (2.) All Joysts on the back of a Chimney ought to be laid with a Trimmer, at 6 Inches distance from the back. (3.) No Joysts ought to bear at a longer Length than 10 Foot. (4.) No Joysts ought to lie less than 8 Inches into the Brick-wall. (5.) Some Carpenters Furr the Joysts, (as they call it,) that is, they lay two rows of Joysts one over another; the undermost of which are fram'd level with the under-side of the Girder, and the uppermost (which lie cross the the lower ones) lie level with the upper side of the Girder.

Ionick Order.

See Column, and Order.

Iron.

1. *What*] Iron is a Metal so universally known, that I neither need, nor shall spend Time in its Description.

2. *Kinds.*] There are several kinds of Iron, as (1.) *English*, which is a coarse sort of Iron, hard and brittle, fit for Fire-bars, and other such coarse Uses. (2) *Swedish*, which is of all sorts the best we use in *England*. It is a fine tough sort of Iron, will best endure the Hammer, and is softest to file, and therefore most coveted by Workmen to work upon. (3.) *Spanish*, this would be as good as *Swedish* Iron, were it not subject to *Read sear*, (as Workmen phrase it,) that is to crack betwixt hot and cold. Therefore when it falls under your Hands, you must tend it more carefully at the Forge. But tho' it be a good, tough, soft Iron, yet, for many uses Workmen refuse it, because 'tis so ill and unevenly wrought in the Bars, that it costs them a great deal of labour to smooth it; but it is good for all great Works that require Welding; as the Bodies of Anvils, Sledges, large Bell-clappers, large Pestles for Mortars, and all thick strong Bars, &c. But 'tis particularly chosen by Anchor-smiths, because it abides the Heat better than other Iron, and when 'tis well wrought, is toughest. (4.) There is some Iron that comes from *Holland*, (tho' in no great Quantity,) but is made in *Germany*. This sort

ſort of Iron is call'd *Dort-ſquares*, (only becauſe it comes to us from thence, and is wrought into Bars of $\frac{3}{4}$ of an Inch ſquare) 'Tis a bad coarſe Iron, and only fit for coarſe Uſes, as Window-bars, Brewers-bars, Fire-bars, *&c.* (5.) There is another ſort of Iron us'd for making of Wire, which of all ſorts is the ſofteſt and tougheſt: But this ſort is not peculiar to any Country, but is indifferently made where any Iron is made, tho' of the worſt ſort; for 'tis the firſt Iron that runs from the Mine-ſtone when 'tis melting, and is only preſerv'd for the making of Wire.

3. *To know good.*] Generally the ſofteſt, and tougheſt Iron is the beſt. Therefore when you chuſe Iron, chuſe ſuch as bows ofteneſt before it breaks, which is an argument of toughneſs, and ſee it breaks ſound within, be of a grey Colour, like broken Lead, and free from ſuch glittering ſpecks as you have ſeen in broken Antimony, and no Flaws or Diviſions in it; for theſe are Arguments that 'tis ſound, and was well wrought at the Mill.

4. *Price when wrought.*] Iron being wrought (by the Smith) into Dogs, Iron bars, Staples large Hooks, Thimbles, and Hinges or Hides, Grates, *&c.* The uſual Price is three Pence Half penny, or 4 *d. per* Pound. But for ſmall and neat Hooks, Hinges, Bolts, Staples, *&c* various, as from 4 *d.* to 8 *d per* Pound.

5. *To make Blue.*] To beautifie Iron with a blue Colour, take a piece of a Grind-ſtone, or Whet-ſtone, and rub hard upon your Work to take off the black Scurff from it; then heat it in the Fire, and as it grows hot, you will ſee the colour change by degrees, coming firſt to a light Gold-colour, then to a darker Gold-colour; and laſtly, to a Blue-colour. But Workmen ſometimes grind Indico and Salad-oyl together, and rub that mixture upon it with a Woolen-rag, while it is heating, and let it cool of it ſelf.

6. *Of Twiſting*] Square and flat Bars of Iron are ſometimes (by Smiths) twiſted for Ornament; which is very eaſily done; for after the Bar is ſquare or flat Forged (and if the Curioſity of the Work require it truly Fil'd,) you muſt take a *Flame-heat*, or if your Work be ſmall, but a *Blood-red-heat*, and then you may twiſt it about, as much, or a little as you pleaſe, either with the Tongs, Vice, or Hand-vice, *&c.*

Juffers.

A Term us'd by ſome Carpenters, for Stuff about 4 or 5 Inches ſquare, and of ſeveral Lengths.

Kerf.

THE Sawn-away Slit in a piece of Timber, or Board, the way made by the Saw, is call'd a Kerf.

Key.

Key-stone.

See *Arch*. Numb. 6.

King-piece.

As *Crown post.*

Knee,

A piece of Timber cut crooked with an Angle, is call'd a *Knee-piece*, or *Knee-rafter.*

Keys,

For Doors, are of various Prices, according to their Size and Workmanship, Master-keys, *per* Piece 2 or 3 *s.*

Larmier.

SEE *Drip.*

Latches,

For Doors, are of various Kinds and Prices. Common Iron-latches, *per* Piece, 6 *d.* if large, 8 *d.* or 10 *d.* long varnish'd Latches, about 10 *d.* *per* piece. Rim'd Latches with a sliding Bolt, 2 *s. per* piece. Spring-latches, *per* piece, 1 *s.* or 1 *s.* 6 *d.*

Laths.

1. *What.*] Laths are long, narrow, thin Slips of Wood, us'd in Tyling and Walling; every one knows what they are without any further Description.

2. *Kinds of*] There are three kinds of Laths, *viz.* Heart of Oak, Sap-laths, and Deal-laths; the two last Sorts are us'd for Ceiling and Partitioning, and the first for Tyling only. Again, as to their Length, they are each of them distinguish'd into three sorts, *viz.* 5 Foot, 4 Foot, and 3 Foot-laths. All these Sorts of Laths are necessary, (especially in repairing of old Buildings) because all Rafters are not spaced alike, nor yet the Proportion strictly observed in every one and the same Roof. See *Tyling*, Numb. 8.

3 *Bundle of*] A bundle of Laths is so many as are bound up together, and is generally called a Hundred of Laths; tho' of the 3 Foot laths there goes 7 Score, (or 140 to the Hundred, or Bundle, and of the 4 Foot laths, 6 Score, but of 5 Foot Laths,) there goes but just 5 Score to the Hundred, or Bundle.

4. *Size of.*] The Statute allows but of two sorts of Laths, one of 5, the other of 4 Foot in Length; of either sort each Lath ought to be in Breadth an Inch and half, and in Thickness half an Inch; but they are commonly less, and are seldom exact, either in their Tale or Measures.

5. *Of Cleaving.*] (1.) Lath-cleavers having cut their Timber into Lengths, they cleave each Piece (with Wedges) into 8, 12, or 16 Pieces, (according to the Bigness of their Timber,) which they call *Bolts*; then they cleave these Bolts (with their Dowl-Ax) by the *Felt-grain*,

Felt-grain, [which is that Grain which is ſeen to run round in Rings at the end of a Tree,] into Sizes for the Breadth of their Laths, and this Work they call *Felting*. Then, laſtly, (with their Chit) they cleave their Laths, into their Thickneſſes, by the *Quarter grain*, which is that Grain which is ſeen to run in ſtraight Lines towards the Pitch. (2) Some Men affirm, That a Foot of Timber will make a Bundle, or Hundred of Laths; but this I know to be a Miſtake, (unleſs the Laths are made very ſlight;) for by ſeveral Experiments, which I have caus'd to be made, I find that a Tun, or 40 Foot of Round-oaken-Timber, will not make above 30 Hundred, or Bundles of Laths. Of which Number above one third part, *viz.* 10 Hundred) will be Sap-laths.

6. *Price of.*] (1.) The common Price for cleaving of Laths, is 5 *d.* or 6 *d.* the Bundle. But I know a Carpenter (in *Suſſex*) that buys a great deal of Timber, and has it cleft into Laths, and he tells me, that he uſes to give but 11 *s. per* Load for the Cleaving of them, reckoning a Load to be 30 Bundles, which is not full 4 *d.* $\frac{1}{2}$ *per* Bundle.

(2.) The Price of Laths muſt needs be various, there being ſo great Diſparity in the Commodity, as to its Goodneſs, Plenty, or Scarcity, *&c.* But the Prices are generally between a Shilling and half a Crown the Bundle: And the common Rate for Heart-laths is about 20 *d. per* Bundle, and Sap-laths are commonly about $\frac{2}{3}$ of the Price of Heart-laths. The Carpenter mention'd above, (in this Number,) tells me, that he uſes to ſell his Laths for 4 *l.* 10 *s.* the Carriage: He reckons a Carriage 60 Bundles, whereof 40 are Heart, and 20 Sap-laths; at which Rate, (reckoning Sap-laths to be $\frac{2}{3}$ of the Price of Heart, he ſold his Heart-laths at 20 *d.* $\frac{1}{4}$ *per* Bundle, and his Sap-laths at 13 $\frac{1}{2}$ *per* Bundle.

7. *Nails allow'd to a Bundle of*] The common Allowance is five Hundred, (at ſix Score to the Hundred, that is 600) Nails to a Bundle of Laths.

8. *How many to a Square.*] Workmen commonly allow a Bundle of Laths to a Square of Tyling, which (if the Diſtances of the Rafters fit the Lengths of the Laths, without any Waſte,) is a ſufficient Allowance; for (then) about 90 *Five-foot*, and 112 *Four-foot Laths*, will compleat a Square of Tyling, (Counter-laths and all) at 7 Inches Gage; and at 8 Inches Gage, a Square will require fewer.

Lathing.

The Price of Lathing, Plaiſtering, Rendring, and Waſhing with Size, is about 10 *d.* 12 *d.* or 14 *d. per* Yard, Materials and Work.

Lead.

1. *What; and its Uſe.*] Lead is a Material us'd in Buildings, well known, and needs no Deſcription. Its chief Uſes

Uſes are for covering for Gutters, for Pipes, and for Glaſs. Covering with Lead is the moſt magnificent, and is generally moſt us'd for the covering of Churches, Princes Palaces, Caſtles, and great Men's Houſes. It's generally laid almoſt flat to walk upon, allowing the Water a little Fall to the Battlements, thence privately to deſcend in Pipes. But in ordinary Tyled Buildings, 'tis chiefly us'd for Gutters to convey the Water from the Houſe into ſome convenient Place.

2. *Sorts of.*] There are three ſorts of Lead, white, black, and Aſh-colour; the White is more perfect and precious than the Black, and the Aſh-colour between both.

3. *Of Caſting Sheet.*] I ſhall here exhibit ſome of my Obſervations on the Method of Caſting Lead into Sheets; as I have often ſeen it done, and have (for Curioſity) caſt ſome Sheets my ſelf. For this Purpoſe there is a Mould provided, which is made ſomething longer than the intended Length of the Sheets, that the End where the Metal runs off from the Mould may be cut off; becauſe 'tis commonly thin, and uneven, or ragged at the End.

This Mould, (which is juſt as broad as the Sheet is to be, muſt ſtand very even, or level in Breadth, and ſomething falling from the End where the Metal is pour'd in, *viz.* About an Inch, or an Inch and half in 16 or 17 Foot.

This Mould (commonly) conſiſts of ſeveral Treſſels, upon which Boards are laid, and nail'd down faſt, and upon theſe, at a due Diſtance, (according to the intended breadth of the Sheets,) the Sharps are fixed. Theſe are two Pieces of well ſeaſon'd Timber, of about 4 Inches Square, and 16, 17, or 18 Foot long, according to the Size of the Sheets. But this Method of fixing down the Sharps, Workmen have found to be inconvenient; and therefore ſome do only fix one of the Sharps firmly, nailing the other but ſlightly, and then they fix ſeveral Pieces firmly to the Boards, without the ſlightly fixed Sharps betwixt which and the Sharp, they drive Wedges, to make the Sharps come nearer together, as they ſee occaſion: For they find by Experience, that the moiſten'd Sand, (when it has lain a while on the Boards,) makes the Board ſwell ſo much, that in ſpight of the Nails the Sharps will be too far a ſunder.

At the upper End of the Mould ſtands the *Pan*, which is a Concave Triangular Priſm, compos'd of two Planks nail'd together at Right Angles to each other, and two Triangular Pieces fitted in betwixt them at the Ends. The length of this Pan is the whole breadth of the Mould wherein they caſt their Sheets, and the breadth of the Planks whereof 'tis compos'd, may be 12 or 14 Inches, or more, according to the Quantity of Lead they have occaſion to put into it, to make a Sheet of, and the thickneſs of the Planks an Inch

Inch and a half. This Pan ſtands with its bottom, (which is a ſharp Edge) on a Form at the end of the Mould, leaning with one ſide againſt it, and on the oppoſite ſide is a handle to lift it up by, to pour out the melted Lead ; and on that ſide of the Pan next the Mould are two Hooks of Iron, to take hold of the Mould, and prevent the Pan's ſlipping when they pour the melted Lead out of it into the Mold.

This Pan is lined on the inſide with moiſten'd Sand, to prevent his being fired with the hot Metal. The Mould is alſo fill'd up (from the upper End towards the lower end, about $\frac{2}{3}$ parts of the way) with Sand ſifted and moiſten'd, and then a Man gets upon it, and treads it all over with his Shooes on, to make it ſettle cloſe to the Mould. This being done, they begin to ſtrike it level with the *Strike*, which is a piece of Board about 5 Inches broad, in the middle of which, and towards the upper Edge is a wooden Pin (about 5 or 6 Inches long, and 1, or 1 $\frac{1}{4}$ Inch Diameter,) to hold it by when they uſe it. The length of this *Strike* is ſomething more than the breadth of the Mould on the in-ſide, and at each end is cut a Notch (on the under-edge,) about two Inches deep ; ſo that when the Strike is us'd, he rides upon the Sharps with thoſe Notches, and the lower edge of the Strike rides about two Inches below the upper ſide of the Sharps.

Then, in leveling the Sand with this Strike, they begin towards the lower end of that part of the Mould that was fill'd, and taking the Handle of the Strike in their Right-hand, and laying their Left-hand upon one end of it, they draw the Sand back into that part of the Mould that was empty. Then they begin again a little nearer to the upper end, and draw the Sand back, (as before) but not ſo far as the empty part of the Mould ; for it is thus level'd at 5 or 6 Places in the length of the Mould ; (if he be 18 Foot long, as that was I made my Obſervation from) ſo that when it is thus level'd the whole length of the Mould, there are as many places that ſeem to be unlevel'd, as there are level'd, by reaſon of the Sand which is a little drawn back. Then the next Operation is to draw all the looſe and hover Sand, (rais'd in the laſt Operation of leveling it) into the empty part of the Mould ; which is done by beginning at the upper end of the Mould, and ſtill as the Sand is drawn back, the level'd part muſt be examin'd, to ſee if there be no Cavities in it; for if there be, a little Sand muſt be put into them, and that muſt be ſettled cloſe and faſt in the Cavities, by lifting up one end of the Strike, (letting the other reſt upon the other Sharp) and rapping upon the looſe Sand, which was put in thoſe Cavities, and ſo it will be ſettled cloſe and faſt.

This

This being perform'd all over the upper $\frac{2}{3}$ parts of the Mould, and all the loose Sand drawn back into the lower $\frac{1}{3}$ part of the Mould *that* is also trod on, and settl'd all over, and level'd in all Respects as the other $\frac{2}{3}$ were; and its loose Sand is drawn off the Mould, down into a Place 2 or 3 Inches below the lower end of the Mould, where the Sand is made into two Concavities to receive the Overplus of the Lead.

The Sand being thus level'd, it is next to be smoothed all over with the *smoothing Plane* (as they call it) which is a thick Plate of polish'd Brass about 9 Inches square, a little turn'd up, on all the 4 Edges; so that the under-side looks something like the Diamond-cut Looking-glasses, on the upper-side, (which is a little Concave, like a Latten pan,) is a brass Handle solder'd on, upon which is a wooden one also, like a Case-smoothing-iron. With this Instrument the Sand is smoothed all over; and where there are any small Cavities, there must be a little Sand put in 'em, (with the two Fore-fingers and Thumb,) and then smoothed down.

The Sand being thus smoothed, the Strike must be made ready, by tacking (that is, slightly nailing) on the Notches, two pieces of an old Felt-hat. (or else by slipping a Case of Leather at each end,) thereby to raise the under-side of the Strike about $\frac{1}{8}$ of an Inch above the Sand, or something more, according as the Sheets are to be in Thickness, which will make a middle siz'd Sheet of about 9 or 10 Pound *per* Foot; as I have observ'd in the casting of Lead for a Plat-form. (But for Hips, and Window-soils, and such Places where it does not lie flat, the Lead need not be above $\frac{1}{10}$ of an Inch thick; but sometimes Plat-form-lead is near $\frac{1}{4}$ of an Inch thick)

Then they Tallow the under-edge of the Strike, and lay him cross the Mould, close by the Pan, to prevent Drops of Lead from spattering into the Mould, before it be ready to pour. Then the Lead being melted, (and the Pan made ready, by being lined with moisten'd Sand, as was said above,) it is laved into the Pan, and when he is full, (or a sufficient Quantity for the present Purpose) then with the end of a piece of Board (two or three Inches broad,) draw off the floating part, or Scum of the Metal round about to the edge of the Pan, and there let it settle upon the Sand, which will thereby prevent the Sand from falling out of the Pan into the Mould, when the Metal is pour'd out.

The Metal being thus prepar'd, and cool enough, (which it will be when it begins to stand with a Shell, or Wall round about on the Sand,) then two Men must take the Pan by the Handle, and pour it into the Mould, and a third Man stands ready with the Strike, (facing of them, and his Right-side to the Mould,) and as soon as they have done pouring

pouring in the Metal, he immediately puts the Strike on the Mould, and runs back the whole length of the Mould, and so draws off the Over-plus of the Lead, into the Cavities made to receive it; and then immediately, (with a Knife) the ragged End is cut off before it is cold.

When the Sheet is a little cool'd, 'tis begun to be rowl'd up, from the upper end downwards, ('tis handled with pieces of old Felt-hats,) and as they rowl it up, they rub off the Sand from it.

When the Sheet is taken off from the Mould, the Sand is immediately rak'd over with the Rake to let it cool, and then if it be too dry, 'tis sprinkled with a little Water; but Care must be taken that none of the Mould be too wet; for if it be, the melted Lead will fly like Shot when it comes upon it. After the Sand is raked, 'tis all turn'd up-side-down with a Spade, and when it has lain a while, 'tis again thrown into $\frac{2}{3}$ parts of the Mould, and settled down by treading, as at first, *&c.* To make it ready for the next Casting, which is commonly in an Hour and a half, or two Hours, if the Furnace heat well.

Thus much I have observed of the Method of casting Lead into Sheets. I have insisted the longer upon it, because I know of none that has written of this so useful a Subject.

4. *Weight of a Foot of Sheet*] Every square Foot of Sheet-lead, (if it be design'd for Gutter, which is commonly run thinner than for Plat-forms,) is reckon'd to weigh 6 or 7 Pound if old, 8 or 9 Pound if new. And every square Foot of Sheet-lead for Plat-forms, is reckon'd to weigh 8, 9, or 10 Pound, if old, and 11 or 12 Pound, if new, and very good.

5. *How much one hundred Weight will cover.*] One hundred Weight of Sheet-lead (at 12 ℔ *per* Foot, (will cover a Square-yard, or 9 Square-foot. And is a lighter covering than Tiles, tho' dearer.

6. *Sheet for Gutters*] Sheet-lead design'd for Gutters, is commonly run thinner than for Plat-forms. And some Plumbers in *London* tell me, That 'tis the best way in laying long Gutters, to make a Drip, (Fall, or Step) about the middle, (of 1, 2, or 3 Inches deep;) for by this Means, say they, the Lead (being cut into two pieces which are shorter,) is not so subject to crack, (by being dilated and contracted with Heat and Cold) as otherwise it is.

7. *Sheet, of laying on in Plat-forms*] I have observed the Method of Plumbers in laying down the Lead in Plat-forms, to be this: Having roll'd open two Sheets, they beat them flat with their Dresser, [which is an Instrument of Wood, of 16, 18, or 20 Inches long, (according as they are of stoutness) and about 3 or 4 Inches broad at the bottom, and in heighth something more, in the

the Form almoſt, of a Parallelopipedon, only the upperſide is rounded off, and at one end the under-ſide is cut away, ſo as to leave a handle running out ſtraight with the top.] Then (with a Line and Chalk, or with a ſtraight Ruler, and a pair of Compaſſes,) they ſtrike a Line about 2 ½ Inches diſtant from one edge of one of the Sheets; this is for the *Standard.* In the ſame manner they ſtrike a Line about 3 ½ Inches diſtant from the edge next to it of the other Sheet; this is for the *Orlop* [The *Standard* is about 2 ½ Inches of a Sheet of Lead, which is ſet up at Right-Angles to the Sheet, all along one edge of it. The *Orlop* is about 3 ½ Inches of the edge, (next to the Stander) of the other Sheet, rais'd up in the ſame manner as the Stander.] Then with their *Pincers*, (which are ſomething different from common Pincers; for theſe have a ſmall Cylinder of Iron, (of about ½ an Inch Diameter, and 3 or 4 Inches long) fixed to one of the Chaps in ſuch a Poſition, that when the Pincers are ſhut, they ſeem to hold it betwixt their Chaps;) they raiſe up the Stander, and Orlop, by putting the ſharp Chap under the Sheet, and the Cylindrical one on the top near the Line, and ſo they bend up the edge of the Sheet, both for the *Stander* and *Orlop* Then they proceed to ſet it in better Order with the *Dreſſer*, with which they make the *Stander* and *Orlop*, as upright and ſtraight as they can, by placing one edge of the *Dreſſer* upon the Line which they ſtruck, and ſtriking hard Blows on the top of him with a Smith's Hand hammer.

Having thus made the *Stander* and *Orlop* as ſtraight as they can, and ſet them up at Right Angles to the Sheet; they bring them together, and proceed to make a Seam of them, by firſt turning the *Orlop*, (which is an Inch broader than the Stander) over the *Stander*, by the Help of the *Dreſſer*, and *Seaming mallet*, [which is an Inſtrument of Holly, or ſome other hard Wood, wrought away from the middle to one end, almoſt to a ſharp edge, and ſo it is likewiſe at the other end, only thoſe Edges ſtand at right Angles to each other, like a Croſs-mattock. And into the middle of it is put a Handle like a Mallet] And then they continue to beat the *Orlop*, and conſtantly work upon him with the Dreſſer, till they have reduced him and the *Stander* into as little Room as they can, by wraping them one in another, till at laſt it ſeems to be a kind of Semicircle, and this is what they call a Seam.

Some Plumbers tell me, That they ſometimes lay Platforms of Lead, without Seams; but then the Joyſts are wrought in hollow, about three Inches broad, and near as deep, in the Foam of a Semi-concave-cylinder, and when they lay the Sheets down, the edge of the firſt

firſt Sheet lies ſo far on the Joyſt, that it comes over beyond the Concavity, and ſo much of the Sheet as lies over the Cavity, is ſet down into it with the Seaming-mallet, and the next Sheet is laid over that, and ſet down into the Chanal alſo; and ſo the Water that comes into thoſe Chanals, runs down into the Gutter.

8 *Mill'd.*] One Mr. *Roberts*, (then Maſter of the Company of Plumbers in *London*,) tells me, That *Mill'd-lead* is of but little Uſe; not only becauſe 'tis ſo very thin; but alſo becauſe by the way of milling it, 'tis ſtretched to that Degree, that when it comes to lie in the hot Sun it ſhrinks, and cracks, and (conſequently) will not keep out the Water. For 'tis, (ſays he,) like Cloath ſtretched on the Clothiers Tenters, which when taken off, naturally inclines to return to its former State. He farther added, That there was ſufficient Proof for what he ſaid; and if any one deſir'd to be ſatisfi'd about it, he might repair to *Greenwich hoſpital* which is covered with Mill'd-lead, and has not been done above 4 or 5 Years, and yet it rains in, almoſt all over the Hoſpital; upon which Account the Maſter and Wardens of the Company of Plumbers were ſent for to the Parliament, who order'd them to go and view this Mill'd lead-work at *Greenwich-hoſpital*, which they did; and when they returned to the Parliament, they all unanimouſly declar'd, That Mill'd-lead was not fit to be us'd: Whereupon the Parliament had Thoughts of putting down the Milling of Lead: But whether they have actually done it, I have not yet heard.

9. *Pipes of.*] I underſtand by diſcourſing with ſome Plumbers in *London*, that they give diſtinct Names to their Leaden pipes, according to their weight at a Yard long, e. g. they have 6 ℔. 8 ℔. 10 ℔. 12 ℔. 14 ℔. 20 ℔. and 28 ℔. Pipes, and if I miſsremember not, one ſize larger; ſo that a Pipe of 6 ℔. to the Yard, they call a 6 lb. Pipe, and ſo of the reſt. I cannot at preſent tell the particular Sizes of all theſe ſorts of Pipes; but if I miſs remember not, the 10 lb. Pipe was about 1 $\frac{1}{4}$ Inch Diameter, from out-ſide to out-ſide.

10. *For Glaſs.*] Some Glaziers tell me, That they uſually allow 50 lb. of Turn'd-lead to 100 Foot of Quarry-glaſs. They call it *Turn'd-lead*, when the *Came* has paſs'd thro' the Vice, and is thereby made with a Groove on each ſide, to go on upon the Glaſs. See *Came*. Their Turn'd-lead for Quarries is commonly about $\frac{3}{16}$. (which is almoſt $\frac{5}{16}$) of an Inch broad; and for large Square-glaſs, their Turn'd-lead is $\frac{8}{16}$, or $\frac{1}{2}$ an Inch broad. So that I find (by Diſcourſe with Glaziers,) they have it of different Sizes, as $\frac{4}{16}$, $\frac{5}{16}$, $\frac{6}{16}$, $\frac{7}{16}$, and $\frac{8}{16}$ of an Inch broad.

I have alſo obſerved, That ſome Glaziers in *London* have three Sizes of Turn'd-lead for Glaſs-windows, *viz.* Of $\frac{7}{16}$,

$\frac{5}{16}$,

$\frac{5}{16}$, and $\frac{4}{16}$ of an Inch broad: The largeſt ſize, they tell me is for large Squares, that of $\frac{5}{16}$ for Quarries, and the $\frac{4}{16}$ for Crocket-work, (or Fret-work, as ſome Glaziers call it,) it being more pliable for that Uſe than broader Lead.

Some Glaziers tell me, They can turn Lead of different ſizes in the ſame Vice, by changing their Cheeks for each ſize, with another pair of Spindles, whoſe Nuts almoſt meet or touch, they turn Lead for Tyers, which when it comes out of the Vice, is almoſt cut aſunder in two Thickneſſes, which they can eaſily rend aſunder. Theſe Tyers are very tough, but they are commonly made too ſlight, and therefore ſome uſe to caſt Tyers, which are ſtouter, but not ſo tough, being more apt to break in winding.

11. *Nails for*] See Nails. N. 11.

12. *Of Soddering*] An ingenious Plumber, (who eſteem'd me as his peculiar Friend) told me the Method of *Paleing*, (as they call it) or ſoddering on of Emboſt Figures on Leaden Work; as, ſuppoſe a Face, or Head in Baſ-relief, were to be pail'd on a Pump-ciſtern for an Ornament to it. To do this, (ſaid he) the Plate where it is to be pal'd on muſt be ſcrap'd very clean, and ſo muſt alſo the back-ſide of the Figure, that it may ſit cloſe with a good Joynt. Then, (ſaid he) place that part of the Ciſtern (where the Figure is to be fix'd) Horizontal, and ſtrew ſome pulveriz'd Rozin on the place where you made your Joynt. Then into the Ciſtern, (juſt under the place where your Figure is to ſtand) ſet a Chaffing-diſh of Coals, (till you ſee the Rozin is changed rediſh, and begins to raiſe in Pimbles, or Bladders) then take a piece of ſoft Sodder, (made of a longiſh Figure) and rub the end of it round about your Figure, and at the ſame time keeping your Figure ſteady in its place, ſo that it may work into the Joynt. And when this is done, your Figure will be well pal'd on, and will be as firm, as if it had been caſt on there

But if your Ciſtern, (or the like) be ſo thin, as that you have reaſon to fear that it will be too hot, and be apt to run, or bend, and yield, before your Figure, (which is on the outſide of it) will be hot enough; you may then lay your Figure on the hot Coals, till it and the place to receive it are both in a good temper for paleing, and then ſet the Figure on its place, and proceed with your Sodder, as before.

By this Method I ſaw him, ſodder on Bottoms to Leaden-ſtands, or Ink-holders.

I alſo ſaw and obſerv'd him, in ſoddering the Leads of a Church, thus to manage it, *viz* When he ſodder'd the Sheets of Lead that are fix'd into the Wall on one edge, and with the other edge lap over the ends of thoſe which are ſeam'd in the Platform, at every other Sheet, in the middle betwixt the ſeams, he ſoddered the Lapping-ſheet down to the

other,

other, thus —: with one corner of the Scraper, [which is an Inſtrument made of a Plate of Steel, in the Form of an Equilateral Triangle, in the middle of which is fix'd an Iron Strig, on the end of which is fixed a Wooden-knob, or Handle. The Plate is flat on the ſide next the Handle, but on the other ſide the edges are ground off with a Bezel like a Chizzel, only very obtuſe.] He firſt marked out, (partly on the edge of the Lapping ſheet, and partly on the other) an Oblong Rectangular Figure, of about five or ſix Inches long, and three or four broad. Then he ſcraped the Metal bright, having firſt, (becauſe it was new Lead,) green'd it, (as they phraſe it) all round about, to prevent the Sodder's taking any where but where they ſcrape it. (This *Greening* is only rubbing it with ſome green Vegetable, it matters not what, he did it with Poor-man's-pepper, that being at hand: He told me, That in the Winter they us'd Cabbage-leaves, or any green thing they could get.) It being thus ſcrap'd, he rub'd it with Tallow, then having an Iron red hot, (which are much like the Irons us'd by Glaziers) he took him with a piece of Felt in his Right-hand, and a piece of Sodder in his Left, and holding it againſt the Iron, 'till it drop'd on the cleanſed place, and when there was enough of it melted, he took a Linen-clout in his Left-hand, and therewith kept the Sodder continually ſhov'd up on the cleanſed place, and at the ſame time work'd it about with the Iron in his Right-hand, thus he did, 'till he thought it was pretty well incorporated with the Lead, and then he made it up into a kind of Swelling form in breadth, and then croſs the breadth of it, he made it into a kind of Seams, with the Point of his Iron. This being done, he took their Knife and a Dreſſer to knock him with, and ſo cut it ſtraight on the ſides and ends, and what he thus cut off, by reaſon of the *Greening* eaſily peel'd off.

Ater the ſame manner he ſodder'd Holes, or Leaks in old Lead, only then he made the Sodder flat, and not ſwelling, (but he made it alſo in little Seams,) neither did he green it before he ſcrap'd it.

13. *Price.*] (1.) *The Priee of Lead in Pigs* (ſays Mr. *Leybourn*) is uncertain, as from 10 to 20 *s.* the hundred weight. I know a Plumber (at *Lewis* in *Suſſex*) who tells me he gives 12 *s.* 6 *d. per* hundred for Lead in Pigs at *London:* Some Glaziers tell me they give but 12 *s. per* hundred, if they buy but half a hundred. Mr. *Wing* tells us, That a Fodder of Lead is 22 $\frac{1}{2}$ hundred weight, (I know not how he reckons; for I am ſure, moſt Authors reckon a Fodder of Lead but 19 $\frac{1}{2}$ hundred,) and is worth from 9 *l.* to 12 *l.* which will caſt 315 Foot of Sheet, at 8 Pound *per* Foot.

(2) *The Price of Sheet-lead.*] Mr. *Leybourn* ſays, That in exchange of old Lead for Sheets new run, there is commonly allow'd

allow'd 3 *s.* in every hundred weight, for Waſte and Workmanſhip. I ſaw Sheet-lead (in 1701) ſold at *Lewis* for 16 *s. per* hundred weight, (they ſometimes ſell it for 17 *s.*) The Tinker who bought it to skirt a Furnace with, ſaid it was good thick Lead. I computed it to weigh about 11 $\frac{1}{4}$ Pound *per* Foot; for there was four pieces of it, each about three Foot long, and fifteen and an half Inches broad, all which weigh'd 174 Pound.

(3.) *The Price of Caſting Sheet-lead.*] The Tinker above mention'd tells me, That Plumbers commonly reckon 4 *s. per* hundred, for caſting old Lead into Sheets; but I apprehend that the Plumber (for this Price) makes good ſo many hundred weight of Sheet-lead, as he receiv'd of old Lead. For Mr. *Leybourn* tell us, That Sheet-lead is caſt out of old Lead, for 3 *s. per* hundred, allowing for Waſte and Workmanſhip: And Mr. *Wing* ſays, That there is about 2 *s.* 6 *d.* (in every hundred) loſs, in caſting old Lead into Sheets: He alſo ſays, That caſting old Lead into Sheets, is worth 1 *s* 6 *d. per* hundred. Yet I know a Plumber that had 3 *s, per* hundred for caſting of Sheet-lead; but then it was weigh'd after it was caſt, and he made very great Wages.

(4.) *The Price of laying on of Sheet-lead in Roofing, &c.*] This (ſays Mr. *Wing*) is worth 15 or 16 *s per* hundred weight, Lead and Workmanſhip. And Mr. *Leybourn* tells us. That covering with Lead is uſually valu'd at 13, 14, or 15 *s. per* Yard Square, (according to the goodneſs of the Lead) or between 7 and 8 Pound the Square of 10 Foot, beſides Sodder.

(5.) *The Price of Sodder,* (ſays Mr. *Leybourn*) is 9 *d.* or 10 *d. per* Pound, as it is allay'd with Lead, and Scal'd: For Tin is 10, 11, or 12 *d. per* Pound neat.

(6.) *The Price of Leaden-pipes* is various, according to their different bigneſs. An ingenious Country-plumber of my Acquaintance tells me, That for Pipes of half an Inch Diameter in the Bore, they have 1 *s.* 4 *d. per* Yard, for $\frac{3}{4}$ Inch Pipe, 1 *s.* 10 *d.* for Inch Pipe, and 1 $\frac{1}{4}$ Inch Pipe, 2 *s* or 2 *s.* 6 *d.* (for, ſays he, they are caſt both in a Mould, only the Inch Pipe has a leſs Bore; and I think he ſaid they were both of a Price; tho' I think, for this Reaſon the Inch Pipe ought to be the deareſt, ſince it contains moſt Lead, and the Work is the ſame in each.) For Pipes of 1 $\frac{1}{2}$ Inch Bore they have 3 *s.* 6 *d per* Yard, and for 3 Inch Pipe, 5 *s.* or 5 *s.* 6 *d.* The *London* Plumbers, (I find) rate their Pipes according to the weight of a Yard in length. Their 10 Pound Pipes are 2 *s.* 2 *d. per* Yard.

(7.) *The Price of Turn'd lead for Glaſs windows,* is various according to its breadth. I know ſome Glaziers in *London* ſell Turn'd-lead of $\frac{7}{8}$ Inch broad, for 18 *s. per* hundred, that of $\frac{7}{8}$ Inch broad for 17 *s. per* hundred.

1. *White, for Painting.*] White-lead is a Colour (well known

known) much us'd in Painting of Gates, *&c* In *London*, 'tis commonly ſold for 2 ½ *d.* or 3 *d per* Pound un-ground. I have alſo known it bought (in *London*) for 5 *d. per* Pound ready ground with Oil.

Ledgers.

See Putlogs.

Lime.

1. *What.*] A Material us'd in Building, (and well known,) made of burnt Stones, commonly of Chalk.

2. *VVhereof, and how made.*] Mr. *Leybourn* tells us, out of *Palladio*, That Stones whereof Lime is made, are either dug out of Hills, or taken out of Rivers: That Lime is the beſt which is made of the hardeſt, ſound, and white Stones, and being burnt, remains a third part lighter than the Stones whereof it is made. All dug Stones are better to make Lime of than gather'd Stones, and from a ſhady and moiſt Pit, than from a dry. All Stones are ſooner or later burnt, according to the Fire which is given them; but ordinarily they are burnt in ſixty Hours.

The ingenious Sir *Henry Wotton*, tells us, That to make Lime (without any Choice) of refuſe Stuff as we commonly do, is an *Engliſh* Error, of no ſmall Moment in our Buildings. Whereas the *Italians* at this day, and much more the *Ancients* did burn their *firmeſt Stone*, and even Fragments of *Marble* where it was plenty, which in time became almoſt *Marble* again for its hardneſs, as appears in their ſtanding Theaters.

There are two kinds of Lime commonly made in *England*, one made of Stone, which is the ſtrongeſt, and the other of Chalk, both being burnt in a Kiln.

The Lime that is made of ſoft Stone, or Chalk, is uſeful for Plaiſtering of Ceilings and Walls within Doors, or on the inſides of Houſes, and that made of hard Stone is fit for Structures, or Buildings, and Plaiſtering without Doors, or on the outſide of Buildings that lie in the Weather: And that which is made of greaſie clammy Stone, is ſtronger than that made of a poor lean Stone, and that which is made of ſpongy Stone, is lighter than that made of firm and cloſe Stone; that is again more commodious for Plaiſtering, this for Building.

Alſo very good Lime may be made of Mill-ſtone, not courſe and ſandy, but fine and greaſie. Likewiſe of all kind of Flints, (but they are hard to burn, except in a Reverberatory Kiln,) except thoſe that are rolled in Water becauſe a great part of its Increaſe goes away by a kind of Glaſs. Alſo the Shells of Fiſh, as of Cockles, Oyſters, *&c.* are good to burn for Lime.

About us in *Suſſex*, Lime is made of hard Chalk, dig'd out of the Hills, and is burnt in Kilns, like Brick-kilns; but with this difference, That they have no Arches in them, but

but only a kind of Bench, or Bank on each side, upon which they lay the largest Stones, and so truss them over, and make an Arch, after the manner of Clamps for Bricks (See *Clamp*) And when they have thus made an Arch with the largest Stones, they fill up the Kiln with the smaller ones

A Mason of my Acquaintance tel's me, That the *Kentish* Lime is far better than that commonly made in *Sussex*: For, (says he) a Gallon of Water will make as much more *Kentish* Lime run, as it will of *Sussex* Lime: So that it should seem (by the Consequence of Discourse,) that *that is* the best Lime which will run with the least Moisture

The ingenious Gentleman, *Walter Burrel*, Esq, of *Cuckfield* in *Sussex*, was the first that introduced the use of Fern for burning of Lime, which serves that purpose as well as Wood, (the Flame thereof being very vehement,) and is far cheaper.

3. *Hundred of- how much.*] In (and about) *London* Lime is commonly sold by the Hundred, which is 25 Bushels, or one hundred Pecks, whence it had its Name.

4. *Load of- how much.*] In the Country, Lime is commonly sold by the Load, which is 32 Bushels. A Load of Lime, (say some) will make Mortar enough for 250 solid Foot of Stone-work. And 8 Bushels of Lime, (heaped measure) is the common allowance to every Thousand of Bricks.

5. *Price of-*] The Price of Lime is various in different Places as from 8 to 12 *s.* the Hundred, says Mr. *Leybourn.* I know that before these late Wars, (which have made Fuel dear and scarce) Lime (in some parts of *Sussex*) was sold for 20, or 21 *s per* Load, 32 Bushels to the Load; but now in some parts of *Sussex* 'tis sold for 24, or 25 *s per* Load, in others for 32 *s.* Yet in some parts of *Sussex*, 'tis still (to my knowledge) sold for 12 *s per* Load at the Kiln, and for about 15 *s.* 6 *d.* laid in 3 or 4 Miles.

Lintels.

1. *What*] *Lintels* (in Stone and Brick Buildings) are the pieces of Timber that lie Horizontally over the tops of Doors and Windows.

2. *Price.*] The Carpenter commonly puts in these by the Foot running measure, at 6 *d. per* Foot, if Oak; 4 *d.* if Fir, Timber and Workmanship. Some Carpenters in the Country (that do not find Timber) tell me, they have 1 *s per* piece for sawing the Timber, and putting them in.

List, and *Listella.*

Is a litte square Moulding, serving to crown or accompany a larger, or on occasion to separate the *Fluteings* of a Column. It is sometimes call'd a *Fillet*, and sometimes a *Square*: It comes from the *Italian* word *Lista*, any kind of *List* or *Selvage.*

Lobby.

As Anti chamber.

Locks

Locks

For Doors are of various kinds; as for Outer doors, called Stock locks; for Chamber-doors, call'd Spring-locks, *&c.* Alſo the ſeveral Inventions in Locks, (I mean in the making and contriving their Wards and Guards,) are almoſt innumerable. And as their kinds are various, ſo are their Prices; I ſhall at preſent mention only ſome of the chief. As Stock-locks plain, from 10 *d.* to 14 *d. per* piece, or more, S-bitted Stock locks with a long Pipe, 1 *s.* 6 *d.* S-bitted and warded Stock-locks very ſtrong, 7 *s.* Braſs-locks from 5 *s.* 6 *d.* to 9 *s.* Braſs-knobed-locks in Iron-caſes, 3 *s.* double Spring'd-locks 1 *s.* Cloſet-door locks 1 *s.* 4 *d.* Pad, (or ſecret) Locks with Slits inſtead of Pipes, 1 *s.* Plate-ſtock-locks, 3*s.* 8 *d.* ſome ditto for half that Price. Plate ſtock-locks in Shute, 4 *s.* 6 *d.* Braſs knob'd-locks in Shute, 6 *s.* 6 *d* Iron-rimb'd-locks very large, 10 *s.* 6 *d.* The Prices of Locks are various, according to their different kinds, ſizes, and variety of Workmanſhip, ſo that 'twere endleſs to mention them all; therefore I ſhall ſay no more of 'em at preſent, only, that there are ſome Locks made of Iron and Braſs of 50, nay 100 *l per* Lock, as Mr *Chamberlain* tells us in his *Preſent State of* England.

Lome,

A ſort of reddiſh Earth, (well known) us'd in Buildings, (when temper'd with Mud Gelly, Straw and Water,) for Plaiſtering of Walls in ordinary Houſes.

I know one Place in *Suſſex* (where being well temper'd with new Horſe-dung,) it is us'd inſtead of Mortar to lay Tiles with, and they tell me it does very well.

Lome, (as 'tis dug out of the Earth) is commonly ſold in ſome parts of *Suſſex,* for 1 *s. per* Court-load, containing about 12 Buſhels.

Lutherns.

As *Dormers.* Alſo, ſee *Windows.* N. — Their Price of making and ſetting up, (and ſawing the Timber) is various, (according to their bigneſs) from 9 to 20 *s. per* Window.

Marble.

1. WHAT.] A hard Stone, beautiful when poliſh'd, but hard to cut; much us'd in adorning of Palaces, and great Men's Houſes, *&c*

2. *Kinds of.*] The kinds of Marble are almoſt innumerable, ſome *white,* ſome *black,* ſome *grey,* ſome *green,* ſome variegated with Veins, and Spots, *&c* It were endleſs to give the particular Names and Deſcriptions of all the kinds of Marble.

3. *Uſe of*] The principal uſe of Marble in Architecture, is for

for Chimney-pieces, Chimney-foot-paces, Window-ftools, Pavements, *&c.*

The Ancients, (as *Pliny* and other Authors tell us) us d to face their Houfes all over with thin Plates of Marble.

4 *Of Polifhing*] An old experienced Mafon tells me, that he has obferv'd Stone-cutters polifh Marbles for Hearths in this manner, *viz.* By laying three or four of 'em in a row, as even as they could, and then with another of thefe Stones fix'd to a broad Beetle, with a Handle put in at Oblique Angles, (and with Sand and Water) by moving this upper Stone too and fro on the lower ones, they wrought off the Strokes of the Ax, and afterwards with *Emmery* and *Putty* they polifh them

I have (alfo) my felf, (at *Lewis* in *Suffex*) feen and obferv'd them polifhing of Marble for tops of Tomb-ftones, which (as I find in my *Adverfaria*) they did in this manner. They block'd up their Stones to be polifhed, fo as they lay Horizontal about 2 ½ Foot high above the Ground ; (I fay they obferv'd to lay them very level) and then they wrought the upper Surface fmooth and even, with a Tool for that purpofe : This Tool was a piece of whole Deal about 18 or 20 Inches long, and 12 Inches broad, and crofs the Grain of the Wood, on the upper fide were nail'd two Ledges, one at each end, and on thefe Ledges was nail'd a Staff or Handle about eight or nine Foot long, *viz.* (long enough to reach the length of the Tomb ftone) alfo at each end on the under-fide was nail'd a Ledge, and between thefe Ledges there was wedged in (with Wooden-wedges) a Hearth-ftone of Marble that was alfo rough and unpolifh'd. Then flinging Water and Sand upon the Tomb-ftone, they wrought upon it, (by drawing the Hearth ftone too and fro) 'till the Hearth-ftone became pretty fmooth, and then they put in another rough Hearth-ftone, and fo they continue to do, till they have wrought the Tomb-ftone pretty even and fmooth. But you are to note, That while the Tomb-ftone and Hearth-ftones are rough, they lay a confiderable weight, (as a Stone, or the like) upon the upper fide of the Tool, to keep it down hard on the Tomb-ftone, but when the Tomb-ftone is pretty fmooth, they make him yet fmoother, by putting into the Tool, (one after another) feveral of thofe hearth-ftones already begun to be polifh'd, and this they continue to do, till they have brought both them and the Tomb ftone to a more polite Surface ; upon thefe they ufe no weight on the back of the Tool, but they ufe Water and Sand, as before. And if they have no Marble-hearth-ftone to polifh, then the Workmen tell me, they put a Purbeck-ftone into the Tool.

5. *Price of* –] *Chimney pieces* of *Egyptian*, or black fteak'd Marble, or of Rance, or Liver-colour'd Marble is worth (of an

an ordinary ſize) 12 or 14 Pound a piece.

Window-ſtools, of white or black Fleak'd-marble, are worth about 2 *s.* 6 *d. per* Foot.

Pavement of black, or white Marble, is worth about 2 *s. per* Foot. Thus Mr *Wing.*

A Stone-cutter in *London* tells me, he ſells *Engliſh* white Marble vein'd with read, *&c*

in { Squares / Slabs } for about { 2 *s.* 6 *d.* / 5 s. } *per* Foot.

for 2 *s.* 6 *d. per* Foot in Squares for *Pavements*, and Slabs of the ſame ſort of Marble, (long enough for a Chimney foot-pace) for 5 *s. per* Foot.

Egyptian Marble, vein'd with variety of Greens, in Slabs, he ſells for 8 *s. per* Foot.

Italian white Marble vein'd for Chimney-foot-paces, he ſells.

Black-marble he ſells ſomewhat cheaper.

Marble-colour.

The Price of Painting ordinary Marble-colour, on new Stuff, is about 1 *s. per* Yard And an old Colour, about 9 *d per* Yard, Colour and Work.

Maſons.

1. *Work*] The ſeveral kinds of Work done by Maſons, (in relation to Building) with their Prices, and Methods of Meaſuring them, *&c* are too many to be comprehended under this ſo general word of *Maſon's-work*, (eſpecially as the word *Maſon* is accepted in the Country) and therefore I ſhall refer them to their Particulars, (as *Walling*, *ſetting of Fronts*, *Healing*, &c.) where they will much more readily be found.

2. *Bill to make.*] See *Bricklayer's Bill.*

Meaſuring.

Of Artificers Work.] See the particular kinds of Work, in their proper Places of the Alphabet; where they will much more readily be found, than under this general word *Meaſuring.*

Membretto.

A Pilaſter that bears up an Arch.

Mezanine.

The ſame as *Entreſole*; which ſee.

Metops.

Is the ſquare Interval between the Triglyphs of the Dorick Frize, which among the Ancients uſed to be adorn'd with the Heads of Beaſts, Baſons, Vaſes, and other Inſtruments uſed in ſacrificing

The Beauty of theſe *Metopes*, conſiſts in their Regularity; that is, in their being perfect Squares: And yet when they are really Square, they appear to be leſs in height than in breadth; which is owing to the Projecture of the little Bandelet wherein they terminate underneath, that hides a ſmall part

part of their Heighth; for this Reaſon M le Clerc is for making the *Metopes* a minute or two more in Heighth, than in breadth; being of Opinion, They ought rather to appear Square, without being ſo, than really be ſquare, without appearing ſo.

Minute.

A Minute is uſually the 60th part of a Module; as a Module is uſually the Diameter of the lower part of a Column It is alſo ſometimes taken for the 12th part of an Ounce. See *Module.*

Mitchels.

Purbeck ſtones for paving, pick'd all of a Size, from 15 Inches ſquare to two Foot. Being ſquar'd, and hew'd ready for Paving a Stone-cutter in *London* tells me, they commonly ſell them at about 2 *s* 10 *d.* *per* Foot.

Model.

An original Pattern which any Man propoſes to imitate: properly (in Architecture) a ſmall Pattern of a Houſe, or the like, (made of Wood, or any other Material) made by a ſmall Scale, wherein an Inch, or half an Inch repreſents a Foot, for the more exactly carrying on a great Deſign. Sometimes the Word is us'd, (tho' improperly) in the ſame Senſe with *Module.*

Modern.

This Word, in its genuine Meaning, is only applicable to ſuch Architecture as partakes partly of the Gothick, retaining ſomewhat of its Delicacy and Solidity; and partly of the Antique, whence it borrows Members and Ornaments, without any Proportion or Judgment.

Module.

A Meaſure made uſe of to regulate the Proportions of the ſeveral Members of a Column In the Dorick Order, a *Module* is half the Diameter of the Body of the Column below: In other Orders 'tis the whole Diameter. A *Module* is commonly ſuppos'd to be divided into 60 equal Parts, call'd *Minutes.* *Module* comes from *Modulus.*

Modilions.

In Italian *Modiglioni* (a ſort of *Cantalivers*) are little inverted *Conſoles* under the *Soffit*, or Bottom of the *Drip*, in the Ionic, Compoſite and Corinthian Cornices, and ought to correſpond to the middle of the Columns. Theſe are particularly affected in the Corinthian Order, where they are always enriched with carv'd Work. In the Ionic and Compoſite they are more ſimple, having ſeldom any Ornaments, excepting ſometimes a ſingle Leaf underneath. In Latin they are call'd *Mutuli.*

Modilion.

Modilion-cornice.

See *Cornice*. N. 9.

Moresk-work.

A kind of Antick-work in Painting and Carving, after the Manner of the Moors, (whence it has its Name) consisting of several Grotesco's, wherein there is no perfect Figure, either of Men, or other Animals, and wherein there is a wild Resemblance of Birds, Beasts, Trees, *&c.* intermingled.

Mortise.

From the French, *Mortaise* the Hole made in one piece of Wood, to receive the Tenon of another Piece. See *Tenon.*

Mortar.

1. *What*] From the French, *Mortier*, a sort of Plaister, commonly made of Lime, and Sand, and Water, used by Masons and Bricklayers, in Building of Walls of Stone and Brick. For plaistering of Walls, they make their Mortar of Lime, and Ox, or Cow-hair, tempered well together with Water, and this is commonly call'd *white Mortar.*

2. *Of making common* —] As for making of common Mortar, and for the Proportions of Lime and Sand to be us'd about it, as many Men are of many Minds, I shall give you their several Sentiments about this Matter.

Vitruvius says, you may put three parts of dug, (or Pit-sand) to one part of Lime, to make Mortar, but (says he) if the Sand be taken out of a River, or out of the Sea, then two parts thereof and one of Lime (He also says, That if to River, or Sea sand, you put a third part of Powder of Tiles, or Bricks, it works the better) But *Vitruvius*'s Proportion of Sand seems too much, tho' he should mean of Lime before 'tis slack'd; for one Bushel of Lime before 'tis slack'd, will be five Pecks after 'tis slack'd.

About *London*, (where for the most part Lime is made of Chalk,) they put about 36 Bushels of Pit-sand to 25 Bushels of Quick-lime, that is, about a Bushel and a half of Sand, to a Bushel of Lime.

Some Workmen in *Sussex* tell me, That they commonly put two of their Court loads (that is about 24 Bushels) of Sand to 1 Load, (that is 32 Bushels) of Lime, which is but three Pecks of Sand to one Bushel of Lime.

Other Workmen in *Sussex* tell me, That their usual Proportion of Lime and Sand, in making of Mortar, is 4 Court-load, (that is about 48 Bushels of Sand to one Load, (or 32 Bushels) of Lime, which is exactly a Bushel and half of Sand to one Bushel of Lime, near the *London* Proportion. But they tell me, 'tis of Stone-lime; for they allow but three Load, (or 36 Bushels) of Sand to one Load, (or 32 Bushels) of run Lime; (for, say they, a Load a Load of run Lime is nothing near ſo much as a Load of Stone- [or quick] Lime) which is but 9 Gallons of Sand to a Buſhel of Lime.

Other Workmen in other parts of *Suſſex*, tell me, that they allow 4 Load (at 18 Buſhels to the Load) of Sand, to one Load (or 32 Buſhels) of Lime, which is 2 $\frac{1}{4}$ Buſhels of Sand to one of Lime.

Another Workman (in *Suſſex*) tells me, that (to his Knowledge) ſome *London* Bricklayers put as much Lime as Sand in their Mortar; eſpecially for Front-work.

A Gentleman in *Suſſex* tells me, That the *London* Bricklayers make their Mortar much more durable than our Country ones; for he told me that at his Brother's Houſe, and at another Gentleman's Houſe (which he nam'd to me,) the Mortar was not ſcal'd at all; but at his own Houſe (which was done by Country-workmen,) it ſcal'd very much, and fell out of the Joynts. But (ſaid he,) the *Londoners* make their *Mortar* by proportioning their Lime and Sand, *viz.* By meaſuring it all; but the Country-workmen, (for the moſt part) make it by gueſs. Now (ſaid he), our Country workmen do not make their *Mort r* fat enough; for they put in too little Lime to their Sand. Nevertheleſs, his Workman told me, That he did put in, as near as he could gueſs (by the Shovels full,) at leaſt twice as much Lime as Sand in his *Mortar*, and took Care to ſift all his Lime and Sand; and yet, (to my Knowledge) ſome of his Walls ſcal'd pretty much, eſpecially thoſe that were done towards the latter end of the Year; tho' (ſaid he) I never made *Morter* ſo fat in my Life before. But indeed, none of his Walls were coped, they were only cover'd with Straw on the top, and Boards or Slabs laid on it to keep it on, which ſometimes were blown off in the Winter, and ſo let in the Wet; which, (ſaid he) was the cauſe of the ſcaling of the Morter; but his Maſter deny'd this, and ſaid, it did ſo where it was never uncover'd all the Winter.

From all theſe various Proportions (of Lime and Sand) above-mentioned, all aſſerted by able Workmen, I think it reaſonable to infer, That the Proportion of *Lime* to *Sand* in making of *Mortar*, ought to be various, according to the goodneſs or badneſs of theſe Materials; and therefore is rather to be regulated by the Judgment of experienced and skilful Workmen in each particular Country, than by any ſtated Proportions So let this ſuffice (at the preſent) for the Proportions of the Materials. I ſhall next ſay ſomething of —

The Method of making of Mortar.] Some Workmen tell me, that 'tis the beſt way not to uſe *Mortar* as ſoon as 'tis made; nor (in making it) to make the Lime run before it is mixt with the Sand, (as ſome will do,) but rather to take the Sand and throw it on the Lime whilſt it is in Stones, before

fore it is run, and ſo to mix it together, and then wet it; by which Means, (ſay they) it will be the ſtronger, and when it has lain a while made before it is us'd, will not be ſo ſubject to blow and bliſter.

Others adviſe to let *Mortar* (when made) lie in a heap two or three Years before 'tis us'd; for ſo (ſay they) 'twill be the ſtronger and better; for the Reaſon of ſo many inſufficient Buildings, (ſay they) is the uſing of the Mortar as ſoon as 'tis made.

Others tell us, (1) That when you ſlack the *Lime*, you muſt take Care to wet it every where a little (but not over-wet it,) and cover with Sand every Laying or Bed of Lime (being about a Buſhel) as you ſlack it; that ſo the Steam, or Spirit of the Lime may be kept in, and not fly away, but mix it ſelf with the Sand, which will make the Mortar much ſtronger, than if you ſlack all your Lime at firſt, and throw on your Sand all together at laſt, as ſome uſe to do. (2.) That you ought to beat all your Mortar with a Beater, 3 or 4 times over before you uſe it; for thereby you break all the Knots of Lime well together, and the Air which the Beater forces into the Mortar at every ſtroak, conduces very much to the Strength thereof. (3) That when you deſign to build well, or uſe ſtrong Mortar for Repairs, you ſhould beat the Mortar well, and let it lie two or three Days, and then beat it well again when 'tis to be us'd. (4.) That in *Summer-time* you ſhould uſe your Mortar as ſoft as you can, but in *Winter* pretty ſtiff, or hard.

Mr. *Worlige* ſays, That if you intend your *Mortar* to be ſtrong, where you cannot have your choice of Lime you may chuſe your Sand and Water; for all Sand that is duſty, makes the Mortar the weaker; and the rounder the Sand, the ſtronger the Mortar, as is uſually obſerv'd in Water-drift Sand; that makes better Mortar than Sand out of the Pit.

Therefore, (ſays he) if you have occaſion for extraordinary *Mortar*, waſh your Sand in a Tub, till the Water, after much ſtirring, come off clear, and mix that with new Lime, and your Mortar will be very ſtrong and durable. And if your Water be foul, dirty or muddy, your Mortar will be the weaker.

He alſo tells us That 'tis a great Error in Maſons, Bricklayers, *&c.* to let the Lime ſlacken and cool before they make up their *Mortar*, and alſo to let their *Mortar* cool and die before they uſe it: Therefore, (ſays he) if you expect your Work to be well done, and long to continue, work up your Lime quick, and but a little at a time, that the Mortar may not lie long before it be uſed. So that you ſee, that in this Point alſo, Men differ in their Sentiments; ſome affirming it beſt to uſe their *Mortar* new, others, after it has lain made ſome time.

An

An old experienced Mason of my Acquaintance, tells me, That being at work at *Eridge-place*, (at my Lord *Abergaveny* s) at *Fant* in *Sussex*, they would have him make use of some Mortar that had been made four Years. But he, (when he came to try it,) told them it was good for nothing, by reason it was so very hard, that there was no tempering of it. Whereupon a Jesuite (residing in the House, and who had been a great Traveller) told him, That to his Knowledge, at several Places beyond Sea, they always kept their *Mortar* 20 Years before they use it ; but then (he saith) they keep it in Cisterns for the Purpose, and always keep it moist. Now, the old Mason (above-mention'd) tells me, he believes this Method may make the Mortar good and tough.

As for the Scaling, (or Crimbling) of Mortar out of the Joynts of Stone and Brick-walls, some Masons tell me, It proceeds from the badness of the Sand, or Lime, or both, as well as from the Season of the Year when the Work is done.

3. *Of making other Kinds of*.] Besides the *common Mortar*, (us'd in laying of Stones, Bricks, and Tiles) above-mention'd, there are several other Kinds, as ——

4. *White Mortar*.] This is used in plaistering of Walls and Ceilings, that are first plaister'd with Lome, and is made of Ox, or Cow-hair, well mixed and temper'd with Lime and Water, (without any Sand :) The common Allowance in making this kind of Mortar) is one Bushel of Hair to six Bushels of Lime. The Hair serves to keep the Mortar from cracking, binding it, and holding it fast together.

5 *Mortar us'd in making of Water-courses*, *Cisterns* &c.] This kind of *Mortar* is very hard and durable, as may be seen at *Rome* at this Day. It is used not only in building of Walls, but also in making of Cisterns to hold Water, and all manner of Water-Works, and also in finishing, or plaistering of Fronts to represent Stone work

And I find two kinds of this Mortar us'd by the Ancients ; both of which are compounded of Lime and Hog's-grease ; but to one is added the Juice of Figs, and to the other Liquid-pitch, and is first wet, or slack'd with Wine, then pounded, or beat with Hog's-grease, and Juice of Figs, or with the same and Pitch ; that which has Pitch in it, is blacker and easily distinguish'd from the other by its Colour, and that which is plaister'd with this kind of Mortar, is done over with Linseed-oil.

6. *For Furnaces*, &c] Some Chymists, in building their Furnaces, make use of a kind of *Mortar* made with red Clay, not too fat, least it be subject to Chinks ; nor too lean, or sandy, least it bind not enough. This Clay is wrought in Water, wherein store of Horse-dung and Chimney-soot has been

been ſteeped and well mingl'd, by which a Salt is communicated to the Water, binding the Clay, and making it fit to abide the Fire.

Some Metaliſts uſe a kind of *Mortar* to plaiſter over the in-ſides of their Veſſels, (for refining of Metals) to keep the Metal from running out: And this kind of *Mortar* is compounded, and made of Quick-lime, and Ox-blood, the Lime being beat to Powder and ſifted and then mix'd with the Blood, and beat with a Beater

The Glaſs-makers in *France* uſe a ſort of *Mortar* (for plaiſtering over the in-ſides of their Furnaces,) made of a ſort of Fuller's-earth, which is gotten from *Beliere* near *Forges*, which is the only Earth in *France* that has the Property of not melting in this exceſſive Heat. And 'tis of this ſame Earth that the Pots are alſo made which will hold the melted Metal for a long time

7. *For Sun Dials.*] An exceeding ſtrong and laſting Mortar to make a Dial-plain on a Wall, may be thus made: Take Lime and Sand, which temper with a ſufficient Quantity of Linſeed-oyl; this ſpread upon the Wall, will harden to the Hardneſs of a Stone, and not decay in many Years. *Note*, If you cannot get Oil, you may temper your Lime and Sand with ſcum'd Milk, (but Oil is better) and this will laſt ſix times as long as the ordinary Plaiſter made of Lime and Hair with Water.

I have known a very ſtrong and tough Mortar (for a Sundial-plain,) made in this manner. To about five or ſix Gallons of Brook-ſand, (which was dry'd on an Oaſt, and ſifted through a fine Splinted-ſieve,) there was put as much, or rather more Sifted-lime, and a Gallon of Boreing (or Gun) Duſt ſifted alſo; all which was wet and temper'd well with 6 or 7 Gallons of Scum'd-milk, and about a Pottle of Linſeed-oil. This was laid on the Wall firſt, well wet with Milk; but the Workman found much Trouble to ſet it ſmooth, by reaſon it dry'd ſo very faſt; but by keeping it often ſprinkl'd with Milk, and ſmoothing it with the Trowel, it at laſt ſet with a very ſmooth and ſhining Surface. But notwithſtanding all his Care, it (as it dry'd) crack'd pretty much; which I fanſie might proceed from the want of Hair in it: It did alſo blow in Bliſters, tho' the Lime were ſifted; and therefore I fanſie, that if the Lime had been prepar'd as it is in *Freſco Painting*, it might have been prevented.

8. *Extraordinary good for Floors, Walls, and Ceilings.*] If you temper Ox-blood, and fine Clay together, and lay the ſame in any Floor, or plaiſter any Wall, or Ceiling with it, it will become a very ſtrong and binding Subſtance, as I have been told, (ſays my Author) by a Gentleman Stranger, who affirm'd to me, That the ſame is of great uſe in *Italy*.

9. *A profitable and cheap kind of*--] A wise, wealthy, and and ancient Soap-boiler, dwelling without *Aldgate*, has (for the better Encouragement of others,) long since erected a fair and stately Edifice of Brick for his own Habitation; upon the good Success whereof he has since built another House of some Charge and good Receipt; the *Mortar* whereof did consist of two Load of waste Soap-Ashes, one one Load of Lime, one Load of Lome, and one Load of *Woolwich* Sand.

So likewise, another Gentleman of the same Faculty, (being likewise of good Credit, and great Experience) has us'd only Lome and Soap-Ashes temper'd and wrought together for *Mortar*; whereby he has laid both the Foundations, Chimneys, and their Tunnels in his Dwelling-house in *Southwark*, and they have endur'd those Storms already past, which have overturn'd many other Tunnels, both new and old, that were built with the ordinary *Mortar*.

It may be, that many Lime-men (and some of those Bricklayers that are in Fee with 'em) may speak against this Practice, and labour (by all possible Means) to discredit it; but there is no *Reason* can hold against *Experience*, nor no *Malice* so great; but *Truth* in her Time will be able to vanquish. And if these three *Tryals* be not thought a competent Number to give Credit to a new Invention, I can (says my Author) back and confirm them with threescore more at the least, which have been already made within the City of *London*, and Suburbs thereof.

True, indeed, this kind of *Mortar* is somewhat rough in the laying, and more sharp and fretting to the Fingers than ordinary *Mortar*, which makes it so much neglected and decry'd by some Workmen: But (says my Author) I could soon remedy these two slender Faults; the first whereof is rather an excellent Quality i. Mortar, than a Fault. Yet for the Good-will I bear to all the excellent uniform Buildings of our Time, I will set down the best Advice that I can in this Case, and such as I dare warrant upon my Credit. And first, concerning the Roughness of this kind of Mortar, Who is so blind, as not to see how to remedy it? (For 'tis rather a Work of Labour than of Skill;) for the Soap-ashes (which are in hard Cakes,) being either ground, or stamped into a fine Powder, before they be mixt with the Sand, will soon be brought to a smooth Temper And here we have no need to fear the Charge that will arise thereby; for I dare undertake, that the Profit of one Day's Labour will answer the Charge of three Men's Wages, in the Difference of Price that will be found betwixt one Load of these Ashes, and one hundred of Lime.

Then, *Secondly*, the *Sharpness* wherewith they offend the Bricklayer's Fingers, may in some sort be avoided by wearing

ing of Gloves, (without which they ſeldom lay any Brick at all) to avoid the like Effects, which they find in Lime.

But for an aſſured help in this caſe, (if the ſharpneſs be ſuch as cannot be endur'd of Workmen) let theſe Aſhes be re-imbibed in Water for ſome reaſonable time, till more of their Salt be extracted from them, and then, (without queſtion) they will find them gentle enough, and much of their fretting Nature taken away from them.

10. *For laying of Tyles*] I know ſeveral Places in *Suſſex*, where for laying of Tiles upon Houſes, *&c.* they make a kind of *Mortar* of Lome, and new Horſe-dung, well temper'd and mix'd together. This ſome Workmen commend for a good, ſtrong, and cheap *Mortar*; and others tell me, That 'tis more agreeable to the Tiles, than the common Mortar made of Lime and Sand; which, ſay they, corrodes and frets the Tiles, cauſing them to ſcale and fly to pieces, which this does not.

I have taken particular notice of one Houſe, where the *Tiles* were laid in this kind of *Mortar* and had been laid about four or five Years, and yet the Mortar did ſtick very well under the Corner-tiles, where it generally lies thickeſt.

11. *For Plaſtering of Fronts of Houſes in imitation of Brick-work*] Some Workmen tell me, That they make *Mortar*, (for this kind of Work) of Powder of Bricks, ſharp Sand, and Lime, and ſome Red-oker. I know a Houſe that is Plaſter'd with this kind of Mortar; it has been done above twenty Years, and yet looks very well, and paſſes (with common Paſſengers) for a Brick Houſe, tho' it be only Timber Plaſter'd over. They have commonly 1 *s per* Yard for doing ſuch Work, only Workmanſhip.

12 *How much allow'd to a Rod of Brick-work, or a Square of Tiling*] Workmen commonly allow a hundred and half, (or $37\frac{1}{2}$ Buſhels) of Lime, and 2 Load, (or 72 Buſhels) of Sand to make Mortar enough for a Rod of Brick-work

And for Tiling, 4 Buſhels of Lime, and 6 or 8 Buſhels of Sand will make Mortar ſufficient to lay 1000 of Tiles, which is about a Square and half. So that a Square of Tiling will take up (for Mortar) about $2\frac{2}{3}$ Buſhels of Lime, and about 5 Buſhels of Sand.

13. *A Caution about*] 'Tis a general Caution in all parts of a Building, that where Stones, or Bricks are contiguous to Timber, they ought to be laid dry, or without Mortar; becauſe Lime and Wood are inſociable, the former very much corroding and decaying the other.

14. *Rough Mortar*, which ſee in *R.*

Moſaique or *Moſaical-work.*

Is a curious kind of Work, conſiſting of ſmall inlaid Pebbles, Cockles, and Shells of ſundry Colours; and (of late) likewiſe with pieces of Glaſs figured with pleaſure.

Moſs.

Moss.

Us'd in Tiling] In some parts of *Sussex* they lay Tiles in Moss instead of Mortar; and when the Workmen get the Moss themselves, they are allow'd 2 *d.* in a Square the more for their Work. But some Workmen contemn this way of Tiling with Moss; because, say they, in windy wet Weather when the Wet, Rain, Snow, or Sleet is driven under the Tiles in the Moss; if there follow a Frost whilst the Tiles are wet, it then Freezes the Moss, and so raises the Tiles out of their Place.

Mouldings.

Under this Name are comprehended all those *Jettings* or *Projectures* beyond the *Naked* of a Wall, Column *&c.* which only serve for Ornament; whether they be square, round, straight or crooked. Of these there are seven Kinds more considerable than the rest, *viz* the *Doucine*, the *Talon* or *Heel*, the *Ovolo* or *Quarter-round*, the *Plinth*, the *Astragal*, the *Denticle*, and the *Cavetto*.

Multiplication

Of Feet and Inches by Feet and Inches.] See *Cross-Multiplication.*

Munions,

In Architecture, are the short upright Posts that divide the several Lights in a Window-frame.

Muring.

An old Term in Architecture, signifying the Raising of Walls. See *Walls.*

Mutule.

Same as *Modillions*: which see. The *Latins* call it *Mutulus.*

Nails.

1. WHat] These are a Material so well known, that (in the general) they need no Description. But the particular Kinds of 'em (which are very numerous) shall be described in the following Articles.

2. *Back — and Bottom.* —] These kinds of Nails are made with flat Shanks, and so as to hold fast, and not open the Grain of the Wood; being proper for Nailing of Boards together for Coolers, for Guts to save Water under the Eves of a House, or for any Liquid Vessels made of Planks, or Boards.

3. *Clamp.*] These are proper to fasten Clamps in Building, and repairing of Ships.

4. *Clasp.*] These are of two sorts, *viz.* (1) *Long,* proper for any fine Building with Firr, or other soft Wood: The clasping of the Head brings them into little compass, and admits of their sinking into the Wood, makes the Work smooth, and will admit a Smoothing-plane to go over them when drove. The sizes are

are 7, 7½, 8, 10, 13, 14, 15, 18, 21, 22, 23, 28, 32, 36, and 40 ℔ *per* Thouſand.

(2.) *Strong*, theſe are fit for Oak, and other hard Woods The ſizes are 15, 18, 28, 32, and 40 ℔. *per* Thouſand.

5. *Clench.*] Theſe are commonly us'd by Boat, Barge, and Lighter Builders, with Boves, and often without: They are proper Nails for any Building with Boards, that muſt be taken down again, becauſe they will drive without ſplitting the Wood, and draw, (or admit of punching out, (if right made) without breaking. The ſorts are too many to be here enumerated, for fine Work they are made with Claſp-heads.

6. *Clout*] Theſe are commonly us'd for nailing on of Clouts to Axle-trees, but are proper to faſten any Iron to Wood; and (if right made) the Heads will hold driving home without flying. The ſizes are 4½, 7, 8, 9, 12, and 15 ℔ *per* Thouſand.

7. *Deck*] Theſe are proper for faſtening of Decks in Ships doubling of Shipping, and Floors laid with Planks They are of two ſorts, Dye-headed, and Claſp-headed. The ſizes are 4, 4½, 5, 5¼, 6, 6½, 7, 8, and 9 Inches long.

8. *Dog.*] Theſe are proper for faſtning of Hinges to Doors, for (if made right) they will hold the Hinge cloſe without the Heads flying off, or withou the help of botching, by putting Leather between the Head and the Hinge. The ſizes are 9, 12, 20, 25, 30, 40, 60, 80, and 120 ℔. *per* Thouſand.

9. *Flat Point*] Theſe are of two ſorts, *viz.* (1.) *Long*, which are much us'd in Shipping, and are very proper where there is occaſion to draw and hold faſt where there is no Conveniency to clench. The ſizes are 7½, 8, 9, 10, 11, 12, 13, 14, 16, 18, 21, 22, 23, 26, 40, 55, 75, and 110 ℔. *per* Thouſand. (2.) *Short*, theſe are fortified with Points to drive into Oak, or other hard Wood and are often us'd to draw the Sheating boards to, very proper where Oak or other hard Wood is us'd. The ſizes are 5, 9, 18, 26, 32, 40, 55, 75, and 110 ℔ *per* Thouſand.

10. *Jobent*] Theſe are commonly us'd to nail thin Plates of Iron to Wood, and to nail on ſmall Hinges for Cubboard-doors *&c* The ſizes are 2 and 3 ℔. a Thouſand.

11. *Lead*] Theſe are commonly us'd to nail Leather, and Canvas to hard Wood. The ſizes are 4¼, 7 and 8 ℔. *per* Thouſand.

12. *Port*] Theſe are commonly us'd to nail Hinges to the Ports of Ships. They muſt be made ſtrong, becauſe they will not admit of being clenched, without being prejudicial to the Lining; and therefore care muſt be taken that they be demanded of ſuch a length, as that they may come near through, (ſo as to take ſufficient hold) and yet not ſo long as to come quite through. The ſizes are 2½, 3, 4, and 5 Inches long.

13. *Pound.*] Theſe are four ſquare in the Shank, and are much

much us'd in *Eſſex*, *Suffolk*, and *Norfolk*; but in few other Countries, except for Paleing The ſizes are 6 *d.* 8 *d.* 10 *d* 20 *d.* and 40 *d.*

14. *Ribbing*] Theſe are commonly us'd to faſten the Ribbing, to keep the Ribs of Ships in their Place in Building; if theſe Nails are made right, they will hold faſt, and draw eaſie, without injuring the Ribbing, or Timbers. They are alſo very uſeful to faſten Timbers to be uſed for a while, and taken down again for further Service. The ſizes are 5, 5 ½, 6, 6 ½, 7, 7 ½, 8, 8 ½, and 9 Inches long.

15. *Roſe*] Theſe Nails are drawn four ſquare in the Shank, and commonly in a round Tool, as all common 2 *d.* Nails are, and moſt commonly 3 *d* and 4 *d.* In ſome Countries they make all their larger ſort of Nails in this ſhape, but their being ſquare drowneth the Iron, and the Nails do not ſhew ſo fair to the Eye, as thoſe laid upon the flat; but if made of tough Iron, they are very ſerviceable. The ſizes are 1 ¾, 2, 2 ½, 2 ¾, 3, 3 ½ 3 ¾, 4, 4 ¼, 4 ¾, 5, 9, 10, 13, 14 16, 17, 18, 24, 26, 28, 30, 32 36, and 40 ℔. *per* Thouſand

16. *Rother*] Theſe are principally to faſten Rother Irons to Ships, and require a full Head, and to be made ſo as to hold faſt in the Wood to the greateſt Degree, without Clenching.

17. *Round head*] Theſe are very proper to faſten on Hinges, or for any other uſe where a neat Head is required; and if made of the beſt tough Iron, as they ought to be, are very uſeful. The ſorts are Tacks, 2 *d.* 3 *d.* 4 *d.* 5 *d.* 6 *d.* and 8 *d.* The ſame tinn'd for Coffin-handles, and fine Hinges.

18. *Scupper.*] Theſe are principally to faſten Leather, and Canvas to Wood, and therefore require a broad Head, that neither may work looſe. The ſizes are 4 ¼, 7, and 8 ℔. a Thouſand.

19. *Sharp.*] Theſe are much us'd in all Countries, eſpecially in the *Weſt Indies*, being made with ſharp Points, the Shank flat, and is a very proper Nail for ordinary Uſes, where ſoft Wood is us'd. The ſizes are 2 ½, 2 ¾, 3, 3 ½, 4, 4 ½, 5, 5 ½, 6, 6 ½, 7 ½, 8, 9, 10, 11, 12, 13, 14, 15, 18, 19, 20, 21, 22, 23, 28, 32, 36, 40, 55, and 75 ℔ *per* Thouſand.

20. *Sheathing*] Theſe are commonly us'd to faſten Sheathing-boards to Ships. The Rule for uſing them, is to have the Nail full three times as long as the Sheating-board is thick, provided the Plank be of a ſufficient thickneſs, which ought to be enquir'd into; for the Sheathing-nail ought not to go through the Plank by half an Inch, leaſt it ſhould make the Ship leaky. The Shank muſt not be ſo ſtrong as to cleave the Board, and the Head muſt be well clapſed, or died, ſo as it may ſink into the Wood, and the Ships ſide left ſmooth. They are alſo a uſeful Nail in doubling of ſmall Ships. The ſizes are 1 ¼, 1 ½, 1 ¾, 2, 2 ¼, 2 ½, 2 ¾, 3, 3 ¼, and 3 ½ Inches long.

21. *Square.*]

21. *Square.*] These are of the same shape as sharp Nails, and is a most useful Nail for Oak, and other hard Wood, as also for nailing up Wall-fruit, the Points being made something stronger than the Points of sharp Nails, which fortifies them to go forward, and not turn back upon a small Opposition, as weaker points will do. The sizes are 2 $\frac{1}{2}$, 2 $\frac{3}{4}$, 3, 4, 4 $\frac{1}{2}$, 5, 5 $\frac{1}{2}$, 6, 6 $\frac{1}{2}$, 7, 8, 9, 10, 11, 12, 13, 14, 15, 16 18, 19, 20, 22, 23, 24 28, 30, 32, 36, 40, 55, and 75 ℔ *per* Thousand.

22. *Tacks.*] The smallest of these are to fasten Paper to Wood, midling for Wool-cards, and Oars, and the larger for Upholsterers, and Pumps. The sizes are 2 $\frac{1}{4}$, 5, 6, 8, 9, 14, and 15 Ounces a Thousand.

There are many more sorts of Nails, which for Brevity, (and because they are not so proper for our present Business) I shall omit

23. *Allowance of in Lathing.*] In Lathing, the common Al lowance of Nails is 500 to a Bundle of 5 Foot-laths, and 600 to a Bundle of 4 Foot-laths, at six Score Nails to the Hundred.

24 *Allowance of in Flooring.*] In laying of Floors 200, (that is 240) Nails is a compleat Allowance for a Square of Flooring.

25. *To Toughen.*] A Neighbour of mine, a Mason, tells me, That —— the Ironmonger at *Rotherbridge* taught a Kinsman of his, (who is also a Mason) to toughen his Nails that were brittle, by heating them hot in the Fire, (in a Fire-shovel, or the like) and putting some Tallow, or Grease to 'em, the first he says is best. This Ironmonger keeps a Nailer at work.

26. *Of driving.*] There is requir'd a pretty Skill in driving a Nail; for if, (when you set the point of a Nail) you be not curious in observing to strike the flat Face of the Hammer perpendicularly down upon the Perpendicular of the Shank the Nail, (unless it have good Entrance) will start a-side, or bow, or break, and then you will be forced to draw it out again with the Claw of the Hammer. Therefore you may see a Reason when you buy a Hammer, to chuse one with a true flat Face.

Perhaps it may not be unacceptable to some Readers, if I here mention a little *Trick* that is sometimes used among some (that would be thought cunning Carpenters) privately to touch the Head of the Nail with a little Ear-wax, and then lay a Wager with a Stranger to the Trick, that he shall not drive that Nail up to the head with so many blows. The Stranger thinks he shall assuredly win, but does assured ly lose; for the Hammer no sooner touches the head of the Nail, but instead of entering the Wood it flies away, or starts aside, notwithstanding his utmost care in striking it down right.

Nave,

Nave,

In Architecture is commonly us'd to signifie the main part, or Body of a Church; which by the nearness of the word may seem to be deriv'd from the *Latin*, *Navis* a Ship; but it may more significantly be deriv'd from the *Greek* *ναὸς*, (that is *Naos*) a Temple

Newel

The upright Post that a pair of Winding-stairs are turn'd about.

Niches, or *Nices*,

The hollow places in a Wall, wherein Statues or Images, are set. If these Images be of white Stone, or Marble, let not the Concavities be colour'd too black; for tho' *contraria juxta se posita magis illucescunt*, be an old Rule, yet 'tis observ'd, that our sight is not well pleas'd with sudden changes, from one Extream to another; therefore let them have rather a duskish Tincture, than an absolute black.

Oak.

1. WHat] This is a sort of Timber well known, and needs no Description. 'Tis one of the pincipal Materials in Building, being strong in all Positions, and may well be trusted in cross and transverse Work; as for Summers, and Girding, or Binding-beams, *&c.*

2. *Of Sawing.*] Oak is worth sawing 2 *s* 8 *d.* *per* hundred, some 3 *s* and upwards to 3 *s.* 6 *d.* *per* hundred, that is the hundred Superficial Feet.

O G. *Ogee*, or *Ogive.*

A sort of Moulding in Architecture, consisting of a round and a hollow, like an S; *Virtruvius* makes it 2 Quarter-circles, *Scammozzi*, and some other Authors make the Arches flatter, by striking them from two Equilateral Triangles. See *Cima.*

Orders,

In Architecture are the different Forms and Proportions of Columns, *&c.* There are five Orders (commonly reckon'd) in Architecture, *viz.* The Tuscan, Dorick, Ionick, Corinthian, and Composite. Of all the Parts, which enter the Composition of a Magnificent Building, the Orders of Columns being the most considerable I shall transcribe from M. *le Clerc* all that is necessary to be said upon this Article

An Order of Columns, is usually understood of a Column, bearing its Entablement; but the Order is scarcely compleat, except the Column be rais'd on a Pedestal.

The Pedestal, Column, and Entablement, are three Compound Parts, each consisting of three others.

The

The Parts of the Pedeſtal, are the Baſe, Die, and the Cornice; thoſe of the Column, the Baſe, Shaft, and the Capital. Thoſe of the Entablement, the Architrave, Frieze, and the Cornice; each of which Parts have their particular Characters and Members; call'd by the general Names of *Mouldings* or *Ornaments*.

The Ancients have given us five ſeveral Orders of Columns; the Tuſcan, Dorick, Ionick, Roman and Corinthian.

The Tuſcan Order is the ſtrongeſt, and the moſt ſimple of all others: Its Name ſhews its Original.

If we believe M. *de Chambray* in his Parallel, this Order ought never to be uſed any where but in Ruſtick, or Country Houſes and Places. And 'tis certain, that in the manner *Vitruvius*, *Palladio*, and ſome others deſcribe it, it ſcarce deſerves to be uſed at all. Methinks, however, in *Vignola*'s manner of Compoſition, it has certain Beauties even in its Simplicity, which add a value to it, and render it worthy to be uſed not only in private Houſes, but alſo in publick Buildings; as in Portico's of Markets, of publick Halls, in Magazines or Granaries of Cities, and even in Palaces, and Seats of Princes and Noblemen, particularly in the lower Apartments, Offices, Stables, (not Equeries, as 'tis tranſlated lately,) *&c.* And in general, in all Places where Strength and Simplicity are required, and where any of the richer and more delicate Orders would be unſuitable.

The Dorick Order is the moſt ancient, and was given us by the *Greeks*. Its Compoſition is Grand and Noble; and the Triglyphs, which make the Ornaments of its Frieze, bearing ſome reſemblance to a Lyre, ſeem to intimate it to have been originally intended for ſome Temple conſecrated to *Apollo*. As we are now-adays furniſhed with richer and more delicate Orders for Temples; the Dorick is moſt properly uſed in the Gates of Cities, in Arſenals and Places of Arms, in Halls of Guards, and other Buildings that have relation to War; where Strength, and a rough, but noble Simplicity are particularly required. In the moſt ancient Monuments of this Order, the Columns are without Baſes, the reaſon of which is not eaſy to aſſign. Monſieur *de Chambray*, in his Parallel, is of opinion with *Vitruvius*, that the Dorick Column not the Ionick, as *le Clerc*'s Tranſlator has it, having been compoſed in imitation of a naked Man, nervous and robuſt as an Hercules, it ought to have no Baſe; imagining a Baſe to be that to a Column, which a Shooe is to a Man. But for my own part, I muſt confeſs, I can't conſider a Column without a Baſe, but in comparing it to a Man, I rather form the Idea of a Man without Feet, than without Shooes. For this reaſon, I am rather of opinion, either that the ancient Architects had not yet thought of adding Baſes to their Columns, or or that they declin'd on purpose to give them any, with design to keep the Pavement clear and unembarrassed with the Angles and Projectures of Bases, which are apt to occasion People in passing by to stumble. This too appears the more probable, in regard the Architects of those Times used to range their Columns exceedingly near one another; so that, had they been furnished with Bases, the Passages between, would have been extreamly narrow and incommodious. And this appears to be the reason why *Vitruvius* orders the *Plinth* of the *Tuscan Column* to be rounded off; that Order, in the manner he describes it, being particularly adapted to the servile Offices of Business and Commerce where Conveniency is always to be consulted before Beauty. Be this as it will, every Man of good Taste will allow, that a Base adds a Grace to a Column; and that it is a very necessary Appendage, in regard it makes it stand the more firmly on its Plan: So that if no Columns are now made without Bases, this ought not to be imputed to the Prejudices of our Architects, as some Admirers of Antiquity will have it, but to their Prudence.

The first Idea of the *Ionick Order* was given by the *Ionians*, who according to *Vitruvius*, compos'd this Column on the Model of a young Lady dressed in her Hair, of an easie and delicate Shape; as the *Dorick* had been form'd on the Model of a strong robust Man. 'Tis said, the Temple of *Diana* at *Ephesus*, the most celebrated Edifice of all Antiquity was of this Order. It may now be used in Buildings of Piety, as in Churches, Courts of Justice, in Apartments of Ladies, and in other Places of Quietude and Peace.

The *Roman Order* is usually call'd the *Composite* in regard its Capital is composed of the principal Parts of the Capitals of all other Orders It has a Quarter-round as the *Tuscan* and *Dorick*; Volutes as the *Ionick*; and a double Row of Leaves underneath, as the *Corinthian* I call it *Roman*, as believing, with many others, that the *Romans* first invented it. Most of our Architects, in compliance with Usage and Custom, place this after the *Corinthian*; doubtless because it was the last that was invented. *Scammozzi* is the only Author who varies from the Rule; but he does it with so much Judgment, that we make no scruple to imitate him.

This Order may be used in every place and on every occasion, where 'tis requir'd that Strength, Richness, and Beauty should be found together.

The *Corinthian Order* is the Noblest, the Richest, and the most Delicate Order of Architecture. This is indeed a Master-piece of Art, for which we are indebted to the City of *Corinth*: It ought always to be used in the most stately and most magnificent Buildings.

These several Orders have been very judiciously composed at various times, in order to to ſuit the various kinds of Buildings, which either Neceſſity or Magnificence ſhould occaſion Men to erect; and theſe are ever made more or leſs ſimple, each in its kind, and more or leſs ſlender, according to the Buildings they are uſed in, and the riches of the Princes, People, or private Perſons who build them. So far ſays, M. *le Clerc*. It may not be amiſs here to add, that the Herb *Acanthus*, or *Branca Urſina*, or *Bears-foot*; whoſe Leaves are repreſented in the *Capital* of the *Corinthian Column*, is ſaid to have been the firſt Occaſion of this *Ornament*; which was invented by *Callimachus*, an ingenious Statuary of *Athens*, upon ſeeing this Plant ſpreading it ſelf around a Basket that had been placed upon the Tomb of a young *Corinthian* Lady, and cover'd over with a Tile. The Leaves he imitated in his *crolls* or *Volutes*, the Basket in the *Vaſe* or *Tambour*, and the Tile or Covering in the *Abacus*. There are two kinds of *Acanthus*, the one wild, and arm'd with Prickles, the other ſmooth, and cultivated in Gardens; the former of which we find repreſented in *Gothick* Buildings, and the latter in thoſe of the *Antique*.

Orlo.

The *Plinth* or *Square* under the Baſe of a Column, or under the Baſe of its Pedeſtal.

Orthography,

Is a Word deriv'd from the *Greek*, *Orthos*, *true* or *right*, and *Grapho*, to *write* or *deſcribe* In Architecture, it ſignifies the Front or (any other) upright Side of a Houſe; or the Draught on Paper of thoſe Parts of a Houſe.

Ovolo,

As *Echinus*.

Over ſpan.

See *Clamp*. N. 2.

Painting.

1. OF *Out-door-work in general.*] Doors, Shop-windows, Window-frames, Pediments, Architraves, Friezes, and Cornices, and all other Timber-works that are expos'd to the Weather, ought at firſt ſetting up be be Prim'd with *Spaniſh brown Spaniſh white* and Red-lead (about a fifth part) to make the other two Colours dry; theſe well ground with Linſeed-oyl, will make an excellent Primer; then afterwards with the ſame Colour, (but much whiter) for a ſecond Primer, and laſtly, with fair White, made of White-lead, and about a fifth part in quantity, (not in weight) of *Spaniſh-white*.

Out-door-work thus colour'd, may be afforded for 3 *d.* or 3 *d.* half-penny, or 4 *d.* the Yard Square, for each time laid over.

2. Of

2. *Of Measuring.*] Painters measure their Work by the Yard superficial, and in taking the Dimensions of their Work, they run a String all over where the Brush goes; for they say, (and 'tis but Reason) we ought to be paid for all where the Brush goes. But sometimes in Rails, and Banisters, they will measure it as if it were flat Measure. I have seen the Experiment try'd, and the Difference would not countervail the Trouble of girting. So that Painters-work is measur'd the same as Joyners, only Painters never reckon Work and half, but work once, twice, or three times, *&c.* done over; or at so much *per* Yard, according to the Work. They always reckon double Work for painting of Window-shutters, if both sides are painted alike; otherwise, according to the value of the Painting. But they reckon Sash-frames by themselves, (at so much *per* Piece, and likewise Mantle-pieces) when there is no Painting about them; but if they stand in the Wainscot, they measure them as plain Work deducting nothing for the Vacancy.

3. *Of Wainscot-colour*] If on new Stuff, is worth about 8 *d. per* Yard, on old Colour about 7 *d.*

4. *Of Walnut-tree Colour.*] It is worth 10 *d.* say some, others say 16 or 18 *d per* Yard.

5. *Of ordinary branch'd Painting.*] Is worth 12, 14, or 16 *d. per* Yard.

6. *Of ordinary Marble-colour*] If on new Stuff, is worth 1 *s. per* Yard on old Colour, 9 *d.*

7. *Of white Colour.*] Is worth 10 *d* or 1 *s. per* Yard.

8 *Of plain Japan, either black or white.*] Is worth 3 *s.* 6 *d.* or 4 *s. per* Yard.

9. *Of Gates, and Outward-doors*] Is worth 3 *d.* or 3 *d.* half penny, or 4 *d per* Yard.

10. *Of Shop windows*] The same as Gates, and Outward-doors.

11. *Of Window-frames.* Is worth from 3 *d.* or 4 *d.* to 6 *d.* or 8 *d.* each Light, according to their Size.

12. *Of Sash-lights.*] Is worth about 1 *d per* Light.

13. *Of Sash-frames.*] Is worth about 1 *s per* Frame.

14 *Of Iron casements.*] Is worth three half-pence, 2 *d.* or 3 *d. per* Casement, according as they are of bigness

15. *Of Iron bars of Windows.*] Is worth 1 *d. per* Bar, or more, if very large

16. *Of Chimney pieces*] Is worth about 2 *s per* Chimney-piece.

17 *Of Pales*] Is worth about 10 *d.* or 12 *d per* Yard.

18. *Colours*] The Colours us'd in Painting, are of several Kinds, as White, and Red-lead, *Spanish* white, and brown, Verdigrease, Smalt, *&c.* Of which see in their proper Places of the Alphabet.

Paleing.

1. *With Cleft pales, Rails and Posts.*] Some Work-men tell me, That for paleing with 3 Rails, Cleft-pales, Rails and Posts

Posts, cleaving, making, and setting up, they have 3 *s.* 6 *d* or 4 *s. per* Rod, Felling the Timber and all. But then their Materials are all laid down to their Hand, so that they have no carrying.

Others tell me they have 2 *s* 6 *d. per* Rod, for (only) making and setting up of Cleft-posts, Rails, and Pales.

2. *VVith saw'd Pales, Rails and Posts.*] Some Workmen tell me they have 1 *s*, 6 *d. per* Rod for making and setting up of Saw'd-posts, Rails, and Pales.

Pales.

1. *Price of Cleaving*] Some Workmen tell me, That they have 2 *s per* Hundred for cleaving of Pales; but others that cleave in Brocks, say they, have but 1 *s.* 8 *d per* Hundred *Note*, A hundred of Pales is various, according to their length: for of 5 Foot Pales, 5 Score Pales is a Hundred, but of 4 Foot there goes 6 Score, and of 3 Foot, 7 Score to the Hundred.

2. *Of the Number a Tun will make.*] This is very uncertain, by reason of the Difference in Timber's cleaving, some cleaving much better, (and less to waste) than other some; yet by comparing several Observations, which I receiv'd from an ingenious Workman, I gather, that a Tun of good cleaving Timber may make three Hundred. (or perhaps something more) of 4 Foot Pales, and a Tun of the like Timber may make 4 Hundred of 3 Foot Pales; the Reason of which is, because Timber generally cleaves better, (and less to waste) in short lengths than in longer.

But the Number of *Sawed-pales* (that may be made of a Tun of Timber) is more certain, than of *Cleft-pales*; for I have found (by the Draught of a Tree, and Calculations.) that a Tun of Timber will make about 400 Foot of Inch-boards; which (if the Timber fit for length) being cut out

into { 5—Foot, 4—Foot, 3—Foot } Pales, will make

{ 80, 100, 133 } Pales, each a Foot broad; which in Paleing

will reach about three Times as far as the like Number of Cleft Pales will do

Palisade, or *Palisado.*

1. *VVhat.*] A sort of slight open Pale, or Fence, set to beautifie a Place, or Walk.

2. *Pales.*] Some Workmen tell me, that making and setting up of Pallisado-pales, (if the Heads are handsomely cut, the Palisades mortis'd through, the Posts at the Corners higher than the rest, and the Rails, Kneeling-rails) is worth 14 *s. per* Rod, Carpenter's Work, and wing.

An ancient and experienced Carpenter informs me, That the Carpenter had 25 *s. per* Rod, (for Timber and Workmanship; for the Palisado-pales at the Bowling-green at *Mount*

Mount Ephraim at *Tunbridge-wells* and likewise for the Palisades at the High-house behind the Bowling-green. This old Carpenter told me, he guess'd the Carpenter's Work of these Palisades to be worth about 10 *s. per* Rod.

I am also inform'd, that the Carpenter had 30 *s. per* Rod for the Palisades at the Walks at *Tunbridge-wells*. I mention these about the Wells, because I suppose them to be well known to most Gentlemen. For there is such Variety in the Workmanship of Palisado-pales, that there can be no certain Price for it by the Rod.

3. *Gates.*] These are as various in the Forms and Fashions as Palisade-pales, and consequently their Prizes are also as various, *viz.* From 6, or 7. to 10, or 12 *s. per* Yard running Measure, at about 6 or 7 Foot high

4. *Of Iron*] Palisado-work of Iron in Gates, or otherways, is from 4 *d per* Pound to 8 *d.* according to the Work.

Pallification.

A Term in *Architecture*, signifying the pileing of the Ground-work, or strengthning of the Ground-work with Piles of Timber driven into the Ground, when they build upon a moist and marshy Soil.

Pantry.

A Room to set Victuals in, a Store-room.

Pan-tiles.

See *Tiles*, Numb. 7.

Parapet.

From the Italian *Parapetto*, a *Save Breast*, is a little Wall, or sometimes a Rail serving either as a *Rest* for the Arm or as an *Inclosure* about a Key, Bridge, Terrass, *&c.*

Parastatæ.

Pilasters which stand alone, not adjoyning to the Wall, and which the French call *Isolees* or *Insulate*; from *Insula* an *Island*, as I take it.

Pargeting

1. *What.*] In *Architecture*, signifies the plaistering of Walls; sometimes 'tis us'd to signifie the Plaister it self

2. *Price.*] Pargeting, or plaistering is of divers Kinds. As (1.) White Lime, and Hair-mortar laid upon bare Walls, at 3 *d* or 4 *d* the Yard. (2.) Upon bare Laths, as in Partitioning, and plain Ceilings, from 8 *d* to 14 *d per* Yard (3.) Rendring the insides of Walls, or doubling Partition walls, at 2 *d.* or 3 *d* the Yard. (4.) Rough-cast upon Heart laths, from 1 *s.* to 3 *s* the Yard square, Workmanship and all Materials. (5) Plaistering upon Brick-work with finishing Mortar, in Imitation of Stone-work, from 1 *s.* to 18 *d.* or 2 *s.* the Yard square. (6.) And the like upon Heart laths, from 18 *d.* to 2 or 3 *s.* the Yard. See more in *Plaistering*.

Parlour.

A fair lower Room design'd principally for the Reception, and

and Entertainment of Company.

Partitions.

1. *Of Framing*]. See *Framing*, Numb. 4.

2. *Of Meaſuring.*] Partitions are commonly meaſur'd by the Square; but they commonly make Deduction for Doors and other Vacancies.

Paſſage.

An Entry, or narrow Room, ſerving only for a Thorough-fair, or Entrance into other Rooms.

Paving.

1. *What.* Is the laying a Floor with Bricks, Tiles, or Stones

2. *With Statute-bricks*] Paving with Statute-bricks, is done at *London* for about 4 *d. per* Yard. But I know ſome Workmen in *Suſſex* that have 5 *d* or 6 *d. per* Yard, into which Price they make ready the Floor for the Work, by clearing out the Earth, and levelling the Floor with a convenient quantity of Sand, (if they lay the Bricks dry, as ſometimes they do) which they ſpread evenly with the Rake; then laying the Bricks level by a Line, they (with a Trowel) put a ſufficient quantity of Sand under each Brick, to raiſe him full as high as (or a little higher than) the Line, and ſo knock him down (level with the Line) with the Handle of their Hammer; which being done they ram in the Sand (on the ſide of, and) againſt the bottom of the Brick with the Handle of their Hammer, to make him lie faſt. Having thus laid the whole Floor, they ſtrew Sand all over the Bricks, to the thickneſs of an Inch, more, or leſs, with a Command to the People of the Houſe, that they let it lie for the ſpace of five or ſix Weeks; now and then ſweeping it too and fro, that thereby, and by their treading on it, it may fill up all the Joynts betwixt the Bricks.

If they lay the Bricks in Mortar, the Price (they ſay) is the ſame as if they were laid dry.

There are ſome Maſons, that having laid the Floor dry, will make a very thin Mortar, which they ſpread all over the Floor ſweeping it too and fro with a Broom, to fill up the Joynts of the Bricks

This kind of paving (with common or Statute bricks) is uſual for Cellars, Waſh-houſes, Sinks, Fire-hearths, and for Halls and Kitchins in common Houſes.

Of theſe kind of Bricks, 32 will pave a Yard ſquare, if laid flat-ways, and 64 if edge ways.

3. *With ſquare Tiles, or (as ſome call them) Paving-bricks.*] The paving with Square tiles is commonly valu'd by the Square, and the dearer, the ſmaller the Tiles are; for theſe kind of Tiles are of ſeveral Sizes, *viz.* 6, 8, 10, and 12 Inches ſquare; their Price from 6 to 20 *s.* the Hundred. In *Suſſex* theſe kind of Tiles, (or as they call them Paving-bricks,) are 9 Inches ſquare, and

and commonly sold at 1 *d. per* Piece, or 8 *s per* Hundred.

If you would know how many of either of these sort of Tiles will pave any Floor, then

Note, That { 36, 21, 16, 13, 9 } Tiles of { 6, 8, 9, 10, 12 } Inches square, will pave a Square Yard.

4. *VVith Flemish bricks.*] The paving with these Bricks is far neater and stronger than common Bricks: They are of a yellowish Colour, and must be laid in Sand. Earth-brick is 6 Inches and a quarter long, 2 Inches and a half broad, and 1 Inch and a quarter thick.

Now, allowing a quarter of an Inch for the Joynt, then 72 of them will pave a Yard square, but if they be set edge-ways, then to pave a Yard square will require 100 Bricks. These Bricks are usually sold at 2 *s* the Hundred, and the Price of laying them is 4 *d* 5 *d*. or 6 *d*. the square Yard.

5. *VVith Rough, or Rag-stone*] This is the cheapest of all Pavements, and is valu'd from 12 *d*. to 15 *d* the Yard.

6. *VVith Free-stone.*] Paving with broad Stone taken out of the Quarries, (commonly call'd Free-stone) and cut into Lengths and Breadths promiscuously, (as they will hold) and in Thickness about 2 or 3 Inches, is usually rated at 6 *d*. 7 *d*. or 8 *d* the Foot square or 4 *s*. 6 *d*. 5 *s*. 3 *d*. or 6 *s*. the Yard square for Stone and Workmanship. This kind of Paving is laid in common Yards, and Passages before Shop doors, and Stalls, *&c.*

But if the Stones be squared all to a size (as sometimes these Stones are cut perfectly square, as Paving-tiles are, but much bigger, as 8 20, and 24 Inches square, and upwards;) then, as they are neater so they are dearer; as 12 *d*. or 14 *d per* Foot, or 9 *s*. or 10 *s*. 6 *d*. *per* Yard. But if the Stones, thus squared and sized be good and well polished, (as they ought to be for Kitchins, Dairies, and neat private Places) then they may be worth 15 or 16 *d*. *per* Foot, or 11 *s*. 3 *d*. or 12 *s*. *per* Yard square.

7 *VVith* Rigate, *or Fire stone*] This kind of Pavement is good for Chimny fire-hearths, Ovens, Stoves, *&c*, and is somewhat dearer than common Purbeck-pavement. For the Price of these Stones, see *Fire-stone*, Numb 2.

8. *VVith Pebble stones, or Bolders.*] Paving with Pebble-stones laid in Gravel, for Materials and Workmanship, may be worth 15 or 18 *d*. the Yard square

9 *With Marble*] Paving with Marble is of all other the most beautiful, of which there are several sorts; as white, black, and grey: Some Pavements, (as in Foot-paces before Chimneys)

neys) are laid all of one ſort, or colour, and in one entire Stone; others of two Colours laid ſquare, or Chequer-ways, the ſide of one by the ſide of the other; others are laid Arrace-wiſe, of two Colours, laid Angle to Angle, and this laſt is the neateſt way; but there may be divers Forms contriv'd to lay them in; as you may ſee in ſeveral Chancels, in the Choir of St. *Paul's*, and in the *Royal Exchange* in *London*, and divers other Places. This kind of Pavement is valu'd from 2 to 3 *s*. the Foot ſquare, and upwards, according as 'tis well laid and poliſh'd. For the Price of Marble, ſee *Marble*, Numb 5.

10. *Diamond.*] Diamond-pavement, (ſays Mr. *Wing*) is worth 3 *d* or 4 *d*. *per* Foot.

11. *Random.*] Random-pavement, (ſays Mr. *Wing*) at the Quarry, is worth 2 *d*. half-penny, or 3 *d*. *per* Foot.

12 *Of Meaſuring*] Paving is commonly meaſur'd by the Yard ſquare. And therefore the Length of any Pavement in Feet and Inches, being multiplied by the Breadth in Feet and Inches, (which, how it is done, ſee *Croſs-multiplication*. Numb. 2.) will produce the Content in Feet; which being divided by 9 (becauſe 9 Square Feet make a ſquare Yard) will give the Content in Yards requir'd.

Pavement.

See *Paving*.

Pediment.

In French *Fronton*, from the Latin *Frons*, the Forehead, is an Ornament that crowns the *Ordonnances*, finiſhes the Fronts of Buildings, and ſerves as a *Decoration* over *Gates*, *Windows*, *Niches*, &c. It is ordinarily of a Triangular Form; but ſometimes makes an Arch of a Circle *Vitruvius* calls it *Faſtigium*. Pediments over Doors, are commonly valu'd at ſo much *per* Piece, dearer or cheaper, according to their Largeneſs, Goodneſs of the Materials, and Curioſity in Workmanſhip.

Pedeſtal.

1. *What.*] Is a ſquare Body, with a Baſe and Cornice, ſerving as a Foot for the Columns to ſtand upon; and having, according to *Vignola*, the third part of the Height of its Column. It is different in different Orders.

2. *Kinds*] There are as many kinds of Pedeſtals as there are Orders of Columns, *viz.* 5. The Tuſcan, Dorick, Ionick, Corinthian, and Compoſite. The Heighth of the Pedeſtal in each Order ought to be a third part of the whole Column, comprehending the Baſe and Capital; and their upper Adjuncts, as Architrave, Frieze, and Cornice a fourth part of the ſaid Pillar. This Rule of ſingular Uſe and Facility,

Facility, I find settled by *Jacobo Baroccio*, and I hold him a more credible Author, (as a Man that most intended this piece of Architecture,) than any that vary from him in those Dimensions, says our famous *English* Architect, Sir *Henry Wotton*

Nevertheless, other Architects differ from him in the Height of the Pedestal I shall at present, (for Brevity sake) only give the Description of the several Orders of Pedestals from *Vitruvius*

3. *Tuscan*] According to *Vitruvius*, the whole Heighth of the Tuscan Column, comprehending the Architrave, Friese, and Cornice, is divided into nine Parts, whereof two go to the Heighth of the Pedestal.

This Pedestal he describes in two different Forms, one of which is plain, having only a Plinth for the Base, and another for the Capital; the Heighth of each of those Plinths is $\frac{1}{6}$ of the whole heighth of the Pedestal; and the Projecture of each of these Plinths is $\frac{1}{6}$ of their heighth.

In the other fashion'd Pedestal which he describes, he also divides the whole heighth of the Pedestal into 6 parts, one of which goes to the Base, and one to the Capital.

Again, he divides the Base into two parts, one of which goes to the Plinth below, and the other to the rest of the Base; and this being sub-divided into 4 parts, three of 'em goes to the Scima-reversa, and the List below it, which is $\frac{1}{2}$ a part, and the other to the List above it.

4 *Dorick*.] The whole of this Column, (comprehending the Architrave, Friese, and Cornice) is by *Vitruvius* divided into 8 parts, whereof two go to the heighth of the Pedestal, which agrees with *Jacobo Baroccio*'s Rule mention'd above, Numb 2.

This Pedestal is (by *Vitruvius*) also described in two different Forms; in both of which the Base and Capital are each $\frac{1}{7}$ of the whole heighth of the Pedestal.

In one of the fashion'd Pedestals, the Base is divided into two parts, whereof one goes to the Plinth below, and the other to the rest of the Base; and this part being sub-divided into two parts, one of them makes the lower *Thorus*; and the other being again sub-divided into three parts, two of them go to the upper *Thorus*, and the other to the List above it.

The Capital of this fashion'd Pedestal is divided into four Parts, whereof the lowermost makes the Astragal, (whose List is $\frac{1}{3}$ of the whole Astragal) and the other three parts goes to the Cimatium, whereof the List at the Top is one of those Parts

In the other fashion'd Pedestal the heighth of the Base is also divided into two parts, whereof the lowermost goes to the Plinth, and the other part being sub-divided into three parts, two of them make the Thorus, and the other part the List above it.

The

The whole heighth of the Capital of this faſhion'd Pedeſtal is divided into five parts, whereof the lowermoſt goes to the Aſtragal, (whoſe Liſt is $\frac{1}{5}$ of the whole,) the next two parts go to the O G; the two parts remaining, being ſub-divided into three parts, the two lowermoſt of them go to the Square, and the other to the *Cimatium*, whoſe Liſt is $\frac{1}{5}$ of the whole.

5 *Ionick.*] The whole heighth of this Column being divided into 14 parts, the height of its Pedeſtal, (according to *Vitruvius*) is three of thoſe Parts.

This Pedeſtal he alſo deſcribes of two different Forms, in each of which, the Baſe and the Capital are each $\frac{1}{5}$ of the whole heighth of the Pedeſtal.

In one of theſe faſhion'd Pedeſtals, he divides the heighth of the Baſe into three parts, whereof the lowermoſt goes to the *Plinth*, the next part goes to the *Scima-reverſa*, with its Liſt at top and bottom, which are each $\frac{1}{8}$ of the whole; the uppermoſt grand Diviſion being ſub-divided into two, the lowermoſt of them goes to the *Caſement*, or *Hollow*, with its Liſt at the top, which is one $\frac{1}{5}$ of the whole; the other part goes to the Thorus, and its Liſt above it, which Liſt is $\frac{1}{3}$ of the whole.

The Capital of this faſhion'd Pedeſtal, is divided into two parts, the lowermoſt of which goes to the *Scima-reverſa*, with its Liſt above and below it: whereof the lower Liſt is $\frac{1}{4}$ of the whole, and the upper Liſt $\frac{1}{8}$ of the Remainder. The other grand Diviſion being ſub-divided into three parts, the two lowermoſt of them go to the Square, and the other to the Cimatium, whereof its Liſt is $\frac{1}{3}$ part of the whole Cimatium.

In the other faſhion'd Pedeſtal, the Baſe is alſo divided into three parts, whereof the lowermoſt goes to the Plinth, the other two grand Diviſions being ſub-divided into five, the three lowermoſt of them go to the *Scima-reverſa*, and the Liſt under it, which Liſt is $\frac{1}{6}$ of the whole; the other two Diviſions being again ſub-divided into three parts, the two lowermoſt of them go to the Thorus and the remaining part to the Liſt above it.

The Capital of this faſhion'd Pedeſtal is divided into two parts, the lowermoſt of which being ſub divided into four parts, the lowermoſt of them goes to the Aſtragal; (whereof its Liſt is $\frac{1}{3}$ part,) the other three of thoſe Sub-diviſions go to the Scima-reverſa, and its Liſt above it, which Liſt is $\frac{1}{8}$ of the whole; the other grand Diviſion being ſub divided into three parts, the two lowermoſt of them go to the Square, and the other part to the Aſtragal, whoſe Liſt is $\frac{1}{3}$ of the whole

6. *Corinthian.*] The whole heighth of this Column being divided into 9 parts, the heighth of its Pedeſtal, (according to *Vitruvius*) is two of thoſe parts.

The whole heighth of the Baſe being divided into five parts,

parts, the two lowermoſt of 'em goes to the Plinth; the Remainder being ſub-divided into 4 Parts, the lowermoſt of 'em goes to the Thorus; the two next parts make the Scima-reverſa, and the Liſt below it, which Liſt is $\frac{1}{5}$ of the whole; the remaining part goes to the Aſtragal, whereof its Liſt is $\frac{1}{3}$ part.

The heighth of the Capital is divided into two parts, the lowermoſt of which being ſub-divided into four parts, the lowermoſt of thoſe go to the O G, the other three Sub-diviſions being again ſub-divided into two parts, the lowermoſt of thoſe goes to the Scotia, or Hollow, and the Liſt above it, (which Liſt is $\frac{1}{3}$ part of the whole;) the remaining part goes to the Boultin. The other grand Diviſion being ſub-divided into three parts, the two lowermoſt of them go to the Corona, and the remaining part to the Cimatum, whoſe Liſt is $\frac{1}{3}$ of the whole.

7. *Compoſite.*] The whole heighth of this Column being divided into 13 parts, the heighth of its Pedeſtal, (according to *Vitruvius*) is three of thoſe parts

The Baſe being divided into 7 parts, two of them go to the Plinth, one to the Thorus, two to the Scima-reverſa, one to the Scotia, and one to the Aſtragal; $\frac{1}{3}$ of the Aſtragal makes the Fillet above the Scotia

The Capital being divided into 7 parts, one of 'em goes to the Aſtragal, two to the Frieſe, one to the Boultin and Liſt under it, two to the Corona, and one to the Cimatium.

Peers, or *Piedroitis.*

A kind of ſquare Pillar, part whereof is hid within the Wall; the only thing wherein it differs from a Pilaſter; being this, that the latter has a Baſe and Capital, which the former has not.

2. *Scantlings, or Size.*] I find the Scantlings of *Stone-peers*, ſet down in an Act of Parliament for the Re-building of the City of *London*, after the late dreadful Fire, (which Scantlings were well conſider'd by able Workmen before they were reduced into an Act) to be as follows, *viz.* In the firſt ſort of Houſes, *Corner peers*, 18 Inches ſquare; *middle*, or *ſingle Peers*, 14 and 12 Inches; *double Peers* between Houſe and Houſe 14 and 18 Inches. In the 2 *d.* and 3 *d* ſort of Houſes, *Corner-peers* 2 Foot 6 Inches ſquare, *middle*, or *ſingle Peers* 18 Inches ſquare, *double Peer* between Houſe and Houſe, 14 and 19 Inches.

3. *Price*] Peers are ſometimes meaſur'd and rated by the Foot running Meaſure; but they are more commonly rated at ſo much *per* piece, dearer or cheaper, according to their Size, Goodneſs of the Stuff, and Curioſity in Workmanſhip.

A pair of Stone-peers with Seat-arches, 4 or 5 Foot wide, and 14 or 16 Foot high, may be worth 40 or 50 Pounds.

A pair of Ruſtick-peers of Stone, may be worth 10, 12, or 14 Pounds, according to their

their heighth and ſubſtance; Plain-peers, 8 or 10 Pounds; Revailed and Pilaſter-peers, from 10 to 14 Pounds a pair.

Pentadoron.

A kind of Bricks ſo call'd. See *Bricks*, N. III. §. II.

Periptere.

In the ancient Architecture, is a Building encompaſs'd round with Columns. The word comes from the *Greek*, *Peri* about, and *Pteron* a Wing.

Perrons.

Are Steps rais'd before the Doors of great Houſes.

Perſian Order.

The *Perſian Order* is that which has Figures of *Perſian Slaves* to ſupport the Entablement, inſtead of Columns, as the *Caryatick Order* has the Figures of Women ſerving for the ſame Purpoſe. The former Columns are uſually like robuſt Men, with long Beards; and ſuch Figures are much fitter to repreſent an unhappy Slavery, than thoſe of Women. The Character of Slavery is expreſt in theſe Figures, either by tying their Hands before or elſe behind their Backs. Columns of this kind may be very properly uſed in a Gallery of Arms, *&c* Princes Palaces; in which caſe they may be made Gigantick, and their Entablature Dorick

We wou'd not, however, inſinuate, that the Figures of Men are always Marks of Slavery: They are frequently uſed as Symbols of Vertues and Vices; of Joy, Strength, Valour, and even of fabulous Deities: As when they are made in the Figure of *Hercules* to ſignifie Strength, of *Mars*, to ſhew Valour, of *Mercury* to repreſent Dexterity, and of *Fauns* or *Satyrs*, to inſpire Mirths and Jollity.

Piazza's.

See *Architrave*, N. 2.

Piedroit.

In *Architecture* is a ſquare Pillar that is partly within the Wall. See *Peer*.

Pillar.

Is a kind of round Column diſengaged from any Wall, and made without any Proportion; being always either too maſſive or too ſlender: Such are the Pillars which ſupport the *Vaults* of *Gothick-Buildings*. In *Latin* they are called *Pilæ*.

Pilaſters.

1. *What.*] In Architecture are a kind of half Columns (ſtanding againſt a Wall) with Baſe and Capital, as Columns have; but differing from Columns in this, that Pilaſters are ſquare, but Columns are round.

2. *Of their Size, and Situation*] Pilaſters muſt not, (ſays Sir *Henry Wotton*) be too tall and ſlender, leaſt they reſemble Pillars; nor too dwarfiſh and groſs, leaſt they imitate Piles, or Peers of Bridges: Smoothneſs

neſs does not ſo naturally become them as a Ruſtick Superficies; for they aim more at State and Strength, than Elegancy.

In private Buildings they ought not to be narrower than one third, nor broader than two thirds of the Vacuity, or Inter-ſpace between Pilaſter and Pilaſter: But to thoſe that ſtand at the Corners, may be allow'd a little more Latitude by Diſcretion for ſtrength of the Angles.

In Theatres, and Amphitheatres, and ſuch weighty Works, *Palladio* obſerves them to have been as broad as the half, and now and then as the whole Vacuity, or Inter-ſpace. He noteth likewiſe, (and others conſent with him.) That their true Proportion ſhould be an exact Square; but (for leſſening of Expence, and enlarging of Room) they are commonly made narrower in Flank than in Front.

Their principal Grace conſiſts in half, or whole Pillars apply'd to 'em; in which caſe it is well noted by Authors, that the Columns may be allow'd ſomewhat above their ordinary length, becauſe they lean to ſo good Supporters. And thus much ſhall ſuffice at the preſent, touching the ſize and Situation of Pilaſters, which is a cheap, a ſtrong, and a noble kind of Structure.

3. *Price*] Theſe are ſometimes meaſur'd and rated by the Foot running Meaſure; but they are more commonly valu'd at ſo much *per* piece, according to the ſize, goodneſs of the Materials, and Curioſity in Workmanſhip.

Pitch.

By this Term Architects underſtand the Angle a Gable-end (and conſequently the whole Roof of a Building) is ſet to. If the length of each Rafter be $\frac{3}{4}$ of the breadth of the Building, then that Roof is ſaid to be *true Pitch*; if the Rafters are longer, 'tis ſaid to be a *high*, or *ſharp pitch'd* Roof; if ſhorter, (which it ſeldom is,) then 'tis ſaid to be a *low*, or *flat pitch'd* Roof.

Pitching.

The ſame as *Paving*. See *Paving*.

Pins for Tiles.

What they are every one knows; they ought to be made of Heart-oak, and to every 1000 of Tiles is uſually allow'd two Gallons of Tile-pins, from 3 *d* to 6 *d*. the Gallon, ſays Mr. *Leybourn* I know not how he reckons, but I am ſure 1000 of Tiles requires but 1000 Pins, which ſome Workmen in *Suſſex* tell me they reckon but 2 *d* or 3 *d*. for they tell me they ſell their Pins for 6 *d*. *per* Gallon; and that they uſe about a Gallon of Pins to a Square and a half of Healing.

Pin-

Pinning.

Of Tiles] Some Workmen in *Suſſex* tell me, That they commonly reckon 8 *d. per* Thouſand, for pinning of Tiles, and finding Pins But for the Workmanſhip only 6 *d. per* Thouſand.

Pilaſter-bricks.

See *Bricks*, N. III. §. 13.

Pipes of Lead.

See *Lead*, N. 9.

Place-bricks.

See *Bricks*, N. III. § 12.

Plain cornice.

See *Cornice*.

Plain-tiles.

See *Tiles*, N. III.

Planchier.

The Ornament to which the Cornice is faſten'd.

Plaiſtering.

1. *Of Walls*.] Some Maſons in *Suſſex* tell me, That for Lathing and Plaiſtering of Walls with Lome on both ſides, they have 3 *d. per* Yard; but if it be done with white Lime, and Hair-mortar on both ſides, then they have 4 *d. per* Yard.

I am informed, That at *Tunbridge-wells* the Maſons will do Plaiſtering of Walls (where they plaiſter over all the Timber) and Ceilings for 2 *s.* 10 *d. per* Square. I know a Gentleman that told me, He had ſuch Work done for 2 *s.* 6 *d. per* Square.

2. *Of Ceilings*.] For Ceilings, our Maſons in *Suſſex*, have (for Lathing, Plaſtering, and Finiſhing) 4 *d. per* Yard. In ſome Countries they make their Ceilings with Reed, Lime and Hair; for which the Workmanſhip is worth 3 *d per* Yard: But if the Workman find all Materials, 'tis worth 5 *d.* or 6 *d per* Yard.

3. *With rough Mortar, or Rough-caſt.*] In ſome parts of *Kent* they commonly *Rough-caſt*, (as they call it) upon old Lome-walls, that is, they give them one Coat (upon the Lome) of *Rough-mortar*, or *Rough caſt*, as they call it, tho' it be commonly ſtruck ſmooth like Lime and Hair. For this Work they have 3 half-pence *per* Yard, only Workmanſhip: But if the Wall be new and Lathed, and Plaſter'd with Lome on both ſides, and a Coat of Rough-mortar on the out-ſide then they have 4 *d. per* Yard only Workmanſhip. But if the Rough-caſting be wrought in Flouriſhes, then they have 8 *d. per* Yard, only Workmanſhip. But if the Workman find all Materials, 'tis worth from 1 *s.* to 3 *s per* Yard, according to the variety and goodneſs of the Work.

4. *On Laths in imitation of Brick*.] I know a Houſe that is plaiſter'd in imitation of Brick-work, the Mortar was made of Powder of Bricks, ſharp Sand, Lime,

Lime, and ſome Red oker. This Houſe has been done this 20 Years, and yet looks very well, and paſſes for a Brick-houſe with common Paſſengers, tho' it be only Timber plaiſter'd over.

Some Workmen tell me, That they have 1 *s. per* Yard for ſuch Work only Workmanſhip.

5. *Of Floors.*] Plaiſter-Floors running, (ſays Mr *Wing*) the Workman finding all, is worth 1 *s.* 4 *d per* Yard, but the working part only is worth 4 *d.* 5 *d.* or 6 *d. per* Yard Plaiſter at the Pits may be had for 4 *s.* or 4 *s.* 6 *d. per* Load, *viz.* 40 C. weight, which will do about 40 Yards of Flooring.

6. *Of White-waſhing.*] White-waſhing with Size upon plaiſter'd Walls, is commonly reckon'd at 2 *d. per* Yard.

7. *Of Meaſuring.*] This kind of Work is commonly done by the Yard ſquare, as *Paving*, which ſee, Numb. 12. But *Note*, That in meaſuring of Partitions, if the Workman find Materials, the Doors and Windows are meaſur'd by themſelves, and deducted from the whole; as is alſo $\frac{1}{7}$ part (of the reſt) for the Quarters in rendring Work: But if the Workman do not find Materials, there is commonly no Allowance made for them the Trouble of cutting and fitting the Laths being equivalent to the void Space left for the Doors and Windows Neither (in Caſe of Workmanſhip only) is there to be any Allowance made (in rendring) for the Quarters, Braces, or Interties, the Work being as much as (if not more than) if it were all plain. See more of Plaiſtering, in the word *Pargetting*.

Platebands.

The Liſts, or Fillets between the Fluteings of the Ionick, Corinthian, and Compoſite Columns. They are (each) in breadth a quarter of the Flute. *Plate bands* are alſo a ſquare Mould ſet at the end of an Architrave of the *Dorick Order.* Plat-band, *Perrault* ſays, is a ſquare Moulding, having leſs Projecture than Height Such are the Faces of an Architrave, and the Plat band of the Modillions of a Cornice. The Plat-band in *Vitruvius*, is ſignified by the words *Faſcia*, *Tænia* and *Corſa*.

Platform.

1. *What*] This word in Architecture is ſometimes us'd to ſignifie the Ichnography, or Draught of the Ground-plot of a Houſe; but more commonly for a broad ſmooth, and open Walk upon the top of any Building. *Platform* is alſo a Row of Beams which ſupport the Timber work of any Roof, and lie on the top of a Wall, where the Entablature ought to be raiſed.

2. *Of Covering with Lead*] See *Lead*, N. 7.

Plat-fond.

A *French* word for the Ceiling or Roof of a Chamber, or other Room, &c. The ſame as *Soffit*

Plaſti.

Plastique, or *Plastick-Art.*

The *Plastique-art,* is a Branch of Architecture that is not only comprehended under Sculpture, but is indeed very Sculpture it self; but with this difference; that the Plaisterer (by his *Plastique art*) makes Figures by Addition, but the Carver by Subtraction; whereupon *Michael Angelo* was wont to say, (somewhat pleasantly) That Sculpture was nothing but a Purgation of Superfluities: For take away from a piece of VVood, or Stone, all that is superfluous, and the remainder is the intended Figure.

Of this *Plastique-Art,* the chief use with us is in the graceful *fretting* of Roofs, (commonly known amongst us by the Name of Fret-work;) but the *Italians* apply it to the Mantlings of Chimneys with great Figures; a cheap piece of Magnificence, and as durable almost within Doors, as harder Forms in the Weather.

Plinth.

Is derived from the *Greek, Plinthos,* (a square Brick) and is in Architecture, a square Piece or Table, under the *Mouldings* of the *Bases* of *Columns* and *Pedestals.*

Plumbery.

An Art belonging to Architecture, it being the Art of working in Lead.

Poitral.

See *Architrave.*

Porphyry.

A fine reddish Marble, streaked with divers Colours. *Pliny,* (*in Hist. Nat. Lib.* 36. *Cap.* 7.) says, This kind of Marble comes out of *Egypt,* where there are large Quarries of it.

Portico.

See *Architrave,* N. 2.

Portland-stone.

A Stone-cutter in *London,* tells me, That they usually sell Slabs of *Portland stone,* (ready polished for Chimney-foot-paces) for 1 *s.* 8 *d. per* Foot superficial. 'Tis a Stone much us'd in Building, and much softer and whiter than *Purbeck.*

Port-nails.

See *Nails,* N. 12.

Portal.

An ancient Term in Architecture, I say an ancient Term, for the thing signifi'd by it is grown out of fashion; it was us'd to signifie a little square corner of a Room, shifted off from the rest of the Room by the Wainscot The word seems to come from the *French, Portail* a *Gate* or *Entrance*; because through it they enter into the Room.

Portail.

The Decoration of the Face or Front of a Church, call'd alſo Frontiſpiece. There are ſome Gothick, as that of *Weſtminſter-Abbey*, &c. and others *Antique*, as in many of the neweſt Churches. The word alſo, upon other occaſions, does ſignifie the principal Gate of a Palace, Caſtle, Pleaſure-houſe and the like.

Portico.

A kind of Gallery rais'd upon Arches, where People walk under Shelter. It has ſometimes a *Soffit*, or Ceiling, but is more commonly vaulted.

Poſts.

1. *What*] Pretty big pieces of Timber, ſtanding upright in a Houſe, *&c.*

2. *Principal*] In Architecture, are the Corner-poſts of a Houſe, *viz.*

3. *Prick poſts*] The Poſts that are Fram'd into Breſſummers, between principal Poſts, for the ſtrengthning the Carcaſs of the Houſe.

4. *Of Preſerving*] An ingenious Gentleman one Mr. *Walter Burrel*, Eſq; of *Cuckfield* in *Suſſex*, deceaſed, uſed to burn (to a Coal on the out-ſide) the ends of all the Poſts which he ſet in the Ground; whereby they will continue a long time without rotting, which otherwiſe would ſuddenly decay.

Poſt and Rail.

See *Fencing*, N. 2. and *Paleing*, N. 1, 2.

Pound-nails.

See *Nails*, N. 13.

Prick-poſts.

See *Poſts*, N. 3.

Priming.

See *Painting*, N. 1.

Principal.

1. *Poſts.*] See *Poſts*, N. 2.
2. *Rafters*] See *Rafters.*

Prices.

Of Work and Materials, ſee the Particulars, that you would know the Price of in their proper Place of the Alphabet.

Profile.

In Architecture is a Draught repreſenting the Breadth, Depth, and Height of a Building, or Fortification, but not the Length; which properly belongs to a *Plan* or *Ground-plot*, ſo that it is in a manner the ſame with a Proſpect of a Place or Building view'd ſideways, and expreſs'd according to the Rules of Perſpective. Alſo, but I think improperly, 'tis taken for the *Contour*, or Out-line of any Member in Architecture, as that of a *Baſe*, a *Cornice*, or the like.

Pſeudo.

Pseudo-Diptere.

See *Diptere.*

Pudlays.

Pieces of Stuff to do the Office of Leavers, or Hand-spikes.

Pulvinata.

A Frieze swelling like a Pillow.

Punchins.

1. *What*] Short pieces of Timber placed under some considerable weight to support it They commonly stand (upright) between the Posts; they are shorter (and slighter) than either Principal posts, or Prick posts Those that stand on each side of a Door, are call'd *Door-punchins.*

2. *Price.*] Carpenters commonly reckon 1 *d.* or three half-pence *per* Foot for putting in of new Punchins.

Purlins.

1. *What*] Those pieces of Timber that lie a-cross the Rafters, on the in side, to keep them from sinking in in the middle of their length.

2. *Size*] By the ACT of Rebuilding the City of *London* all *Purlins* in length from 15 Foot 6 Inches, to 18 Foot 6 Inches, ought to be in their Square 9 Inches, and 8 Inches. And all in length from 18 Foot 6 Inches, to 21 Foot 6 Inches, ought to be in their Square 12 Inches, and 9 Inches.

Purbeck stone.

1. *What*] 'Tis a hard greyish Stone, almost like *Sussex Petties.* They are us'd for Pavements.

2. *Price*] A Stone-cutter in *London* tells me, That they commonly sell Purbeck-slabs, (ready polished for Chimney-foot-paces) for 2 *s per* Foot. And Purbeck paveing of promiscuous Sizes only hew'd and squar'd, they sell for 7 *d per* Foot Also *Mitchels* they value at about 1 *s.* 10 *d. per* Foot. See *Mitchels.*

Putlogs.

Pieces of Timber, or short Poles, (about 7 Foot long,) us'd by Masons in Building of Scaffolds to work on. The Putlogs are these pieces which lie horizontal to the Building, one end lying into it, and the other end resting on the *Ledgers*; which are those pieces that lie parallel to the side of the Building.

Pycnostyle.

This Term is us'd when the Columns are ranged so close to one another, that the *Intercolumnation* does not exceed a *Diameter* and a half.

Pyramid.

From the *Greek, Pyr,* Fire or Flame, this being pointed like that. It is a solid Body, whose Base is either Square, Triangular, or Polygonous, and which from

from that *Base* diminishes continually to its *Vertex* or *Top*.

Pyling.

The Ground for Foundations, See *Foundations*, N. 2. §. 5.

Quarry.

1. OF *Stone*] a Place whence Stones are dug out.

2. *Of Glass*] A piece of Glass cut in a Diamond-form. Quarries of Glass are of two kinds, *viz.* square, and long; and these again are of different sizes, as 8's, 10's, 12's, 15's, 18's, and 20's, [that is, 8 Quarries of 8's make a Foot of Glass, and so does 10 Quarries of 10's, 12 of 12's, *&c*] But all Quarries, (of what size soever) are cut of one sort of Angle for the Square Quarries, and another for the long Quarries: The Acute Angle of the square Quarries being 77 Degrees, and 19 Minutes; and the Acute Angle of the long Quarries 67 Degrees, and 22 Minutes. See more, *Glazing*, N.

Quarters.

In Architecture, all those slight upright pieces between the Punchins and Posts, (which serve to Lath upon) are call'd *Quarters.* They are of two kinds, *single* and *double*: Single Quarters are sawn Stuff, two Inches thick, and four Inches broad. The double Quarters are sawn to four Inches square. 'Tis a Rule in Architecture, that no Quarters be placed at greater distance than 14 Inches.

Quartering.

In Architecture, signifies the putting in of Quarters. Sometimes 'tis us'd to signifie the Quarters themselves.

Quarter-head.

See *Brads*, N. 6, 7.

Quarter-round.

By this Name the Workmen call any *Moulding* whose *Contour* is a Circle, or approaching to a Circle; using this Term where ever the Architects use that of Egg, or Ovolo. See *Echinus*.

Quirk,

In Architecture, signifies a piece taken out of any regular Ground-plot, or Floor. As if the Ground-plot were a Square, or an Oblong, and a piece be taken out of one corner of it, for a Court, or Yard, that piece so taken out is call'd a *Quirk*.

Quins, or *Quoins*.

The corners of Brick, or Stone walls. Also the Stones in the corners of Brick-Buildings. If these Stones stick without the Brick-work, (their edges being cypher'd off) they are call'd Rustick-quoins. The Rustick-quoins, at two Foot, one Face, and one Foot the other, are valu'd from 1 *s.* to

1 *s.*

1 *s.* 4 *d. per* Quoin, Stone and Workmanship.

Quadrels.

A sort of artificial Stones, (so call'd from their Form, they being square) made of a chalky, whitish and pliable Earth, and dry'd in the Shade. They were two Years in drying, and were much us'd by ancient *Italian* Architects.

Rafters.

1. WHAT] Rafters are those pieces of Timber, which (standing by pairs on the Reson,) meet in an Angle at the top, and compose the Roof of a Building.

2. *Scantling, or Size*] In an ACT of Parliament for Re-building the City of *London*, the following *Scantlings*, (which were well consulted by able Workmen, before they were reduced to an Act) are set down, as fitted for all Edifices, great or small, *viz.*

3. *Distance*] 'Tis a Rule in Architecture, that no Rafters be laid at greater distance from each other than twelve Inches.

Principal Rafters, in length	from	to	must be broad at the Foot.	Top.	and thick.
	F. In	F. In	Inches.	Inches.	Inches.
	12. 6	14 6	8	5	6
	14. 6	18 6	9	7	7
	18. 6	21. 6	10	8	8
	21 6	24. 6	12	9	8½
	24. 6	26. 6	13	9	9

Single Rafters in length { 6 Foot 6 Inches, / 8 Foot, / 9 Foot, } must have in their Square { 4, and 3 / 4½, and 3¾ / 5, and 4 } Inches.

Rag-

Rag-ſtone.

See *Paving*, N. 5.

Rails,

In Architecture, are us'd in various Senſes; as for thoſe pices that lie Horizontally *between* the *Pannels* of Wainſcot, and *over*, and *under* them. Alſo for thoſe pieces that lie over, and under Balliſters, in Balconies, Stairs-caſes, *&c.* Alſo for thoſe pieces of Timber that lie Horizontally from Poſt to Poſt, in Fecning with Pales, or without.

Rails and Balliſters.

1. *Price of making.*] Mr. *Wing* tells us, That Rails and Balliſters on Balconies, or about the Platform of great Houſes, are worth (only Workmonſhip) 4 *s. per* Yard, running Meaſure.

2. *Painted, of Meaſuring*] See *Painting*, N. 2.

Raiſer,

A Board lie on edge under the fore-ſide of a Step.

Raiſing-pieces.

Are pieces that lie under the Beams, upon Brick or Timber by the ſide of the Houſe.

Random-pavement.

See *Paving*, N. 11.

Range.

The ſide of any Work that runs ſtraight, without breaking into Angles, is ſaid to *Range*, or *Run-range*: Thus the Rails and Pannels of one ſtraight ſide of Wainſcotting, is ſaid to *Run-range*. See more in the Note in *Glazing*, N. 3. §. 2.

Redſear.

See *Iron*, N. 2.

Regula,

As *Orlo*.

Rendering.

See *Pargeting*, N. 2.

Repoſitory.

A Store-houſe, or Place to keep things in; more peculiarly by Architects 'tis us'd to ſignifie ſuch Places as are built for the laying up of Rarities, either in Painting, or other Arts.

Return.

The Side that falls away from the fore-ſide of any ſtraight Work, is called the *Return*.

Reul.

See *Parapet*.

Ribbing-nails.

See *Nails*, N. 14.

Rides,

Rides,

Or Hinges for Doors, *&c.* are commonly ſold for 4 *d per* Pound. See Iron, Numb. 4.

Ridge.

The meeting of the Rafters on the top of the Houſe, is call'd the Ridge.

Ridge-tiles.

See *Tiles,* Num. 4.

Rigate-ſtones.

See *Fire-ſtone.*

Rondel.

See *Capital,* Numb. 2.

Roofing.

1. *Price*] Roofing in ordinary Buildings is worth 7 or 8 *s. per* Square, but in great Buildings, 10 or 11 *s per* Square. See Pl. *Framing,* Numb. 5.

2. *Of Meaſuring*] Roofing is commonly meaſur'd by the Square, as Flooring. See *Flooring,* Numb. 5.

Roof.

The Covering of a Houſe; but the Word is us'd in Carpentry, for the Timber-work of the Covering.

Roſe-nails.

See *Nail*, Numb. 15.

Roſe.

Roſe is an Ornament cut in the Spaces which are between the *Modillions* under the Plat-fonds of Cornices, and in the middle of each Face of the *Abacus,* in the *Corinthian* and *Compoſite* Capitals.

Rother-nails.

See *Nails,* Numb. 16.

Rough-Stone.

As *Rag ſtone.*

Rough-caſting.

See *Plaiſtering* Numb. 3.

Rough-mortar.

In many Places of *Kent,* where they rough-caſt their Houſes, they make their Mortar (which they call *Rough-mortar)* of a ſort of *Sand,* which when 'tis mixt with the *Lime,* makes it look as red as Blood; but to theſe they put Powder of *Cinders* which changes it to a kind of blueiſh Colour; See Pl. *Mortar.* Num. 11.

Round-heads.

See *Nails,* Numb. 17.

Ruſtick.

A Manner of Building, rather in Imitation of Nature, than according to the Rules of Art. The Columns are encompaſſed

compaſſed with frequent Cinctures.

Ruſtick peers.

See *Peers*, Numb. 3.

Ruſtick-Quoins.

See *Quoins.*

Sagitta.

IN Italian, *Saetta* (an Arrow) ſignifies what we call the Key piece of an Arch.

Salon.

Is a kind of Hall, in the Middle of a Houſe, or at the head of a Gallery, or a large Apartment, which ought to have a Simmetry on all its Sides; and as its Height uſually takes in two *Stories* with two Rows of Windows, the bottom of its Plat-fond ought to be arch'd; as is practiſed in ſome of the Palaces in *Italy.*

Samel, or *Sandel Bricks.*

See *Bricks*, Num. III. §. 14

Sand.

1. *Kinds.*] What it is, every one knows. Its Uſe (in *Architecture*) is in making of Mortar. There are three Sorts of Sand, *viz* Pit-ſand, River-ſand, and Sea-ſand: Pit ſand is of all the beſt, and of all Pit-ſand that which is whiteſt, is (by long Experience found to be) the worſt. Of all River-ſand, that which is found in the Falls of Water is the beſt, becauſe it is moſt purged. The Sea-ſand is the worſt of all.

The Pit-ſand, becauſe it is fat and tough, is therefore us'd in Walls and Vaults. The River-ſand is very good for rough-caſting of Walls.

All Sand is good in its kind, if being ſqueez'd and handl'd it crackles; and if being put upon a white Cloath, it neither ſtains nor makes it foul.

That Sand is bad, which mingl'd with Water, makes it dirty and muddy, and which has been a long time in the Air; becauſe it will retain much Earth and rotten Humour: And therefore ſome Maſons will waſh their Sand before they uſe it.

2. *Price.*] Sand, at *London*, is commonly ſold for 3 *s per* Load, 36 Buſhels to the Load. In ſome parts of *Suſſex* 'tis ſold for 1 *s* 6 *d. per* Load, at 12 Buſhels to the Load. In other parts of *Suſſex* 'tis ſold at 2 *s.* 6 *d. per.* Load, at 18 Buſhels to the Load.

Saffita, or rather *Soffita*, or *Soffit.*

An *Italian* Word (from *Subfixum*, in *Latin*) a ſort of Cieling. In ordinary Buildings, it is taken for the Boards over the tops of Windows, oppoſite to the Window-boards at the bottom. In great Edifices it ſignifies the Ceiling or Wainſcot of any Apartment form'd by Croſs-Beams, or flying

flying Cornices, and having the ſquare Pannels of its Compartments enrich'd with Sculpture, Painting and Guilding, as we may obſerve in the Baſiliques and Palaces of *Italy*. This is what the Latins mean by *Lacunar*, and *Laquear*, with only this Difference, That *Lacunar* ſignifies any Soffit, whoſe Pannels are call'd *Lacus*, Lakes; and *Laquear* is uſed to expreſs that form'd by Compartments, inter-mingled with Plat-bands, in manner of a Nooſe, in Latin *Laqueus*. The bottom of the Drip is by ſome call'd Soffit

Saſh-lights.

See *Painting*, Num. 12.

Saſh frames.

See *Painting*, Num. 13.

Sawyers-work.

See ——

Sawing.

1. *By the Hundred.*] Sawyers do moſt commonly work by the Hundred, that is by the Hundred Superficial Feet, (how meaſur'd, ſee Numb. 12.) For which they have various Prices, not only in different Places, but alſo for different kinds of Timber; as may be ſeen in the following Numbers.

2. *Of Oak*]. The ſawing of Oak is in ſome Places 2 *s*. 8 *d*. in others 3 *s*. in others 3 *s*. 6 *d*. the Hundred.

3. *Of Elm.*] The ſawing of Elm, is in ſome Places 3 *s*. the Hundred, commonly about the Price of Oak.

4. *Of Aſh, and Beech*] The ſawing of Aſh, and Beech, is generally worth 6 *d*. in the Hundred more than Oak, or Elm. In ſome Places 'tis 3 *s*. in others 3 *s*. 6 *d*. in others 4 *s per* Hundred.

5. *By the Load.*] Sawyers do ſometimes work by the Load, *viz*. So much for cutting out a Load, (or 50 Foot) of Timber; the Price various, according as what the Timber is cut to. But the common Price is for Ship-planks of two Inches thick, 10 *s*. the Load. And for Building

Timber	large Size, 6 *s*. or 6 *s*. 6 *d*.	the Load.
	large Size 7 *s*.	
	ſmall Size, 7 *s*. 6 *d*. or 8 *s*.	

An old experienced Sawyer tells me, That *ſawing by the Load* is commonly thus agreed for, *viz*. They have all their Sizes ſet down which they are to cut; and they will cut none ſmaller, neither will they ſlab any, unleſs they are paid for it by Meaſure, over and above what they are to have by the Load. They never cut any thing leſs than Rafters, which are about four and five Inches, and which is generally the ſmalleſt Timber in a Frame, except Quarters, and Window-

Window-ſtuff, which they generally cut by the Hundred. If the Carpenter will have any Pieces clear'd by Slabbing, after they have cut them off to their Size, they will (alſo) be paid by Meaſure for it.

They generally prick off their Sizes from the outer Edges, and what is left in the middle they lay by till they can fit it to ſome other Size, when it is wanted.

This Sawyer tells me, That the Carpenter has a great deal of hewing of out-ſide pieces, when 'tis ſaw'd by the Load: He alſo ſays, That ſawing by the Load is commonly good Work for the Sawyer. The Truth is, (as he confeſſes) it waſteth a great deal of Timber, it being hew'd away to Chucks.

The loweſt Rate that any (about us in *Suſſex*) is cut for by the Load is 6 *s.* and then it muſt be very large Scantlings, or elſe they will have 7 *s.* which indeed is the common Price for ſawing a good large ſiz'd Timber-frame. But if the Timber-frame be ſmall and ſlight, as they commonly build at *Tunbridge-wells* they will have 7 *s.* 6 *d.* or 8 *s. per* Load

6. *Of Ship planks.*] An old experienced Sawyer tells me. That they ſometimes cut Ship-planks by the Load for 10 *s. per* Load: But then the Size of their Planks are (if I remember right) two Inches in Thickneſs.

Sometimes (he ſays) they ſaw them by the Hundred, and then they have 3 *s per* Hundred, and 2 *d.* for petting of every Log. But if there be nothing allow'd for petting the Logs, then they reckon ſo many Carves as there are Pieces, which is one Carf more than there really is.

He alſo ſays, that they commonly cut Planks from 1 ½ Inch, to three Inches thick; but they are never paid for breaking Work, till it comes to a two Foot Carf

7. *Of Compaſs-work*] For ſawing of Compaſs-work, (as Mill wheels, Furnace wheels, Forge-wheels, Rafters for Compaſs-roofs, *&c*) Some Sawyers tell me they have 2 *d. per* Foot.

8. *Of Bevil-work*] Some Sawyers tell me, that in ſawing of Bevil-work, (as Hips, and Sleepers, *&c* Poſts, *&c.* in Bevil-frames; as alſo Poſts or Punchins in Polygonal Turrets, *&c.* Alſo Cant-rails, *&c.*) they work by the hundred, but they always reckon a Carf and half; that is, they reckon ½ as many more Feet of Sawing as there is.

9. *Of Furnace-bellows.*] Theſe they cut by the Foot, Lineal Meaſure, at 1 *s. per* Foot.

10. *Forge bellows.*] Theſe they cut by the Foot Lineal Meaſure, at 4 *d.* or 6 *d. per* Foot.

11. *Of Ground-guts*] Theſe they alſo cut by the Foot, Lineal Meaſure; if ſmall, for 1 *d per* Foot, but if 15 Inches deep, then 1 ½ *d.* if 18 Inches, 2 *d. per* Foot.

12. *Of Meaſuring*] Sawyers, (when they work by Meaſure) generally Meaſure their Work by the Foot Superficial.

There

There is no difficulty in taking the Dimensions; for they reckon the Depth of the Carf for the Breadth; and the length for the length. The breadth, (or depth) and length of a Carf being taken, and multiply'd together, (as is taught in *Cross-multiplication*, Num. 2.) gives the Area, or superficial Content of that Carf.

Having thus found the Number of Feet in one Carf; multiply it by the Number of Carves of the same depth and length; and so you have the Area of them all.

Note, (1.) That having thus cast up their whole Work in Feet; they are paid for it by the *Hundred*, [that is 100 Feet] at various Rates, of which see above, Num. 1. 2. 3. and 4.

(2.) That if the Carf be but 6 Inches (or be less than 6 Inches) in depth, they have a Custom to be paid for Carf and half; (as they phrase it) that is, for half so much more as it comes to by Measure. The reason they urge for this Custom is, their Trouble in often linding, and removing their Timbers.

3. That for *Breaking-work*, [that is cutting a Log through the middle,] and *Slabbing*, [that is, cutting off the out-side Pieces,] if the Carf be more than 12, or 13 Inches deep, they are paid by the Foot, Lineal Measure, at various Prizes, according to the different Depth of the Carf, *viz.* at

Inches deep.	per Foot.
15	1*d.*
18	1½
20	2.
22	2½
24	3
26	3½
28	4
30	4½
32	5.
34	5½
36	6

(4) That in some Places 'tis the Custom to allow the Sawyer but one *Breaking-carf* in a Log, tho' there be never so many deep Carves in the Log: But some Sawyers claim it as a Custom to have half *Breaking-work*, and the other half *Hundred-work*; as if they have four deep Carves, then they will have two Breaking-works, and the other two Hundred-work.

Scantling.

The Size that any Timber is design'd to be cut to.

Scenography.

(From the Greek *Skene*, a Tent, or Tabernacle, and *Grapho*, to write or describe,) is a Model, or Description of the Front and Sides of a House, or the Art of rightly contriving Draughts in Architecture.

Scheam.

Scheam.

See *Arches*, Numb. 6.

Scima,

As *Cima*.

Scima-tum,-tium.

As *Cimatum*.

Scima-recta.

See *Capital*, Numb. 3.

Scima-reversa.

An O-G, with the hollow downwards, See O-G.

Scotia.

A Member of Architecture 'Tis a hollow like a Semicircle. It is particularly plac'd in the Bases of Columns between the Thorus and the Astragal, and sometimes 'tis put under the Drip, in the Cornice of the Dorick Order. *Scotia*, from the Greek *Skotos*, Darkness, signifies, says M. *Perrault*, a hollow, *obscure* Moulding, between the *Tores* of the Base of a Column. It is also call'd by some, the Concave Member, and by others *Trochilus*, from the Greek *Trochilos*, a Pulley, which it resembles, as to Form.

Scribe,

A Term us'd by Joyners, when they are to fit one side of a piece of Stuff against the side of some other piece of Stuff and the side of the Piece they are to fit it to is not regular: To make these two pieces of Stuff joyn close together all the way, they *scribe* it, (as they phrase it) thus; they lay the piece of Stuff (they intend to scribe) close against the other piece of Stuff they intend to scribe to, and open their Compasses to the widest Distance, these two Pieces of Stuff bear off each other: Then (the Compasses moving stiff in their Joynt, they bear the Point of one of their Shanks against the side they intend to scribe to, and with the Point of the other Shank they draw a Line upon the Stuff to be scribed; thus the Points of the Compasses remaining unmoved and your Hand carried evenly along by the side of the piece to be scribed to, that Line scribed upon the piece intended to be scribed, shall be parallel to the irregular side intended to be scribed to: And if you work away your Stuff exactly to that Line, when those Pieces are put together, they shall seem a Joynt.

Sculpture,

The Art of carving in Wood, or Stone. See *Plastick Art*.

Scupper nails.

See *Nails*. Numb. 18.

Seasoning of Timber.

See *Timber*, Num. IV.

Scwer,

Sewers,

In *Architecture*, are Conduits or Conveyances for the Suillage and Filth of a Houſe; which, how baſe ſoever they are in Uſe, yet for the Health of the Inhabitants, they are as neceſſary and conſiderable, as (perhaps) any thing about a Houſe.

Concerning theſe, I find in our Authors, this Counſel, That Art ſhould imitate Nature in thoſe ignoble Conveyances, and ſeparate them from ſight, (where there wants a running Water) into the moſt remote, and loweſt, and thickeſt part of the Foundation, with ſecret Vents paſſing up through the Walls (like a Tunnel) to the wide Air aloft; which all *Italian* Artizans commend for the diſcharge of noiſome Vapours, though elſewhere, (to my Knowledge) little practis'd.

Sells.

1. *What.*] Sells in *Architecture* are of two kinds, *viz. Groundſells*, [which are the loweſt pieces of Timber in a Timber-building, on which the whole Superſtructure is erected,] and *Window ſells*, (ſometimes call'd Window-ſoils,) which are the bottom pieces in a Window-frame

2. *The Price of putting in*] The putting in of Groundſells in a Houſe, is commonly rated at 3 *d*. or 4 *d per* Foot, only Workmanſhip.

Setting.

See *Pitching*.

Setting of Fronts.

See *Fronts*, Numb. 2.

Shaky, or *Shaken.*

Such Stuff as is crack'd, either with the Heat of the Sun, or the Drought of the Wind, is call'd ſhaky, or ſhaken Stuff.

Sharp-nails.

See *Nails*, Num. 19.

Sheathing-nails.

See *Nails*, Num. 20.

Sheet-lead.

See *Lead*, Num. 3, 4, 5, 6, 7.

Shides.

The ſame as ——

Shingles.

1. *What.*] Theſe are ſmall pieces of Wood, or quarter'd Oaken-boards, *ſaw'd* to a certain Scantling; but they are more uſually *cleft* to about an Inch thick at one end, and made like Wedges about 4 or 5 Inches broad, and 8 or 9, (and in ſome places 12) Inches long. They are us'd to cover Houſes with, (but more commonly Churches and Steeples) inſtead of Tiles, or Slates.

This

This kind of Covering is very chargeable, and ſeldom us'd but in covering the Roofs of Churches, and Pyramidal Steeples. Nevertheleſs, where Tiles are ſcarce, and you would have your Houſe but lightly cover'd; Shingles are to be prefer'd before Thatch; and if they are made of good Oak, and cleft out, (not ſaw'd) and then well ſeaſon'd in the Water and Sun, they become a ſure light, and durable Co vering.

2. *Price of.*] Some Workmen tell me, That Shingles are ſometimes ſold for 20 *s. per* Thouſand, but then are very bad Ware; for if they are good they are worth 30 *s. per* Thouſand; nay, they tell me, They have known 40 *s. per* Thouſand given for Shingles to lay upon Steeples; for thoſe that lie ſo high, anu hang ſo perpendicular, ought to be of the beſt ſort.

3. *Price of Cleaving and Making.*] Several Workmen tell me, That the common price of Cleaving and making of Shingles, is 10 *s. per* Thouſand.

4. *How many made of a Tun of Timber.*] Some Workmen tell me, That a Tun of Timber will make 3000 Shingles.

5. *Of laying on.*] For covering with theſe, the Building muſt be firſt well cover'd all over with Boards; which being done, the Shingles are faſten'd to thoſe Boards; with 4 *d* 5 *d.* or 6 *d.* Nails, in every Courſe, at a certain Gage, *viz.* At 3 ½ Inches, or 4 Inches, from under one another; for they commonly make 3 Waters, (as they phraſe it,) that is, they commonly hang 3 Shingles in heighth, in the length of one; ſo that if the Shingles are 12 Inches long, they are laid at 4 Inches Gage.

In breaking of Joynt, they do not obſerve to make one Joint over the middle of the Shingle below; but they ſometimes break Joynt an Inch, an Inch and a half, ore two Inches, according to the breadth of the Shangles; for they (eſpecially if they are cleft) are not all exactly of a ſize.

6. *Price of laying on.*] For laying them on upon Spire ſteeples, where the Work is high and troubleſome, they have (commonly) 20 *s. per* Thouſand; but on low Work, (as upon Houſes and the like,) they will cleave, and make, and lay them on for that Money: Or if they only lay them on upon Houſes, they will do it for 10 *s. per* Thouſand Some Workmen tell me, That for dreſſing and laying on of Shingles upon Churches and Steeples, they have (commonly) 18 *s. per* Thouſand.

7. *Price of dreſſing old ones.*] For dreſſing of old Shingles, [that is new hewing them, and cutting of the ragged lower ends] Workmen tell me they have, (I think) 6 *s. per* Thouſand.

8. *How many will cover a Square.*] If the Shingles are four Inches broad, and laid at four Inches Gage, 81 Shingles will cover a ſquare Yard; and conſequently 900 will cover a Square, (or 100 Superficial Feet) of Healing: But, becauſe

caufe Shingles feldom hold to be all four Inches broad; therefore they commonly allow 1000 to a Square, and of Nails as many.

Shingling,

The laying on of Shingles, See *Shingles.*

Shinlog.

See *Bricks,* N. 5.

Shop-windows.

Thefe may be afforded at the fame rate as plain or batton'd-doors, befides the Iron-work, as Bolts, Staples, Hinges, Locks, Keys, Latches, Chains, *&c.* See *Doors,* N. 4.

Shreadings.

The fame as *Furrings.*

Silery.

As *Cilery.*

Skewback.

See Arches, N. 7.

Skirting-boards.

The narrow Boards fitted round the under-fide of Wainfcot againft the Floor.

Slabs.

The out-fide fappy Planks, or Boards fawn off from the fides of Timber.

Slating.

1. *What.*] Slating is the Covering of Houfes with Slate.

This kind of Covering is very neat, efpecially the *Blue-flate*; as for the other kind of Slate, (known in fome Places by the Name of *Horfham-ftone,*) See *Horfham-ftone.*

The Blue-flate, cut into long Squares, or Efcallops, fhews very handfome, and is commonly us'd in Covering of Summer, and Banquetting-houfes in Gardens; it being a very light and lafting Covering.

But as this kind of Covering is very handfome, fo alfo 'tis very chargeable; for Roofs cover'd with Slate, muft be (firft) boarded over, the Slates hang'd on Tacks, and laid with finer Mortar than Tiles.

But if thefe Slates be rudely cut, and carelefly laid, (in refpect of Form) it is then accounted a cheaper Covering than with plain Tiles; efpecially in thofe Countries where the Earth affords plenty of them.

2. *Price of.*] This kind of Covering is valu'd by fome from 3 *s.* to 6 *s.* the Yard fquare, or by the Square of 10 Foot, (that is 100 Feet) from 30 *s.* to 3 Pounds, or more in fome places.

3. *Price of Pointing of Slates.*] The Pointing of Slates, [that is hewing them, and making them fit for the Work,] is worth, (fays Mr. *Wing*) about 12, or 13 *d. per* Square.

4. *Price*

4. *Price of Slates*] Slates at the Pits are worth, (ſays Mr. *Wing*,) 12 or 14 *s. per* Thouſand, which will nearly do 36 ſquare Yards.

5. *Of Meaſuring.*] Slating is in ſome places meaſur'd by the Rod of 18 Foot Square, which contains 324 ſuperficial Feet, or 36 ſquare Yards.

In meaſuring this ſort of Work, where there are Gutters or Valleys, there is commonly an Allowance, which is to take the length of the Roof all along upon the Ridge; which makes the Gutters double Meaſure, *viz.* as much more as really it is; which in ſome places is allow'd, and in others not; which depends upon the Cuſtom of the Place.

Slates.

See *Slating*, N. 3. 4.

Sleeper,

In Architecture is the Oblique Rafter that lies in a Gutter. See *Hip.* N. 1.

Slipper.

The ſame as *Plinth.*

Sluces.

Vents, or Drains for Water. See *Alder*, N. 3.

Smiths.

1. *Work*] Smith's-work in relation to Architecture, is of divers kinds, as *making of Caſements*; (for which ſee *Caſements* N 2.) Pallifado-work in Gates or other ways, [See Paliſado, N 4.) For making Dogs, Bars, large Hooks, Thimbles, Hinges, Staples, Grates, *&c.* they have in ſome places 3 *d* ½. in others 4 *d per* Pound. But for ſmall and neat Hooks, Hinges, Staples, *&c.* they have from 4 *d.* to 8 *d per* Pound. For Iron Balconies, 5 *d* the Pound.

2. *Bill to make.*] A *Smith's Bill* ſhould be made after this manner.

Mr Zaca-

Mr. Zachariah Zinthos, *of* London, *his Bill of Materials had of, and Work done by,* Samuel Smith, 1724.

	l.	*s.*	*d.*
June 24. For 8 large Casements, weighing 80 ℔. at 6 *d per* Pound.	2	0	0
July 2. For 10 small Casements, weighing 60 ℔ at 6 *d. per* Pound	1	10	0
12. For 10 pair of Hooks and Rides for Doors weighing 60 ℔ at 4 *d. per* Pound.	1	0	0
Sept. 10. For 2 great Bars for the Chimney, weighing 40 ℔. at 4 *d. per* Pound.	0	13	4
Octo. 13. For 3 Bars for Doors, weighing 30 ℔. at 4 *d. per* Pound.	0	10	0
30. For 4 Dogs, weighing 25 ℔. at 4 *d per* Pound.	0	8	4
Nov. 3. For 3 great Bolts for Doors, weighing 4 ℔ ½. at 4 *d per* Pound.	0	1	6
Summ	6	3	2

Soffita.

See *Saffita.*

Soils.

See *Sells.*

Socle, or *Zocle.*

The same as *Plinth.* See *Zoccolo.*

Solder, or *Sodder.*

1. *What*] There are several kinds of Solder; but that which more immediately relates to our present Business is Solder for Lead, which is made of Lead and half as much Block-tin. This is for Plumbers use; for Glaziers use it may be somewhat finer.

2. *Price of.*] This is sold from 8 *d.* to 10 *d. per* Pound, according to its fineness.

3. *To know if fine enough for the Glazier's use*] Some Glaziers tell me, That to know whether their Solder be fine enough for their use; they take a piece of it, and bend it too and fro near their Ear; for if it be of a fit temper it will crackle like Nits.

Solive.

From the *Latin, Solum* a Floor, signifies a Joist Rafter, or piece of Wood either slit or saw'd, wherewith the Builders lay their Ceilings. These are made of different Thicknesses, according as their Lengths require; and their Distances from each other are usually

 equal

equal to their Depths. In *Latin* we call 'em *Tignæ*.

Sommering.

See *Arches*, N. 7.

Spira.

Is properly *Latin* for the Fold of a Serpent laid at Reſt, or the Coil of a Cable Rope, *&c.* In Architecture 'tis ſometimes uſed for the Baſe of a Column, this making a Figure not unlike thoſe.

Splaying.

Of Windows and Doors. See *Bricklayers*, N. 2.

Springs.

For Caſements.] Some Smiths tell me, They have 6 *d. per* piece for Springs for Caſements, of the common or ordinary Faſhion. But I have ſeen ſome Springs for Caſements that were ſomething extraordinary in their Workmanſhip; they being a kind of double Springs, which ſeem'd as if a Right and Left-hand Spring had been join'd together for about three or four Inches from the Shoulder, where they were driven into the Timber; but at the end where the Scrolls were at leaſt two Inches aſunder, they had a Scroll turn'd both upwards and downwards in each Spring; ſo that each Spring ſeem'd like two Springs turn'd back to back. The Smith that made theſe Springs, told me, that he had 1 *s. per* piece for 'em.

Square.

A certain Meaſure, (made uſe of in Meaſuring ſeveral Artificers Works) conſiſting of 100 ſuperficial Feet.

Square-nails.

See *Nails*, N. 21.

Stairs

1. *What.*] Stairs are the Steps whereby we aſcend and deſcend from one Story of a Houſe to another.

2. *Dimenſions of.*] Several Writers of Architecture, have laid down ſeveral, and different Rules for the heighth, breadth, and length of Stairs, or Steps, and that according to the ſeveral Capacities of the Stair-caſes. But (in general) they forbid more than ſix, for the heighth of each Stair; and more than eighteen, and leſs than twelve Inches for the breadth, and more than 16, and leſs than ſix Foot for the length of each Stair.

But here we muſt underſtand, that they mean theſe Meaſures ſhould be obſerv'd only in large and ſumptuous Buildings: For in common and ordinary Houſes, they may be ſomething higher, and narrower, and much ſhorter; yet in theſe they ought not to exceed ſeven, or (at moſt) eight Inches in heighth; for if they do, they will be difficult to aſcend. Neither ought they to be leſs than nine or ten Inches in breadth; nor ought their length

length to be less than three Feet.

To reduce this Doctrine (of the Dimensions of Stairs) to some Natural, or at least Mathematical Ground, *Vitruvius*, (as we see, *Lib.* 9, *Cap.* 2.) borrows those Proportions that make the sides of a Rectangular Triangle, which the Ancient School did express (in lowest Terms,) by the Numbers 3, 4 and 5. That is 3 for the Perpendicular, from the *Stair-head* to the Ground; four for the *Ground-line* it self or Recession from the Wall, (says Sir *Henry Wotton*) and five for the whole Inclination, or Slope in the Ascent. But this Rule is so far from being follow'd in our modern Buildings, that the contrary is rather practis'd; for by this Rule, the lower the Stairs are, the narrower they ought to be; and if a Stair be but six Inches high, he must (according to this Rule) be but eight Inches broad; whereas in this case we seldom make 'em less than a Foot broad. And if we should make Stairs so low as four Inches, (for such the Ancient Architects make mention of,) they must (by this Rule) be but five $\frac{1}{3}$ Inches broad; which certainly is too narrow for any Stair.

3. *Of making*] Tho' we have laid down Rules (in the foregoing Number) for the heighth and breadth of Stairs; yet Workmen are not to be so strictly ty'd to those Rules, as not to vary in the least from 'em: For they must still observe to make all the Stairs of the same Stair-case of an equal heighth and breadth: To do which they must first consider the heighth of the Room, as also the width, or compass they have to carry up the Stairs in.

Then to find the heighth of each particular Stair, they ought first to propose the heighth of each Stair, and by that proposed heighth divide the whole heighth of the Room; which done, the Quotient will shew the number of Stairs: But if the Division fall not out exact, but that there be a Remainder; then (in this case) take the Quotient, (not regarding the Remainder) for the Number of Stairs, and by that number divide the whole heighth of the Room; so the Quotient shall give you the exact heighth of each Stair.

Example. Suppose the whole heighth of the Room be nine Foot three Inches, and suppose you design'd to make each Stair six Inches high, turn the whole heighth of the Room into Inches, 'twill be 111 Inches, which divide by six, the Quotient will be 18, and three remaining; therefore take 18 for the number of Stairs, and by it divide 111, the Quotient will be $6\frac{3}{18}$ Inches, or $6\frac{1}{6}$ Inches, which must be the exact heighth of each Stair.

Then, to find the breadth of each Stair, divide the width, or compass, (that you have to carry them up in) by the number of Stairs, and the Quotient will shew you the exact breadth of each Stair.

Stair-

Stair-case.

1. *What.*] *A Stair-case* is ſometimes taken to ſignifie the Incloſure of a pair of Stairs; whether it be with Walls, or with Walls and Rails, and Baliſters, *&c*. And ſometimes 'tis taken for the whole Frame of a pair of Stairs.

2. Of *making*] To make a compleat *Stair caſe*, is a curious piece of Architecture: The vulgar Cautions about it are theſe, ſaith Sir *Henry Wotton* in his *Elements of Architecture*.

That it have a liberal *Light*, againſt all Caſualty of Slips and Falls.

That the ſpace over-head be large and airy, which the *Italians* uſe to call *Un bel Sfogolo*, as it were, good Ventilation, becauſe a Man ſpends much breath in mounting.

That the Half paces, be well diſtributed at competent diſtances, for repoſing on the way.

That to avoid Encounters, and beſides to gratifie the Beholder, the whole Stair-caſe have no niggard Latitude, that is, for the principal Aſcent, at leaſt ten Foot in Royal Buildings.

That the Breadth of every ſingle Step or Stair be never leſs than one Foot, nor more than 18 Inches.

That they exceed by no means half a Foot in their heighth or thickneſs, for our Legs do labour more in Elevation than in Diſtention.

That the Steps be laid, (adds he) where they join *con un tantino diſcarpa i e*. ſomewhat ſlopeing, that ſo the Foot may in a ſort both aſcend and deſcend together, which tho' obſerv'd by few, is a ſecret and delicate Deception of the Pains in mounting. But this Doctrine of Stair-caſes ought to be regulated in proportion to the Quality of the Building: For a great Stair-caſe in a little Houſe would be as improper, as a little Stair-caſe in a great Houſe; both of them equally Ridiculous.

There is to be great care taken in the well placing the Stair-caſe; for there is not a little Difficulty to find a place convenient, ſo as the Stairs may be diſtributed without Prejudice, or Hindrance to the reſt of the Building.

III. *Kinds*] There are many kinds of Stair-caſes; for in ſome the Stairs are made *ſtraight* in others, *winding*, in others, mixt of both. Of *Straight-ſtairs*, ſome fly directly forward, others are Square, others Triangular; others are call'd *French* Flights. Of *Winding-ſtairs*, (which in general are call'd *Spiral*, or Cockle-ſtairs,) ſome are *ſquare*, ſome *Circular*, or *Round*, and ſome *Eliptical*, or *Oval*; and theſe again are various; for ſome wind about a *Solid*, others about an *open Newel* Stairs mixt of *ſtraight* and *winding*, are alſo of various kinds; ſome are call'd *Dog-leg'd*, others there are that both wind about a Solid-Newel, and fly about a ſquare Open-Newel. I ſhall particularly, (tho

(tho' briefly) deſcribe all theſe ſeveral kinds, in the following Numbers.

IV. *Straight-ſtairs*] Theſe are ſuch as always fly, and never wind, and therefore are by ſome call'd *Flyers.* Of theſe there are ſeveral kinds, as

1. *Direct flyers*, or *Plain-flyers.* Theſe fly directly from one Floor to another, without turning to the right or left, and are ſeldom us'd, unleſs it be for Garret, or Cellar-ſtairs in ordinary Houſes.

2. *Square-flyers*] Theſe fly round the ſides of a Square-Newel, either ſolid, or open, (ſo that there two kinds of 'em,) and at every corner of the Newel, there is a ſquare Half-pace, that takes up ¼ of a Circle. So they fly from one Half-pace to another; and the length of the Stars is Perpendicular to the ſide of the Newel.

3 *Triangular flyers*] Theſe fly round by the ſides of a Triangular Newel, either ſolid or open, (ſo that there are alſo two kinds of theſe,) and at each corner of the Newel there is a Trapezial Half pace, that takes up 120 Degrees. (or ⅓) of a Circle. So they fly from one Half-pace to another; and the length of the Stairs is Perpendicular to the ſide of the Newel.

Palladio tells us, That *Triangular ſtairs* are to be ſeen in ſome ancient Edifices; and of this ſort, (ſays he) are thoſe of the Cupola of St. *Maria Rotunda*, which are open in the middle, and receive Light from above. Thoſe alſo at *Sancto Apoſtolo* in the ſame City, are of the ſame kind.

4. *French-flyers*] Theſe kind of Stairs, firſt fly directly forward, 'till they come within the length of a Stair of the Wall, and then they have a ſquare Half-pace; from which you immediately, (without any Stairs between) aſcend to another Half-pace; and from this ſecond Half-pace the Stairs fly directy back again, parallel to the firſt flight.

V. *Winding ſtairs*] Theſe are ſuch as always wind, and never fly: There are many kinds of theſe Stairs; for ſome wind round a Circle, others round an Ellipſis, or Oval, others round a Square, and others round an Equilateral Triangle: And of each of theſe, ſome wind round a ſolid Newel, and others round an open, or hollow Newel. Again, ſome are ſet upon Columns, and ſome Stairs are double, and ſome are Quadruple. I ſhall deſcribe each of theſe in the following Numbers.

1. *Circular winding ſtairs.*] Theſe are of four kinds. Firſt, Such as wind about a ſolid Newel, and the fore-edge of each Stair is a right-line pointing to the Centre of the Newel. Theſe are common in Church-Steeples, and great old Stone-houſes. Secondly, Such as wind round an open Newel, and the fore-ſide of each Stair is a right Line pointing to the Centre of the Newel. Of this kind are thoſe in the Monument of *London* Thirdly, Such as wind round a ſolid Newel, but

but the fore-ſide of each Sair is an Arch (of a large) Circle, that points quite by the Centre, (and near to the Circumference) of the Newel In theſe, the Stairs are much longer than in the common Winding-ſtairs. Of theſe there may be two kinds: For their Ichnography being drawn, the Stairs may be contriv'd to be either Concave, or Convex on the fore ſide. Fourthly, there are other Stairs, in all Reſpects like thoſe laſt deſcrib'd, only they have an open Newel These kind of Stairs are ſaid to be invented by *Mark Anthony Barbaro*, a Gentleman of *Venice*

Any of theſe Kinds of Winding-ſtairs, take up leſs room than any other kind of Stairs whatſoever

In Stairs that wind round a Solid-Newel, Architects make

the Diameter of the Newel { $\frac{1}{5}$, or / $\frac{1}{4}$, or / $\frac{1}{3}$, or / $\frac{3}{7}$ } of the Diameter of

the whole Stair-caſe; according as the Stair-caſe is in bigneſs; for if the Stair-caſe be very ſmall, they make the Newel but $\frac{1}{5}$ of its whole Diameter; and if very large then $\frac{3}{7}$: and ſo proportionably of the reſt.

In Stairs that wind round an open Newel, *Palladio* tells us, the Newel muſt be $\frac{1}{2}$ the Diameter of the whole Stair-caſe. But I ſee no Reaſon why theſe open Newels ought not to be proportion'd to the Size of the Stair-caſe, as well as the ſolid ones.

Then, as to the Number of Stairs in one Revolution, *Palladio* tells us,

That if the Stair-caſe be { 6, or 7, / 8, / 9, or 10 / 18, } Foot Diameter, then

there may be { 12 / 16 / 20 / 24 } Stairs in one Revolution about the Newel.

2. *Elliptical-winding-ſtairs.*] Of theſe there are two kinds; one winding round a Solid, and one round an open Newel. They are much of the Nature of Circular Stairs, only in thoſe, the Newel is a Circle, but in theſe an Ellipſis, or Oval. Theſe kind of Stairs are very handſome and pleaſant, (ſays *Palladio*) becauſe all the Windows and Doors are commodiouſly placed in the middle and head of the Oval. I have made one of theſe, (ſays he,) with an open Newel

Newel at the Monaſtery of *Charity* at *Venice.*

3. *Square-winding-ſtairs.*] Theſe wind round a Square-Newel, either ſolid, or open; (and therefore are of two kinds) and the fore ſide of each Stair is a right Line pointing to the Centre of the Newel.

4. *Triangular winding ſtairs.*] Theſe wind round a Triangular-Newel, and the fore-ſide of each Stair is a right Line pointing to the Centre of the Newel. And becauſe the Newel may be either ſolid, or open, therefore there are two kinds of them.

5. *Columnated-window-ſtairs.*] *Palladio* mentions a pair of Stairs belonging to the *Portico's* of *Pompey* at *Rome*, that were ſet upon Columns, that the Light (which they receiv'd from above) might diſtribute it ſelf to all Parts alike. Such another pair were made by *Bramante* (an excellent Architect in his Time) at *Belvedere*, the Pope's Palace.

6 *Double-winding-ſtairs*] *Scammozzi* mentions a Stair caſe of this Form, made by *Piedro del Bergo*, and *Jehan Coſſin* at *Sciamburg* in *France* in the King's Palace. They are ſo contriv'd, that two Perſons, one aſcending, and the other deſcending, ſhall not come at one another. Mr. *Grew* (in his *Muſæum Regalis Societatis*,) gives us the Deſcription of a Model of this kind of Stair-caſe (which Model is kept by the Royal Society, in *Greſham-College*,) thus; The Foot of one of theſe Stair-caſes (ſays he) is oppoſite to that of the other; and both make a Parallel Aſcent, and within the ſame Cylinder. The Newel in the middle is hollow, and built with long Apertures to convey Light from Candles placed at the bottom, and on the ſides of the Newel into both the Caſes.

7. *Quadruple winding ſtairs.*] *Palladio* mentions a Stair-caſe of this Form, which King *Francis* the Firſt caus'd to be made in the Caſtle of *Chambor* near *Bloyſe*: It conſiſts of four Stair-caſes (carried up together) which have four Entrances, *viz.* one to each; and go up one over another in ſuch manner, that being made in the middle of the Building, the 4 may ſerve for 4 Apartments; ſo that the Inhabitants of one need not go up and down the Stairs of the other; and becauſe 'tis open in the middle, they all ſee each other go up and down without any Hindrance to one another.

VI. *Mixt Stairs*] Theſe are ſuch as do both fly and wind; and therefore are by ſome call'd by the general Name of *Flyers* and *Winders.* There are ſeveral Kinds of them. As

1. *Dog-legg'd-ſtairs.*] Theſe firſt fly directly forward then wind a Semicircle, and then fly directly back again, parallel to the firſt Flight

2. *Square Flyers and Winders*] Theſe have a ſquare Newel, either ſolid, or open; (and therefore are of two kinds,) they fly by the ſides of the Newel,

Newel, and wind (a quarter of a Circle) at each Corner.

3. *Solid, and open Newel'd-flyers, and Winders.*] These are of two kinds For some do first wind (a quarter of a Circle) about a solid Newel then fly by the side of a square open Newel, then wind by a solid Newel again, then fly again, as before, and so alternately. Others fly first, and then wind, and then fly again, and so alternately.

Let this suffice at present for the various kinds of Stair-Cases. However, in the mean time, the bare Description of these several kinds of Stairs, together with what has been said above, N. V. §. 1. and in *Stairs*, Num 3. may be a pretty good Guide to the ingenious that have a mind to make any of these kind of Stairs.

VII *Price of Stair cases.*] The Price of Stair-cases is various, according to their various Kinds, Sizes, and Curiosity of Workmanship. They are sometimes rated at so much *per* piece; and sometimes at so much *per* Stair.

An ordinary pair of Stairs with Flyers and Winders, of about six Foot, and four Foot, made of Elm-boards, are accounted to be worth 2 *s.* 6 *d.* or 2 *s.* 8 *d per* Stair, the Workman finding all Materials, as Boards, Nails, *&c.* But if the Materials are found by the Owner, then 9 *d.* or 10 *d.* *per* Stair, is a good Allowance for the Workmanship.

But for Stair-cases that have an open Newel, with a Landing-place at every sixth or eighth Stair, being about three Foot all the way: These Stairs, with Rails, Ballasters, String-boards, Posts, Balls, Pendants, and such other Ornaments may very well be worth 4 *s.* 6 *d.* 5 *s.* or 6 *s. per* Stair.

Stanchions.

The same as *Punchins.*

Staples.

What they are every one knows. For their Price, see *Iron,* Num. 4. and Smith's Work, Num 1.

Steening of Wells.

See *Bricks,* Num. III. §. 1.

Steps.

The same as *Stairs.*

Stiles.

In Joinery, the upright pieces that go from the bottom to the top in any Wainscot, or the like, are call'd Stiles.

Stillatory.

The Room that a Still, or Limbeck is set up in, for distilling Strong-waters, *&c.*

Stilobatum.

The Body of the Pedestal of any Column.

Stock bricks.

See *Bricks*, N. III. §. 15.

Stones.

1. *Their Kinds.*] There are ſeveral kinds of Stone; as *Marble Fire ſtone*, *Purbeck ſtone*, *Rag-ſtone*, *Alabaſter*, *Free-ſtone*, and *Common-ſtone*; of all which, except the two laſt, I have already treated in their proper Places of the Alphabet. As for *Free-ſtone*; there is a ſort of Stone commonly dug in the *Peninſula* of *Portland* in *Dorſet-ſhire*, (and commonly known by the Name of *Portland-ſtone*,) that is much us'd in Building; it being much ſofter and whiter than *Purbeck-ſtone*, and is commonly rais'd out of the Quarries in bigger Blocks than *Purbeck ſtone*. This *Portland ſtone* is by ſome Authors call'd Free ſtone, tho' there is a ſort of Stone found in *Oxfordſhire*, that is call'd *Free-ſtone*: And ſome call *Rigate*, or *Fire-ſtone*, *Free-ſtone*.

Common-ſtone needs no Deſcription; it being that which is commonly us'd, and found almoſt every where; and is that of which I ſhall principally ſpeak in the following Numbers of this Word.

2. *Of their Nature.*] If I had Leiſure (ſays the Honourable Eſq; *Boyle*.) I could eaſily ſhew you, That ways (hitherto unus'd) may be found out, (as I have partly try'd) to examine the *Nature* and *Goodneſs* of *Marble*, *Alabaſter*, and other Stones. A competent Knowledge of the Sap that is to be found in Stones imploy'd for Building, is of ſo much Importance, that the experienced Maſter-workmen have confeſſed to me, That the ſame ſort of Stone, and taken out of the ſame Quarry, if dug at one Seaſon, will moulder away in a very few Winters; whereas dug at another Seaſon, it will brave the Weather for very many Years, not to ſay Ages.

Again, (ſays the ſame ingenious Author in another place,) Experienc'd Maſons tell us, That as there are ſome ſorts of Stone that will decay in few Years; ſo there are others that will not attain their full Hardneſs in 30, or 40, or a much longer time.

Again, (ſays the ſame Author) there are in ſome Places Quarries of ſolid and uſeful Stone, which is employ'd about ſome ſtately Buildings I have ſeen, and which yet is of ſuch a Nature, (wherein divers other ſorts of Stone are ſaid to reſemble it) that tho' being dug at a certain Seaſon of the Year, it proves good and durable, as in thoſe Structures newly mention'd; yet imploy'd at a wrong time, it makes but ruinous Buildings; as even the chief of thoſe Perſons, whoſe Profeſſion makes him more converſant with it, has himſelf acknowledged (to me) to have found by ſad Experience.

3. *Of drawing.*] An ancient and experienced Maſon of my Acquaintance, tells me, That common Stones have a cleaving

ving Grain, (as they lie in the Quarry,) and a breaking one; the first, (he says) runs parallel with the Horizon; the other is perpendicular to it. The Method which he uses in *drawing of Stones*, [that is, getting them out of the Quarry,] is thus. Having uncoped it, [that is, taken off the Earth from the Stone,] they observe (by the Grain) where the Stone will cleave, and there they drive in a good many Wedges, till they have cleft him off from the rest of the Rock; and having thus loosen'd him, they next proceed to *break him*, which they thus perform; they applying their Rule to him at both Ends, mark out the Breadth they would have him, (e. g. suppose 10 or 12 Inches, or more, according to the Use they design the Stones for;) and by these Marks they strike a Line with the Corner of their Stone-axe; and by this Line they cut a little Channel with their Stone-axe, and in this Channel they set 6 or 8 Iron-wedges, (supposing the Stone to be but 3 or 4 Foot long;) which they drive very carefully with soft and gentle Strokes, keeping them all forward together, and not one before another, least it break the Stone a-cross, and not by the length of the Channel. Yet, he says, that this Method of driving the Wedges, is not always to be observ'd; for sometimes a Stone is not through the whole length of an equal Solidity, but is in some places softer, and in others harder; this they find, (and observe) in cutting their Channel; and those Wedges that stick in those softer places, they venture to drive a little faster than the others. And this, he says, he has found by long Experience, to be the best way of breaking Stones.

Having thus broken them in length, which by this Method they can do to any Size within less than an Inch; (which is near enough for rough Stones) they next apply a Square to the straight side, and striking a Line, they proceed to break them in breadth, in the same manner, as before in length; also now they size them for the length, as before for the breadth.

By this Method of drawing of Stones, he says one Load of Stones will do as much Walling as a Load and half of such Stones, as in drawing are broken at random; for in this last Case, one Stone has commonly a very acute Angle, another a very obtuse one; whence it comes to pass, that they require abundantly more scapting, and wast much more of the Stones, than when drawn by the Method above mention'd.

The same ingenious old Man tells us, That some *London* Stone-cutters have told him, that hard Stones have not a Cleaving-grain as the soft ones (in our Contrey) have: And therefore when they are minded to break up a Stone in such Quarries, they have great heavy Stone axes, with which they work down a deep Channel in the Stone, into which

Channel

Channel (at the top) they lay two Iron-bars, (such as Smiths have from the Forge to work out,) and between these Bars they drive their Iron-wedges to break off the Stone; for their Wedges will not go where there is not a Channel made for them, as they will in soft Stones.

Some in drawing of Stone make use of Gun-powder; concerning which, take the following Account (in his own Words) from the Honourable Esq; *Boyle*. It has long been, and still is in many places (says he) a Matter of much Trouble and Expence, as well of Time as Money, to cut out of Rocks of Alabaster and Marble, great Pieces to be afterwards squar'd, or cut into other Shapes; but what by help of divers Tools and Instruments, cannot in some Quarries be effected without much Time and Toil, is in other places easily and readily perform'd, by making with a fit Instrument a small Perforation into the Rock, which may reach a pretty way into the Body of it, and have such a Thickness of the Rock over it, as is thought convenient to be blown up at one time; for at the further End of this Perforation, there is plac'd a convenient quantity of Gun-powder, and then all the rest of the Cavity being fill'd with Stones, and Rubbish strongly ram'd in (except a little place that is left for a Train,) the Powder, (by the help of that Train) being fir'd, (and the impetuous Flame being hindred from expanding it self downwards, by Reason of the newly mention'd Obstacle, concurring with its own tending another way, displays its Force against the upper parts of the Rock, which in making it self a Passage, it cracks the Rock into several pieces, most of them not too unweildy to be manag'd by the Workmen. And by this way of blowing up of Rocks, a little vary'd and improv'd, some ingenious Acquaintance of ours, imploy'd by the Publick, to make vast Piles, have lately, (as I receiv'd the Account of themselves) blown up, or scatter'd with a few Barrels of Powder, many Hundred, not to say Thousand, Tuns of common Rock.

4. *Load of Stone, how much*] Some Masons tell me, That 25 Foot of Stone make a Load. But (it's to be observ'd) they do not mean 25 solid Feet, but superficial measur'd on the *Face* of the Stones, and not on any of the Beds.

For a clearer Understanding of this, it must be noted, That every squared Stone has 6 Plains, or Sides, *viz.* The upper, and under Bed, the Face, and the Back, and the 2 Heads, or Ends. Of these 6 Plains, those 2 opposite ones that are the cleaving way of the Stone, (and which in the Quarry lay parallel to the Horizon,) are call'd the *Beds*; and of the best of the 4 Plains that are perpendicular to these, (and consequently are the breaking way of the Stone,) they make the *Face*, and the Plain

Plain opposite to the *Face*, (and, which commonly goes rough as it comes from the Quarry) they call the *Back* of a Stone; and the other 2 perpendicular Plains are call'd the Heads, or Ends.

5. *Cord of Stone, how much.*] In some parts of *Kent*, Stones are sold by the Cord, consisting of 27 solid Feet, *viz.* 3 Feet long, 3 broad, and 3 high.

6. *How much Walling a Load of Stones will do*] An old and experienced Mason, tells me, That a Load of Stones will build about 20 Foot of 18 Inch Wall; this he reckons a Medium, the Extreams he reckons 15 and 25.

7. *Soft Stones, how wrought smooth*] An old experienced Mason, tells me, that some Stones are too soft to bear a good Edge; for when they are scap'd, and wrought smooth, their Edges crumble off; and therefore (in this Case) to make them smooth, they proceed thus: After they are scap'd, they take an old Card, (such as Wool is carded with) and with it they work out the Strokes of the Axe, then they bring it to a better liking, by rubbing it with a piece of the same Stone. And thus our Country-Masons manage all soft Stones.

8 *Price of drawing and carrying of Stones*] The old Mason mention'd above, Num. 3 tells me, That he has 3 *s.* the Load for drawing of Stones, after the Method mention'd, Num. the 3d. and for the Carriage of a Load, (tho' it be not above $\frac{1}{2}$ a Mile) he has 2 *s.* the Load.

Another Mason tells me, That he has drawn Stones for 9 *d.* the Load; but then they lay almost level with the Ground, and requir'd but very little uncopeing. He also told me, That another Mason, which he nam'd to me, (and whom I also knew,) used to draw Stones for 1 *d. per* Foot.

Also a *Sussex* Gentleman of my Acquaintance, tells me, That he can have very good Stones drawn for 2 *s.* 6 *d. per* Cord, and have them carry'd almost a Mile for 3 *s.* 6 *d. per* Cord.

But as the Price of drawing Stones is various in different places, according to the different Manners of drawing them, and according to the different Circumstances of *Difficulty*, or *Easiness* of drawing them, *&c.* So also is the Price of carrying them very various in different Places, according to the Custom of those Places See more concerning this Matter, in the word *Ashlar*.

9. *Price of Scapling Stones*] Several Masons tell me, That they commonly give 5 *s.* for scapling 100 Foot of Stones; this is Journey-man's Wages, out of which (they say) the Master has but small Profit. They also tell me, That they reckon 50 Foot a Day's Work, tho' some Workmen will do 60 Foot in a Day: But (it must be observ'd) the Measure is superficial, and they measure only the *Face* of the Stone tho' they scaple 5 sides to each Stone, *viz.* A *Face*, two *Beds*, and

and two *Ends*; ſo the *back* goes rough as it came out of the Quarry. But in Scapling, they always, (if they can conveniently,) chuſe *that* for the *Face* of the Stone which will be moſt for their Advantage.

Stone-work.

Of *Meaſuring.*] In ſome parts of *Suſſex*, Maſons have a Cuſtom to meaſure their Stone-work thus; they apply one end of a Line to the top of the Copeing, and ſo carry it along the ſlant of the Copeing, and preſs it under the Toothing (if any be,) and from thence they carry it to the Water, or Ground-table, (if any ſuch be in the Wall) where they preſs it in likewiſe, and then carry it over the Table to the bottom of the Foundation; and this Dimenſion, thus taken, they account for the heighth; which multiply'd into the length, gives the Content.

But (I think) in moſt places they are not ſo nice, as to take the heighth by a Line, but are contented with the perpendicular heighth.

Stove.

A Hot-houſe, or Room. *Palladio* obſerves, That the Ancients us'd to warm their Rooms, with certain ſecret Pipes that came through the Walls, conveying Heat, (as I conceive it, ſays Sir *Henry Wotton*) to ſeveral parts of the Houſe from one common Furnace. Whether this were a Cuſtom, or a Delicacy, (ſays Sir *Henry Wotton*) it was certainly both for Thrift and Uſe, far beyond the *German* Stoves.

Strait.

A Term us'd by Bricklayers, it is half, (or more, or leſs than half) a Tile in breadth, and the whole length. They are commonly us'd at the Gable-ends, where they are laid at every other Courſe, to cauſe the Tiles to *break Joint*, as they phraſe it; that is, that the Joynts of one (Courſe) may not anſwer exactly to the Joints of the next Courſe, either above, or below it.

Straight-arch.

See *Arch.* Num. 7.

Structure.

See *Building.*

Struts.

See *Dragon-beams.*

Stuff.

The Wood that Joyners work upon, they call in general *Stuff.*

Stretchers.

See *Arch*, Num. 7.

Stylobatæ.

The ſame as Pedeſtals; a Greek Word from *Stylos* a Pillar

Pillar and *Basis*, the Base or Foot thereof.

Substruction.

See *Foundation*, Num. 2. §. 7.

Summer-tree.

A Beam full of Mortises for the Ends of Joists to lie in. See *Bress-Summers* and *Girders*.

Supercilium.

As List.

Symmetry.

Is the Conveniency that runs between the *parts* (of a Building) and the *whole*.

Systole.

Is that manner of placing Columns, where the Space between the two *Fusts* consists of two Diameters or four Modules.

Table, or *Glass*.

SEE *Case of Glass*.

Tabern,

A Cellar.

Tacks.

See *Nails*. Numb. 22.

Talon.

Same as *Heel*; a sort of Moulding in Architecture; See *Heel*. *Talon* is the French Word for *Heel*.

Taper.

All sorts of Stuff, or Work that is smaller at one end than the other, and diminishes gradually from the biggest end, is said to be *taper*.

Tarrace, or *Tarras*.

An open Walk, or Gallary Also a flat Roof on a House Also a kind of course Plaister, durable in the Weather.

Tassels.

Pieces of Board that lie under the ends of the Mantle-tree.

Teeth.

As *Dentils*.

Templets.

See *House*, N. 4.

Tenia.

As *List*.

Thermes or *Termes*.

A *French* word from *Terminus*, the *Roman* God of Boundaries or Land-marks, which they us'd to represent in a humane Figure, with half a Body, as if it proceeded out of a Sheath or Case. These they fixed in the Earth as Land-marks. In Architecture they serve as a kind of Symbolical Column.

To

To give 'em a Figure proper to represent a delicate Column, their Arms are lopt off, and their Body does not appear below the Girdle: These *Termini* are very proper in the Decorations of a Theatre, as also in Pieces of Architecture *de Treillage*, (as *le Clerc* calls it) crail'd Work kind.

These *Termini* have this in common with the Cariates (or Cariatick Columns) that they should never be brought to match with the common Columns: This Advantage, however, they have in particular. That a Man may give them what degree of Delicacy he pleases, by lengthning out their Sheath, and raising the Figures to any heighth desired. By this means they'll be made to suit gay airy Architecture, such as Cabinets, Sallons, and Arbours of crail'd Work, especially require. 'Tis not reasonable, in my Opinion (adds *le Clerc*) to reduce the Figures of Angels into *Termini*: tho' we see it has been formerly done in Places of Distinction.

Tennon.

A square end of a piece of Timber fitted into a Mortess. See *Mortess*.

Terrass.

As *Thrrace*.

Tetradoron.

A kind of Brick so called. See *Brick*, N. III. §. 18.

Thack-tiles.

As *Plain-tiles*. See *Tiles* N. III.

Thatching.

1. *What.*] Thatching is the covering the Roof of a House or Barn, with Straw or Reed.

2. *With Straw.*] *Thatch* (says *Worlidge*,) is a common Covering in many Places, yet is some to be preferr'd before other some; the best which I have seen, (says he) is that which is call'd *Helm*, that is long and stiff Wheat-Straw, (with the Ears cut off,) bound up in Bundles unbruis'd; which well laid, lies thin, lasts long, and is much neater than the common way.

Thatchers commonly allow about two good Load of Straw for five Square of Thatching, or one Load to 2 ½ square.

A Thatcher of my Acquaintance, tells me. That one *Rubble* a Mason of *Rootham* in *Kent*, proffer'd (for a small matter) to teach him how to Thatch a Roof so, that no Mouse nor Rat should come into it: But he was not so thoughtful then, as to get the Receipt of him, tho' it would have been of no small use to him; for the *Rootham Mason* said, he knew a Thatcher that had 4 *d. per* square more for doing it so. It is a thing worth inquiring after.

In some parts of *Kent* they use no Withs to bind on their Thatching-rods, but (instead thereof) they use *Rope-yarn*, (as they call it) which is a single Strand-

Strand-line, about the ſize of a Penny-Cord; it is Pitched with Pitch, according as ſome do their Well ropes. A *Kentiſh* Thatcher told me, that one Pound of it (which coſts 2 *d.*) will do about a ſquare of Thatching. He had about 18 Pound of it for 18 ſquare and 90 Foot of Thatching on a Barn; and I think he had but 40 Pound for 48 Square and 88 Foot: He tells me, 'Tis more durable than Withs; for they when are grown ſear, will fly and break; but this will not. See P *Withs*.

3. *With Reed.*] In ſome parts of *Suſſex* and *Kent*, they Thatch with Reed inſtead of Straw Some Workmen tell me, That this kind of Thatching will indure 40, 50, or 60 Years. They alſo tell me, That Reed is ſold by the Thouſand, *viz* A Thouſand handfuls, each handful being about 8, 9, or 10 Inches in Circumference, bound up in a lattle Band; a Thouſand of which will coſt 15 or 16 *s.* and will cover about 3 Square of Roofing. For laying of which they have 4 *s. per* Square.

4. *Price of.*] Common Thatching is done in ſome Places for 2 *s.* 6 *d. per* ſquare; but in other Places they have 2*s* 8 *d* and in others 3 *s: per* ſquare And for Thatching with Reed they have 4 *s. per* ſquare.

5. *Of Meaſuring*] Thatching is meaſur'd by the ſquare as Tiling: And in ſome Places they are allow'd ſo many Feet more as the Corners and Gables are Feet in length In other Places they are allow'd (only) ſo many half Feet more to the whole, as the Gable heads are Feet in length; and the Reaſon they urge for this Cuſtom, is, becauſe they have more trouble in turning the Straw (at the Gables) that it may be cut as it is at the Eaves If one ſide of a Roof (only) be Thatched and not the other; they (then) take their Dimenſions over the Ridging, as far as the new Straw goes.

Thimbles.

See *Iron*, N. 5.

Thorough framing.

See *Framing*, N. 7.

Thorough-lighted.

Rooms are ſaid to be Thorough-lighted when they have Windows at both ends.

Tiles.

See *Tiles*.

Timber.

1. *What.*] All thoſe kinds of Trees, which being cut down and ſeaſon'd, are uſeful for the Carpenter. Joyner, or other wooden Tradeſmen to work upon, are call'd *Timber* when they are cut down, and *Timber Trees* when they are growing.

2. *Kinds.*] There are many kinds of Timber; it were tedious to mention 'em all. I shall content my ſelf at preſent,

ſent, briefly to ſhew the moſt common uſes, and of the moſt common kinds of Timber, as I find it ſet down in Mr. *Evelin*'s *Sylva*, and Mr. *Worlidge*'s *Syſtema Agriculturæ*. As follows

1. *Oak*] The ſeveral uſes of Oaken-timber for Buildings, and other Mechanick Uſes, is ſo univerſally known, that 'twere needleſs to enumerate them. To endure all Seaſons of the Weather, there is no Wood comparable to it; as for Pales, Shingles, Poſts, Rails, Boards, *&c.* For Water-works alſo 'tis ſecond to none; eſpecially where it lies obvious to the Air as well as the Water, there is no Wood like it

2. *Elm.*] If the Elm be fell'd between *November* and *February*, it will be all Spine, or Heart, or very little Sap, and is of moſt ſingular uſe (in the Water) where it lies always wet, and alſo where it may be always dry. It is alſo of great uſe for its toughneſs and there fore us'd by Wheel-wrights, Mill wrights, *&c* It is alſo good to make Dreſſers, and Planks to chop on, becauſe it will not break away in Chips like other Timber.

3 *Beech*] Its uſe is principally for the Turner, Joyner, Upholſterer, and ſuch like Mechanick Operations, the Wood being of a clean, white, and fine Grain, and not apt to rend, or ſlit: Yet it is ſome times us'd. (eſpecially of late Years) for Building-timber. And if it lie always wet, (as for Ground-guts, and the like) 'tis thought to endure longer than Oak will in that Caſe.

4. *Aſh.*] The uſe of Ash is almoſt univerſal, good for Building, or any other uſe where it may lie dry; ſerves the Occaſions of the Carpenter, Plough-wright, Wheelwright, Cart-wright, Cooper, Turner, *&c.* For Garden Uſes alſo, no Wood exceeds it; as for Ladders, Hop-poles, Paliſade-hedges, *&c.* It ſerves alſo at Sea, for Oars, Hand-ſpikes, *&c.*

5. *Fir*] This kind of Timber is commonly known by the name of *Deal*, and is of late much us'd in Building, eſpecially within Doors, for Stairs, Floors, Wainſcot, and moſt ornamental Works

6 *Walnut-tree.*] This Timber is of univerſal uſe, (unleſs for outward Edifices,) none better for the Joyner's uſe, it being of a more curious brown colour than Beech, and not ſo ſubject to the Worms.

7. *Cheſnut-tree.*] This Timber is (next to Oak) one of the moſt ſought after by the Joyner and Carpenter, and of very long laſting, as appears by many ancient Houſes and Barns built of it about *Graveſend* in *Kent*.

8. *Service tree.*] This Timber is uſeful for the Joyner, it being of a very delicate Grain, and is fit for divers Curioſities: It alſo yields Beams of a conſiderable bigneſs for Buildings.

9. *Poplar, Abel and Aſpen*] Theſe kinds of Timber differ but little from one another; and of late they are often us'd inſtead of *Fir*; they look as well, and are tougher and ſtonger.

10. *Alder.*

10. *Alder.*] This is useful for Ladder and Scaffold-poles, as also for Sewers, or Pipes to convey Water; for if it lie always wet, it will harden like a very Stone; but where it is sometimes wet sometimes dry, it rots immediately.

11. *Lime tree.*] I have known, (says my Author,) excellent Ladders made of Lime-tree-poles, and of a very great length.

III. *Time of Felling.*] The Time of the Year for this Work is not usually till about the end of *April*, (at which Season the Barks does commonly rise freely, and if there be any quantity of Timber fell'd, the Statute obliges us to fell it *then*, the Bark being necessary for the *Tanner*) But the Opinions and Practice of Men have been very different concerning the best time to fell Timber: *Vitruvius* is for an Autumnal Fall; others advise *December* and *January*: *Cato* was of Opinion, That Trees should have first born their Fruit, or at, least should not be Fell'd 'till the Fruit was full ripe, which agrees with that of the Architect: And tho' Timber unbarked be indeed most obnoxious to the Worm, yet we find the wild Oak, and many other sorts Fell'd over late, (and when the Sap begins to be proud) to be very subject to the Worm; whereas being cut about Midwinter, it neither casts, rifts, nor twines; because the Cold of the Winter does both dry and consolidate: Happy therefore were it for our Timber, if some real *Invention* of *Tanning* without so much Bark, (as the Honourable, Mr. *Charles Howard*, has most ingeniously offer'd) were become universal; that Trees being more early Fell'd, the Timber might be the better season'd, and condition'd for its various Uses.

Then for the Age of the Moon it has been religiously observ'd; and that *Diana*'s Precedency in *Sylvis* was not so much celebrated to credit the *Fictions* of the *Poets*, as for the Dominion of that moist Planet, and her Influenee over Timber: For my part, I am not so much inclin'd to these *Criticisms*, as to Fell Timber altogether at the Pleasure of this mutable Lady; however there is doubtless some regard to be had, *Nec frustra signorum obitus speculamur & ortus*.

The old Rules are these: Fell Timber in the *Decrease*, or 4 Days after the New Moon; some say in the *last Quarter*: *Pliny* says, (if possible) in the very Article of the Change; which hapning, (says he, in the last Day of the Winter Solstice, *that Timber* will prove immortal: *Columella* says) from the 20th, to the 30th Day: *Cato*, four Days after the Full: *Vegetius*, from the 15th to the 25th, for Ship-timber, but never in the Increase, Trees then most abounding with Moisture, which is the only Source of Putrefaction.

Then for the Temper and Time of the Day; the *Wind* low, neither *East* nor *West*; neither in *Frosty*, *Wet*, or *Dewy*

Wea-

Weather; and therefore never in a Forenoon.

Lastly, Touching the Species; Fell *Fir* when it begins to spring; not only because it will then best quit its Coat and Strip; but for that they hold it will never decay in Water; which whatsoever *Theophrastus* deduces from the old Bridge made of this Material, (cut at this Season,) over a certain River in *Arcadia*, is hardly sufficient to satisfie our Curiosity. *Elm* (says Mr. *Worlidge*) is to be Fell'd between *November* and *January*; for then, (says he) it will be all *Heart*, or at least will have but very little Sap. And this he also says is the only Season for Felling of *Ash*

Some Authors advise in Felling of Timber, to cut it but into the Pith, and so let it stand 'till it be dry, because, (say they) by drops there will pass away that Moisture which would cause Putrefaction.

IV. *Of Seasoning*] Timber being Fell'd, and Sawn, is next to be season'd; for doing of which, some advise, That it be laid up very dry in an airy place, yet out of the Wind, or Sun; at least, (say others) it ought to be free from the Extremities of the Sun, Wind, and Rain; and that it may not cleave, but dry equally, you may daub it over with Cow-dung. Let it not stand upright, but lay it along one piece upon another, interposing some short Blocks between them, to preserve them from a certain Mouldiness, which they usually contract while they sweat, and which frequently produces a kind of Fungus, especially if there be any sappy Parts remaining.

Others advise to lay Boards, Planks, *&c.* In some Pool, or Running-stream for a few days, to extract the Sap from 'em, and afterwards to dry 'em in the Sun, or Air; for by so doing, (say they) they will neither chap, cast, nor cleave: (Mr. *Evelin* particularly commends this way of Seasoning of Fir) against *shrinking* there is no Remedy.

Some again commend *Buryings* in the Earth, others in Wheat; and there be Seasonings of the Fire, as for the scorching and hardening of Piles, which are to stand either in the Water, or the Earth. Thus do all the Elements contribute to the Art of *Seasoning of Timber*.

Sir *Hugh Plat* informs us, That the Venetians use to burn and scorch their Timber in the flaming Fire, continually turning it round with an Engine, 'till they have gotten upon it a hard, black, coally Crust; and the Secret carries with it great probability; for that the Wood is brought by it to such a hardness and dryness, *ut cum omnis putrefactio incipiat ab humido*, nor Earth nor Water can penetrate it. I my self, (says Esq; *Evelin*,) remember to have seen Charcoals dug out of the Ground, amongst the Ruins of ancient Buildings, which have in all Probability lain cover'd with Earth above 1500 Years.

V. Of

V. *Of Preserving.*] When Timber, or Boards are well seasoned, or dry'd in the Sun, or Air, and fix'd in their Places, and what Labour you intend is bestow'd upon 'em; the Use of Linseed-oil, Tar, or such like Oleaginous Matter, tends much to their Preservation and Duration. *Hesiod* prescribes to hang your Instruments in the Smoak, to make them strong and lasting; *temonem in fumo poneres:* Surely then the Oil of Smoak, (or the vegetable Oil, by some other Means obtain'd,) must needs be effectual in the Preservation of Timber. Also *Virgil* advises the same, *Et suspensa focis exploret Robora fumus*, says he.

The Practice of the *Hollanders* is worth our Notice, who, for the Preservation of their Gates, Port-cullis's, Draw-bridges, Sluces, and other Timbers expos'd to the perpetual Injuries of the Weather, Coat them over with a Mixture of Pitch and Tar; upon which they strew small pieces of *Cockle*, and other *Shells*, beaten almost to Powder, and mingl'd with *Sea sand*; which incrusts, and arms it after an incredible manner, against all the Assaults of Wind and Weather.

When Timber is fell'd before the Sap is perfectly at rest, (says Esq; *Evelin*,) it is very subject to the Worm; but to prevent, or cure this in Timber, I recommend the following Secret, as most approved.

Let common yellow Sulphur be put into a Cucurbit Glass, upon which pour so much of the strongest *Aquafortis*, as may cover it three Fingers deep: distil this to driness, which is done by two or three Rectifications: Let the Sulphur remaining at the bottom, (being of a blackish, or sad Red-colour) be laid on a Marble, or put into a Glass, where it will easily dissolve into Oil: With this anoint what Timber is either infected with Worms, or to be preserved from them. It is a great and excellent *Arcanum* for tingeing the Wood of no unpleasant Colour, by no Art to be washed out; and such a Preservative of all manner of *Woods*, nay, of many other Things also, as *Ropes*, *Cables*, *Fishing-nets*, *Masts*, or *Ships*, &c. That it defends them from Putrefaction, either in *Waters*, under, or above the *Earth*, in *Snow*, *Ice*, *Air*, *Winter*, or *Summer*, &c.

'Twere superfluous to describe the process of making the *Aquafortis*; it shall suffice to let you know, That our common *Copperas* makes this *Aquafortis* well enough for our purpose, being drawn over by a *Retort*: And for *Sulphur*, the *Island* of St. *Christophers* yields enough, (which hardly needs any refining) to furnish the whole World. This *Secret* (for the *curious*) I thought fit not to omit, tho' a more compendious way may serve the turn, three or four Anointings with *Linseed-oil*, has prov'd very effectual: It was experimented in a *Walnut-tree table*, where it destoy'd Millions of Worms immediately, and is to be practis'd for *Tables*, *Tubes*, *Mathematical*

matical Instruments, Boxes, Bedsteads, Chairs, &c. *Oyl of Walnuts* will doubtless do the same, is sweeter and better Varnish; but above all is commended *Oyl of Cedar*, or that of Juniper.

For *Posts*, and the like, that stand in the Ground, *the burning the out sides* (of those ends that are to stand in the Ground) *to a Coal*, is a great Preservative of them. I have already, (in the fore-going Number) mention'd the Practice of the *Venetians* in a like Case, mention'd by Sir *Hugh Plat*; to which he adds, That a *Kentish* Knight of his Acquaintance, did use to burn (in this manner) the ends of his Posts, for Railing, or Pailing: And this was likewise practis'd with good Success by a *Sussex* Gentleman, *Walter Burrel* of *Cuckfield*, Esq;. And this Practice was probably deduced, from the *Observations* made by several that digged in the Earth, who have found *Charcoal*, which they conjectur'd might have lain there about 100 Years, (nay, Esq; *Evelin* says 1500 Years; see above in the fore-going Number), and yet was not in the least inclin'd to Putrefaction, but was very firm and solid; which plainly demonstrates, That *Timber* thus calcin'd, will resist Putrefaction much longer than it can do without it.

This of burning the ends of Posts, is also practised in *Germany*, as appears by the Abstract of a Letter, written by *David Von-der-beck* a *German Philosopher*, and *Physitian* at *Minden*, to Dr. *Langelot*, registred in the *Philosophical Transactions*, Num. 92. Page 585, in these Words: Hence also, they slightly *burn* the Ends of *Timber* to be set in the Ground, that so by the Fusion made by Fire, the *Volatile Salts*, (which by accession of the Moisture of the Earth, would easily be consum'd, to the Corruption of the Timber) may catch, and fix one another.

VI. *Of closing the Chops, or Clefts in green Timber*] Green-timber is very apt to split and cleave after 'tis wrought into Form, which in fine Buildings is a great Eye-sore. But to close the *Chops*, and *Clefts* in *Green Timber*, I find this Expedient, To anoint and supple it with the *Fat* of *Powder'd-beef Broth*, with which it must be well soak'd, the Chasms fill'd with Spunges dip'd into it; this to be done twice over. Some Carpenters make use of *Grease* and *Saw-dust* mingl'd; but the first is so good a way, (says my Author) that I have seen *Wind-shock timber* so exquisitely clos'd, as not to be discerned where the Defects were. This must be us'd when the Timber is green.

VII. *Of Measuring*.] Timber is commonly measur'd and sold by the *Tun*, or *Load*, which is a solid Measure, containing 40 or 50 solid Feet, *viz* 40 Feet of round Timber, and 50 Feet of hewn Timber is call'd a *Tun*, or *Load*; which Denomination, (I conceive) it receives from the Supposition, that 40 Feet of round Timber, or 50 Feet of hewn Timber weighs about

about a *Tun Weight*, (*i. e.* 20 Hundred,) which is commonly accounted a Cart-load. Now

For measuring of round Timber the Custom is, to gird the Tree about in the middle of the length, and folding the Line twice (to take a quarter of it,) they account that for the true *side of the square*; then for the *length*, 'tis counted from the But-end of the Tree, so far up as the Tree will hold *half a Foot girt*, (as they phrase it) *i. e.* The Line half a Foot, when twice folded.

The Dimensions thus taken, the Timber may be measur'd either by multiplying the side of the square in it self, and that Product by the length, by the Method of *Cross-Multiplication*, (See *Cross-Multiplication*) or more easily and speedily by *Gunter*'s *Line*, by extending the Compasses from 12 to the side of the Square in Inches; for that Extent turn'd twice (the same way) from the length in Feet, will reach to the Content in Feet.

If the Tree have any great Boughs, which are Timber, (as they phrase it) *i. e.* which will hold half a foot Girt; they commonly measure them, and add them to the whole: The Solidity of the whole being thus found, they divide it by 40, which brings it into Tuns.

But (it is noted:) If round Timber be measur'd in order for Sale; they commonly (for *Oak*) cast away an Inch out of the Square for the Bark; [*i. e.* if a Tree be 10 Inches Square, they measure him as if he were but 9] but for *Ash*, *Elm*, and *Beech*, an Inch is too much to be allow'd for the Bark. (2.) That this way of taking $\frac{1}{4}$ of the Circumference for the true Square, is erroneous, and always gives the Solidity less than the Truth, by about a fifth part

For measuring hewn or squar'd Timber their Custom is to find the middle of the length of the Tree, and there to measure the breadth of him, by clapping two Rules, or other straight Things, to the sides of the Tree, and measuring the Distance between them, and in the same manner they measure the breadth the other way; which, if they are unequal, they add them together, and take half their Sum, which they account the true side of the Square.

The Dimensions thus taken, it is measur'd in the same manner as round Timber. So the Content being found in Feet, they divide it by 50 to bring it into Tuns.

But 'tis to be noted. (1) That if the Timber be unequal sided, this Method of taking the Dimensions, always gives the Content more than the Truth, and the greater is the Difference of the Sides, the greater is the Error. (2.) That tho' the Method of taking the Dimensions, both of square, and round Timber, are both erroneous, yet Custom has made them currant.

VIII. *Price of Felling and Hewing.*] Carpenters about us in *Sussex*, and *Kent*, have about 1 *s* or 1 *s.* 2 *d. per* Load for *felling*

felling of Timber, and about 3 *s per* Load for *hewing*.

IX. *How much to a Square of Framing.*] Mr. *Leybourn* tells us, That 20 Foot of ſolid Timber, (cut into convenient Scantlings) will compleat a *Square*. (*i. e.* 100 ſuperficial Feet) of Framing in any Building, great or ſmall. I mean, (ſays he,) of the Carcaſs, *viz.* the outſide Frame, Partitions, Roof, and Floors.

X.—*Buildings, of facing with Brick.*] See *Facing*; alſo ſee *Brick*, Num. 9.

Tinea, or rather *Tænia*.

As Liſt. See *Plat-band*.

Tondino,

As *Aſtragal*.

Top-beam,

As *Coller-beam*.

Torus.

A large round Moulding in the Baſes of Columns; the Word comes from the Latin *Torus* a Bed; the Figure of this Moulding being not unlike that of the Edge of a Quill.

Torſels,

As *Taſſels*.

Trabeation.

Same as *Entablement*, which ſee.

Trammel.

An Iron moving Inſtrument in Chimneys, whereon they hang the Pot over the Fire.

Tranſom.

1. *What.*] The piece that is fram'd acroſs a double *Light-window*.

2. *Windows.*] Tranſom-windows in great Buildings, are worth making (ſays Mr. *Wing*) 1 *s* 9 *d. per* Light, or 7 *s. per* Window.

Traverſe.

A Term in Joynery, ſignifying to plain a Board, (or the like) acroſs the Grain.

Traverſe-tile.

See *Tile*, Num. 10.

Triglyph.

A Term of *Architecture*. The Word is originally Greek, and ſignifies a *hollow Graving like three Furrows, or Gutters*. In *Architecture*, Triglyphs are thoſe kind of Stops, (in the Dorick Freeze) between the Metops. See *Metops*.

Trim.

When Workmen fit a piece into other Work, they ſay they *trim* in a piece.

Trimmers.

In *Architecture*, are thoſe pieces of Timber fram'd at right

right Angles to the Joysts, against the Ways for Chimneys, and Well-holes for Stairs.

Trochilus.

See *Capital*, Num. 4.

Trunk.

From the Latin *Truncus*, signifies the Fust or Shaft of a Column, and the Die of a Pedestal.

Turn'd-lead.

See *Lead*, Num. 10.

Tuscan-Order.

See *Column* and *Order*.

Tusk.

A Bevel Shoulder, made to strengthen the Tenon of the Joyst which is let into the Girder.

Tiles.

I. *What.* What they are every one knows: Yet Bishop *Wilkins* defines them to be a sort of Artificial Stones, (of a laminated Figure,) us'd about the Roofs and Pavements of Buildings.

They are made of Clay, kneaded together, then squeez'd flat in a Mould, and then bak'd in a Kiln.

II. *Kinds of.*] There are many kinds of Tiles, and those known by several Names; as *Plain*, *Thack*, *Ridge*, *Roof*, *Crease*, *Gutter*, *Pan*, *Crooked*, *Flemish*, *Corner*, *Hip*, *Dorman*, *Dormar*, *Scallop*, *Astragal*, *Traverse*, *Paving*, and *Dutch Tiles*: Of which I shall treat in the following Numbers.

III. { *Plain*— or *Thack*— } Of which I shall give

1. *Their Description*] They are the common or ordinary Tiles (of an Oblong Figure,) us'd about covering of Houses, *&c.*

2. *Their Dimensions.*] By the Statute of the 17th of *Edw. Cap.* 4th Plain Tiles ought to be in length 10 ½ Inches, in breadth 6 ¼ Inches, and in thickness half an Inch and half a quarter at the least. But by Observation, I find our *Sussex* Tiles to be of different Dimensions; for some I find to be 10 Inches long, 6 ¼ broad, and ⅝ of an Inch thick. Others I find to be but 9 ½ Inches long, 5 ¾ broad, and about ½ an Inch thick.

3. *Their Weight.*] Mr. *Leybourn* says, That one plain Tile weighs about 2 Pounds and ½; whence 100 of them will weigh 250 Pounds, and 1000 of 'em will weigh 2500 Pound. But by my Observations, one of the largest Size of those I measur'd, *viz.* (those of 10 Inches long) will weigh but about 2 Pound 3 Ounces, so that 100 of 'em will weigh about 220 Pounds, and 1000 of 'em about 2200 Pounds. And one of the other Size that I measur'd, weigh'd about 2 Pounds; so that 100 of 'em will weigh about 200, and 1000 of them about 2000 Pounds.

4. *Their*

4. *Their Price.*] They are in ſome places dearer, and in others cheaper, according to the Scarcity, or Plenty of the Earth whereof they are made, and of the Wood wherewith they are burnt. Mr. *Wing* ſays, They are from 25 to 30 *s.* the Thouſand in *Rutland-ſhire*; Mr. *Leybourn* ſays 25 *s.* the Thouſand in *London*; but about us in *Suſſex* they are ſold from 15 to 17 *s.* the Thouſand.

IV. { *Ridge*— / *Roof* — / or / *Creaſe*— } Of theſe I ſhall alſo give

1. *Their Deſcription.*] Theſe are ſuch Tiles as are us'd to cover the Ridge of a Houſe; they being made circular breadth-wiſe, like a half Cylinder.

2. *Their Dimenſions.*] Theſe, by the fore-mention'd Statute, ſhould be in length 13 Inches, and in thickneſs the ſame with plain Tiles. I have meaſur'd ſome of theſe, and found one of them to be 13 Inches long, about 16 broad by the Compaſs on the outſide, and in breadth (from ſide to ſide) on the inſide about 11 Inches, ſome not above 9 or 10 Inches.

3. *Their Weight*] I weigh'd of theſe kind of Tiles, and found him to weigh about 8 Pounds $\frac{3}{4}$. Whence 100 of 'em will weigh about 875 Pounds, and 1000 about 8750 Pounds

4. *Their Price.*] In ſome Places, ſays Mr. *Leybourn*, 5, 6, or 7 of theſe Tiles are allowed into every Thouſand of plain Tiles; but if bought by themſelves, they are ſold from 20 to 25 *s. per* Hundred. About us in *Suſſex*, they are ſold at 2 *d. per* piece, or 16 *s.* the Hundred.

V. { *Hip*—] / *Corner*—] } Of theſe I ſhall alſo give

1. *Their Deſcription.*] Theſe are to lie on the Hips, or Corners of Roofs. As to their Form, they are at firſt made flat like plain Tiles, but of a Quadrangular Figure, whoſe two *ſides* are right Lines, and two *ends* Arches of a Circle, one end being a little Concave, and the other Convex, which Convex End is about 7 times as broad as the Concave End; ſo that they would be of a Triangular Figure, were not one Corner taken off. Then before they are burnt, they are bent (upon a Mould) in their breadth, after the Manner of *Ridge-Tiles*. They have a Hole at their narrow end to nail them on by, and are laid with their narrow Ends upwards.

2. *Their Dimenſions*] By the Statute above-mention'd, (Num. III §. 2) The Tiles ought to be 10 $\frac{1}{2}$ Inches long, with convenient thickneſs and breadth. I have meaſur'd ſome of 'em, and find them to be in length 10 Inches, in breadth (according to their Compaſs) at the narrow end two Inches, and at the broad end 14 Inches; and the Right-lined breadth at the broad end, about 11 Inches.

3 *Their*

3. *Their Weight.*] I found the weight of one of these Tiles to be about 3 Pounds, and 3 or 4 Ounces; See Page, Num. 6. §. 3.

4. *Their Price.*] They are usually sold, (says Mr. *Leybourn*) at Three half-pence, or 2 *d. per* Tyle, or from 10 to 15 *s. per* hundred. About us in *Sussex*, they are usually sold for Three-half-pence a piece, or 12 *s* the Hundred.

VI. *Gutters.*] Of these I shall also give

1. *Their Description*] These are to lie in Gutters, or Valleys in cross Buildings. They are made like Corner-tiles, only the corners of the broad end are turn'd back again with two Wings; so that the broad end resembles the upper part of the Character for the *Sign Libra.* These have no holes in them, but are laid (with their broad ends upwards, and) without nailing at all.

2. *Their Dimensions.*] I suppose *these* are made in the same Mould as *corner Tiles*, for they have the same Dimensions on the out (or Convex) side. Their Wings (mentioned in the foregoing Section.) are each about four Inches broad, and 8 Inches long, pointing out short of the narrow end, about two Inches.

3. *Their Weight.*] These. (for the Reason mention'd in the foregoing Section.) are of the same weight with *corner Tiles.* So that 100 of either of these kinds of Tiles will weigh about 321, or 322 Pounds, and 1000 of them will weigh about 3210, or 3220 Pounds.

4. *Their Price*] They are of the same Price as corner Tiles, see above, Num. 5. § 4.

VII. { *Pan*— / *Crooked*— / or / *Flemish*— } Of these I shall give

1. *Their Description.*] They are us d in covering of Sheds, Lean-to's, and all kind of flat roof'd Buildings. They are in the Form of an oblong Parallelogram, as plain Tiles; but they are bent (breadthwise,) forward and backward in the Form of an S, only one of the Arches is at least three times as big as the other; which biggest Arch, or Hollow of the Tile is always laid uppermost, and the lesser Arch, or Hollow of another Tile, lies over the edge of the great Hollow of the former Tile. They have no Holes for Pins, but hang (on the Laths) by a knot of their own Earth.

2. *Their Dimensions.*] They are usually in length 14 $\frac{1}{2}$ Inches, and in breadth 10 $\frac{1}{2}$ Inches.

3. *Their Price.*] The Price of these Tiles in most places is about 7 or 8 *s* the Hundred.

VIII. { *Dormar*— / or / *Dorman*— } Of these I shall give

1. *Their Description*] These Tyles consist of a plain Tyle, and a triangular piece of a plain Tile standing up at right Angles to one side of the plain Tyle, and this triangular Piece at

at the broad end is about the breadth of the plain Tile; and ſwept with an Arch of a Circle from the other end, which other end terminates in a point, or has no breadth; and of theſe kind of Tiles there are two ſorts; for in ſome the triangular piece ſtands on the right, in others on the left ſide of the plain Tiles; and of each of theſe there are again two ſorts, for ſome have a whole plain Tile, others but half a plain Tile; but of all theſe ſorts, the plain Tile has two holes (for the Pins) at that end where the broad end of the Triangular piece ſtands.

2. *Their Uſe*] They are uſed to be laid in the Gutters betwixt the Roof and the Cheeks, or ſides of the Dormars, the plain Tile part lying upon the Roof, and the Triangular part ſtanding perpendicularly by the Cheek of the Dormar. They are excellent to keep out the Wet in thoſe Places, which 'tis very difficult to do without either *them*, or ſome Sheet-lead. Theſe Tiles are much us'd in ſome parts of *Suſſex*, the Bricklayers not caring to do any Healing (where there are Dormers) without 'em; tho' to my Knowledge, in ſome parts of *Kent*, they know not what they are; and I believe they are ignorant of them alſo in moſt other parts of *England*; For I never ſaw any Author that ſo much as mention'd them.

3. *Their Dimenſions.*] As to their Dimenſions, the plain Tile part is of the ſame Dimenſions as a plain Tile, both as to its length and breadth; the triangular Part is of the ſame length, and its breadth at one end 7 Inches, and the other Nothing.

4. *Their Weight.*] I have weigh'd one of theſe Tiles, and found him to weigh about 4 $\frac{1}{2}$ Pounds; whence 100 of them will weigh about 450, and 1000 about 4500 Pounds: This was a whole one, as $\frac{1}{2}$ one weigh'd 3 ℔. 2 ℥.

5. *Their Price.*] They are commonly ſold at Three-halfpence, or 2 *d. per* piece, or 12 or 16 *s.* the Hundred.

IX { *Scallop*— or *Aſtragal*— } Of theſe I ſhall give

Their Deſcription.] Theſe are in all Reſpects like plain Tiles, only their lower Ends are in the form of an *Aſtragal*, *viz.* a Semicircle with a Square on each ſide. They are in ſome places us'd for Weather-tyling, and look very handſome

I have not yet learn'd their *Price*, *Weight*, or *Dimenſions*; but I think the latter is the ſame as plain Tiles.

X. *Travers*] Theſe Tiles are (by our common Bricklayers) call'd *Travis*, or *Travas Tiles*; but I ſuppoſe it ſhould rather be *Travers Tiles*; for the word *Travers* is perfect *French*, ſignifying *Irregularity*; and theſe Tiles, which they call *Travers Tiles*, are only irregular plain Tiles, *viz.* Such as have the Pin-holes broken out, or one of the lower Corners broken off Theſe they lay (with the broken ends upwards) upon Rafters,

Rafters, where pinn'd Tiles cannot hang

XI. *Paving.*] These are by some call'd *Paving Bricks*; see Bricks, Num. 3. §. 10.

XII *Dutch.*] Of these I shall give

1. *Their Description*] Of these there are two Kinds, which I shall distinguish by the Appellations of ancient and modern: The ancient Dutch Tiles were us'd for Chimney-foot-paces: They were painted with some antick Figures, and sometimes with the Postures of Soldiers, *&c.* And sometimes with Compartments, and in them some irregular Flourishes; but in general they are nothing so well done, (nor with so lively Colours) as the Modern ones. The modern Dutch Tiles are commonly us'd instead of Chimney-cornerstones, (being plaister'd up in the Jambs, (see Corner-stones.) These Tiles seem to be better glazed, and those that are painted, (for some are only white,) are done with more curious Figures, and more lively Colours than the ancient ones: But both these sorts seem to be made of the same whitish Clay as our white glazed Earthen Ware. The modern ones are commonly painted with Birds, Flowers, *&c.* and sometimes with Histories out of the *New Testament.*

2. *Their Dimensions.*] Those which I call Ancient Dutch Tiles, are 5 $\frac{1}{4}$ Inches square, and about $\frac{1}{4}$ of an Inch thick. The modern Dutch Tiles are 6 $\frac{1}{2}$ Inches square, and $\frac{3}{4}$ of an Inch thick.

3. *Their Weight*] I have weigh'd some of both these sorts of Tiles, and I found one of

the {Ancient / Modern} sort to weigh {1 $\frac{1}{4}$ Pound, / 1 ℔. 3 Ounces} whence 100 of them will weigh {125 ℔, and 1000, 1250 / 169 ℔, and 1000, 1690} Pounds

XIII. *Method of Making and Burning.*] Tiles, (says Mr. *Leybourn*) are made of better Earth than Brick earth, and something near the Potters Earth. According to the Statute of 17 *Edw* 4. *Cap.* 4. Earth for Tiles should be cast up before the first of *November*, shired and turned before the first of *February*, and not made into Tiles before the first of *March*, and should likewise be tried and sever'd from Stones, Marle, and Chalk.

In *Sussex* and *Kent*, Tiles are commonly made of a kind of Clay: As to the particular Method of making them, I must omit it, as foreign to my Purpose. But for the Method of burning them, see *Bricks*, Num. 5. where you will find it at large.

XIV. *Price of Making and Burning.*] For making 1000 of plain Tiles, (says Mr. *Leybourn*) 2 *s.* or 2 *s.* 6 *d.* is the usual Price: But I know not how, or where he means; for an experi-

experienced Workman tells me, That for casting the Clay, and shireing it, and making it into Tiles, and burning them, they have 6 *s. per* 1000.

XV. *How many will cover a Square*] This is various, according to the width they gage for the Laths : At 6 *Inches Gage*, about 800 will cover a Square ; at 6 ½ *Inch Gage*, 740 Tiles will cover a Square ; at 7 *Inch Gage*, 690 ; at 7 ½ *Inch Gage* 640, and at 8 *Inch Gage* 600 Tiles will cover a Square, or 100 superficial Feet. These Numbers suppose the breadth of the Tiles to be 6 Inches ; for (if they are Statute Tiles) they will be thereabouts when they are burnt, allowing ¼ of an Inch for their shrinking with burning. If your Tiles are broader than 6 Inches, then fewer will cover a Square, if they are narrower there must be more.

Tiling.

1. *What*] By *Tiling*, is meant the covering the Roof of a Building with Tiles.

2. *Of Measuring*] Tiling is measur'd by the *Square* of 10 Foot, *i. e.* 100 superficial Feet. And in taking their Dimensions, they measure to the middle of the *Gutters*, *Corners*, and *Ridge tiles* ; and having cast up the Area, they have a Custom to make an *Addition* for all *hollow Ware*, (as they call *Ridge-tiles*, *Corner*, *Gutter*, and *Dormar-tiles*,) and this *Addition* (I think) is in *London* one superficial Foot for every such Foot of such hollow Ware But I am sure, in some parts of *Spain*, 'tis the Custom to reckon one superficial Foot for every such Tile ; 100 of which they reckon one Square of Work, and add it to the Area before found.

3. *Price of*—] Tiling is commonly done by the Square, which in *new Work*, (says Mr. *Leybourn*) and the Workman finding all Materials, as Tiles, Mortar, Laths, and Nails, is usually valu'd at 30, or 32 *s. per* Square. (Mr. *Hatton* reckons but 28 *s. per* Square.) And for ripping of old Work, and new Covering, and making good the old, they reckon 12 or 14 *s.* the Square, according as they find the old Tiling.

But for Workmanship only they reckon for *new Work* 5 *s. per* Square at *London*, in the Country various. Mr. *Wing* says, 3 *s.* in *Rutland*, in some places, says he, 2 *s.* 6 *d.* In several parts of *Sussex*, I know 'tis commonly done for 3 *s. per* Square, and I am inform'd (at second hand) that in some parts of *Kent* they do it for 2 *s* 6 *d. per* Square ; but then their Tiles are large, and they lath wide, at 8 Inches Gages, and pin but half their Tiles, the other half they lay Travers Tiles.

And for Ripping, and Healing again, (only Workmanship) our *Sussex* Bricklayers reckon 3 *s.* 6 *d. per* Square, and if they Counter-lath it, then 3 *s.* 9 *d* or 4 *s.* But in some parts of *Kent*, they Rip, and Heal, and Counter-lath, for 3 *s per* Square, which is very cheap ; but then 'tis suppos'd their Work is done accordingly.

4. *Laths and Nails to a Square of*—] For the number of Laths and Nails, commonly allow'd to

to a Square of Tyling. See Laths, N 8, and Nails, N. 23.

5. *Mortar to a Square of*—] Mr. *Leybourn* says, That about a quarter as much Mortar as is allow'd to a Rod of Brick-work, will do for a Square of Tyling. See Mortar, N. 12.

6. *Pins to a Square of*—] Mr. *Leybourn* says, They usually allow a Peck of *Tyle pinns* (from 2 s. to 4 s. the Bushel) to every Thousand of Tyles; but surely this must be a Mistake, for an Experienced Workman tells me, He uses but about a Peck of Pins to three Square of Healing, which at seven Inch Gage, (the size he commonly Gages) is more than enough for 2000 Tyles. And I think this Work-man told me he reckon'd *Tyle pinns* at 6 *d. per* Gallon.

7 *Without Mortar*—] Some lay Tyles without Mortar, or any thing else, laying them dry as they come from the Kiln. Others lay them in a kind of Mortar made with Lome and Horse-dung, (See Mortar, N. 20.) In some parts of *Kent* they have a way of laying Tyles in *Moss*; when the Workmen get the Moss themselves, they are allow'd 2 *d.* in a Square the more for their Work. But an old Workman of theirs condemns this way of Tyling with Moss; for he tells me, That in windy wet Weather, when the Rain, Snow, or Sleet is driven under the Tyles (in the Moss) if there follow a Frost while the Moss is wet, it then freezes and raises the Tyles out of their Places.

8. *With Pan-tyles.*] These Tyles are for the most part laid dry without any Mortar; yet sometimes pointed within side.

The Laths whereon they hang, are 10 or 12 Foot long, an Inch and an half broad, and an Inch thick. They are usually sold at 2 *d.* or 3 *d.* the Lath, or at 10 or 13 *s.* the Hundred.

The Gage for nailing on these Laths (with 4 *d* Nails) is ten Inches and an half, and the breadth of a Tile when laid eight Inches; whence about 170 Tiles will cover a Square, (or 100 Foot) of this kind of Tiling.

A great Covering with these, spends but little Mortar (if pointed) and but little Time in laying. Mr. *Wing* reckons it worth about 1 *s.* 8 *d. per* Square, Workmanship.

9. *Of its Weight.*] See *Horsham-stone*, N. 4.

Valleys.

THE Gutters over the Sleepers in the Roof of a Building. See *Gutters.*

Vestible.

Among the Ancients was a large open Space before the Door, or at the Entry of a House which they call'd *Atrium Populatum & Vestibulum*, being dedicated, as *Martinus* tells us, to the Goddess *Vesta*, whence he will have the word derived, *i. e. Vestæ stabulum*, it being

being uſual for People to ſtop here, before they went within Doors. The word may likewiſe be derived from the *Latin Veſtis*, a Garment, and *ambulare* to walk; becauſe the Veſtible in the modern Houſes being an open Place at the bottom of a large Stair-caſe, ſerving as a Thorough-fare to the ſeveral Parts of the the Houſe, 'tis here that the Robes are firſt let fall in Viſits of Ceremony. *Veſtible* is alſo ſometimes uſed to ſignifie a little kind of Anti-chamber before the Entrance of an ordinary Apartment.

Vault.

Is a Piece of Maſonry, arch'd without-ſide, and ſupported in the Air by the Artful placing of the Stones which form it; its principal Uſe being for a Cover or Shelter. The chief Vaults in a Building are call'd Maſter-Vaults, to diſtinguiſh them from the other leſs conſiderable ones, which only ſerve to cover Gates, Windows, Paſſages, *&c.* Double Vaults are thoſe which are built over another, to make the Beauty and Decoration of the Inſide, conſiſtent with that of the Outſide: A Chaſm or Vacancy being left between the Convexity of the one, and the Concavity of the other; Inſtances of which we have in the Dome of St. *Peter*'s at *Rome*, St. *Paul*'s in *London*, and in that of the *Invalides* of *Paris*.

Under-pinning.

1. *What.*] By this *Term* is meant, the bringing it up with Stone under the Ground-ſells of a Building. Sometimes it ſignifies the Work it ſelf, when done.

2. *Price of*] In ſeveral parts of *Suſſex*, I know the uſual Price (for the Workmanſhip only) is 1 *d. per* Foot Superficial. In ſome Parts of *Kent* they have three Half-pence *per* Foot. In ſome places 'tis the Cuſtom (in Meaſuring it) to take in half the Sell into their Meaſure.

Volute.

From *Volvo* to fold, is one of the principal Ornaments of the Ionick and Compoſite Capital repreſenting a kind of Bark wreath'd or twiſted into a Spiral Scroll. There are eight angular *Volutes* in the *Corinthian Capital*, and theſe are accompanied with eight other little ones, call'd *Helices*.

Urn.

Comes from the *Latin*, *Urna*, a Veſſel to draw Water in, and ſignifies a low wide Vaſe, ſerving as a Crowning over Balluſtrades, and as an Attribute to Rivers, River-Gods, *&c.* in the Grotto's and Fountains in Gardens. A Funeral *Urn* is a kind of cover'd Vaſe enrich'd with Sculpture, and ſerving as the Crowning or Finiſhing of a Tomb, a Column, Pyramid, or or other Funeral Monument; made in imitation of the Ancients, who deposited the Ashes of their deceased Friends in this kind of *Urn*.

Voussoirs.

The Stones that form the Arch: A *French* Word.

Wainscot,

THE Pannel'd Work round (against the Walls of) a Room.

Wainscotting.

1. *What.*] the making, and setting up of Wainscot is called *Wainscotting*.

2. A *Note* in—] Some Joyners, (as I am inform'd) put Charcoal behind the Pannels of their Wainscot, to prevent the Sweating of Stone and Brick-walls from unglueing the Joynts of the Pannels, which otherwise, (especially in some places) 'tis very apt to do; and others make use of Wool in the same manner, and for the same purpose; yet neither of these ways will prevent their unglueing in some Houses: But the most effectual way to prevent it, is by priming over the Back-sides of the Joynts well with *White-lead*, *Spanish-brown*, and *Linseed-oyl*.

3. *Of Measuring.*] Wainscot is generally measur'd by the Yard square, *i. e.* nine Superficial Feet. Their Custom is to take the Dimensions with a String, pressing it into the Mouldings; for they say, (and 'tis but Reason) we ought to be paid for all where the Plain goes.

Therefore when Joyners would take the Dimensions of a Room they have Wainscotted; they take up a Line on the top of the corner of the Room, and as they carry it down to the bottom, they press it (with their Fingers) into all the Mouldings; this they account the *breadth* and (they measure) the Circumference of the Room from the *length*: Some Joyners will measure this also with a String, but others do not. The Dimensions being thus taken in the Feet, they multiply the length by the breadth, and the Product is the Content in Feet; which being divided by 9, the Quotient is the Content in Yards, But.

Note, (1.) That you must make Deduction for all *Window Lights*, and measure the *Window boards*, *Cheeks*, and *Sapheta's* by themselves.

(2) That for *Window-shutters*, *Doors*, and such things as are wrought on both sides, they reckon *Work and half*; for indeed the Work is half more.

(3.) That *Cornices*, *Bases* and *Sub-bases* are sometimes measur'd by the Foot, Lineal Measure so also are *Freezes*, *Architraves*, and *Chimney pieces* measur'd; unless agreed for by the Great.

(4) *Price of*—] The Price of *Wainscotting* is various, according to the Variety of *Stuff* and *Workmanship*..

Wain-

Wainſcotting with *Norway Oak*, the Workman finding Stuff, is worth 6 or 7 *s. per* Yard. The Workmanſhip only is about 2 *s.* in *London*, in *Rutland* 3 *s.* 6 *d.* or 4 *s. per* Yard; and if the Mouldings are large, 5 *s.* ſays Mr. *Wing.*

Plain-ſquare Wainſcotting, (the Workman finding Deal) is worth 3 *s* or 3 *s.* 6 *d. per* Yard. For only Workmanſhip about 1 *s. per* Yard.

Ordinary Biſection Wainſcotting, (the Workman finding Deal) is worth in *London* 3 *s.* 6 *d*; in the Country, 4 *s.* 6 *d. per* Yard. The Workmanſhip only about 1 *s.* 6 *d. per* Yard.

Large Biſection-work is worth 6 or 7 *s. per* Yard of Dantzick Stuff.

5. *Of Painting of Wainſcot*] See *Painting.*

Walls.

I. *What.*] By this Term in Architecture is meant the *Incloſures* of whole Houſes, or particular Rooms; as alſo of Gardens, Orchards, *&c.* if made of Brick or Stone. Walls are either entire and continual, or intermitted; and the Intermiſſions are either Pillars or Pylaſters

II *Kinds of*—]There are ſeveral kinds of Walls, diſtinguiſhable by different Names, according to the ſubſtance whereof they are made, as *Plaſter'd* or *Mud-walls*, *Brick-walls*, *Stone-walls*, *Flint*, or *Boulder-walls*, and *Boarded-walls*; of all which I ſhall diſcourſe in the following Numbers.

III. *Plaiſter'd*, or *Mud* —.] Theſe kind of Walls are common in Timber Buildings, eſpecially of ordinary Buildings; for ſometimes the Walls are made of Brick betwixt the Timber: But this is accounted no good way; becauſe the Mortar corrodes and decays the Timber.

Theſe Mud-walls, (as they are call'd in ſome places) are thus made. The Walls being quarter'd and lathed between the Timber, (or ſometimes lathed over all) are Plaſter'd with *Lome*, (See *Lome*, alſo, ſee *Mortar*, N. 8. and 11.) which being almoſt dry is Plaiſter'd over again with *white Mortar*, (See *Mortar*, N. 4.)

This kind of Work is commonly meaſur'd by the Yard. For the Price of it, See *Pargeting*, N. 2. and *Plaiſtering*, N. 1.

IV. *Brick*—.] Here I ſhall ſay ſomething.

1. *Of Building them*] And here are ſeveral things to be conſider'd and taken notice of; as firſt, That all *Walls* ought to be moſt exactly Perpendicular to the Ground-work; for the right Angle (thereon depending) is the true cauſe of all *Stability*, both in Artificial and Natural Poſition, a Man likewiſe ſtanding firmeſt when he ſtands uprighteſt.

Secondly, That the maſſieſt and heavieſt Materials be the loweſt, as fitter to bear than to be born.

Thirdly, That the *Walls* as they riſe, diminiſh (proportionally) in thickneſs, for eaſe both of Weight and Expence.

Forthly,

Fourthly, That certain Courſes, or Ledges of more ſtrength than the reſt, be interlay'd, like Bones, to ſuſtain the Fabrick from total Ruin, if the under-parts ſhould decay.

Fifthly, That (all along) care be taken in laying the Bricks, concerning which, ſee *Bricks*, N. 8.

Sixthly, That the Angles be firmly bound, which are the Nerves of the whole Edifice; and therefore are commonly fortify'd by the *Italians*, even in their Brick-buildings, on each ſide of the Corners, with well ſquared Stone, yielding both Strength and Grace.

Seventhly, In working up the Walls of a Building, do not work any Wall above three Foot high before you work up the next adjoyning Wall, that ſo you may joyn them together, and make good Bond in the Work: For 'tis an ill Cuſtom among ſome Bricklayers, to carry, or work up a whole Story of the Party-walls, before they work up the Fronts, or other Work adjoining, that ſhould be bonded, or work'd up together with 'em, which occaſions cracks and ſetlings in the Walls.

Eighthly, That if you build (a Houſe) in the City of *London*, you muſt make all your Walls of ſuch thickneſſes, as the *Act of Parliament* for Re-building of the ſaid City enjoyns; (which *Act* you may ſee in *Houſe*. N. 4.) but in other places you may uſe your Diſcretion; yet for ſome Directions in this Matter. See *Houſe*, N. 3

Ninthly, It may be worth your Notice, that a Wall of a Brick and half thick, with the Joynt, will be in thickneſs 14 Inches, or very near; whence 150, or 160 Bricks will lay a Yard Square meaſur'd upon the Face of the Building, and to the Square of 10 Foot (which is 100 ſquare Feet) are uſually allow'd 1700, or 1800 Bricks, and 4600, or 5000 Bricks will compleatly lay, erect, or build one Rod, Pole, or Perch ſquare; which Rod, Pole, or Perch, (for by all theſe Names 'tis call'd) contains in length, (according to the Statute) 16 ½ Feet; whoſe Square is 272 ¼ Feet, ſuperficial Meaſure, which is 30 Yards and a quarter.

But tho' I have here laid down the number of Bricks for each of theſe Squares, yet theſe Numbers are not to be rely'd on as abſolutely exact; for no exactneſs can be diſcover'd as to this Particular, and that for ſeveral Reaſons: For tho' the Bricks were all made in the ſame Mould, and burnt in the ſame Kiln, or Clamp; yet the Nature, or Quality of the Earth whereof they are made, (whereby ſome ſhrink more than other ſome) and the Bricklayers Hand and Mortar, may cauſe a conſiderable Variation, and beſides ſome Bricks are warp'd in burning, (whereby they will not lie ſo cloſe in the Work) ſome miſcarry, (or are broken) in every Load, or 500 Bricks, and the Tally, or Tale is (for the moſt part, if not look'd after) too little: And beſides all theſe Uncertainties, when Bricks are dear, and Lime cheap, the Workman

man (by the Great) will uſe more Mortar, and make the ampler Joynts, which is much worſe for the Building.

Tenthly, It may be alſo noted, That (when all Materials are ready) a Workman with his Labourer will lay in one Day 1000 Bricks, and ſome 12 or 1500.

Eleventhly, All Brick-work, according to theſe Rules, is ſuppos'd to be one Brick and half thick, which is the Standard Thickneſs. If they are thicker, or thinner, they muſt be reduc'd to that thickneſs, as ſhall be ſhewn how in the next Section of this Number.

I *Of Meaſuring them*] Brick-layers moſt commonly Meaſure their Walls by the Rod ſquare, each Rod, Pole, or Perch, (for by all theſe Names 'tis call'd) being (by the Statute) 16 ½ foot long; ſo that a ſquare Rod contains 272 ¼ Superficial Feet.

Therefore having taken the Dimenſions, (*viz.* the length, and heighth) of a Wall in Feet, multiply the length by the heighth, (See Croſs-multiplication, N. 2.) and divide the Product by 272 ¼, and the Quotient ſhews the number of ſquare Rods in the Superficies of that Wall. But it being troubleſome to divide by 272 ¼, Workmen commonly have a Cuſtom to divide 272 only, which gives the Content ſomething more than the Truth, which notwithſtanding they take for it.

Having thus found the Area, or Content of the whole Superficies of a Wall, they next conſider its Thickneſs; for they have a certain Standard Thickneſs, to which they reduce all their Walls, and this Standard is *one Brick and a half thick*, as they phraſe it, (*i. e.* the length of one Brick, and the breadth of another, ſo that a Wall of three Bricks (length) thick of the ſame heighth and length with another of 1 ½ Brick thick, the former will contain twice as many ſquare Rods as the latter.

Now, to reduce any Wall to this Standard thickneſs, take this plain and eaſie Rule: Say, as 3 is to the thickneſs of the VVall in half Bricks, [that is in the breadth of Bricks, the breadth of a Brick being always half its length,] ſo is the Area before found, to the Area at their Standard thickneſs of 1 ½ Brick.

Thus, If the Wall be all of one thickneſs from the Foundation to the top, it is eaſily reduc'd to the Standard Thickneſs of 1 ½ Brick. But if the Wall be of different thickneſſes (as in Brick Houſes they commonly are, being made thickeſt below, and thinner at every Story;) then the beſt way is to meaſure every different thickneſs by it ſelf and reduce it to the Standard thickneſs; then add all theſe ſeveral Area's into one Sum, out of which deduct the Doors and Windows (meaſur'd by themſelves,) and ſo the Remainder will be the true Area, or Content of the whole Wall.

See more (concerning meaſuring of Brick walls.) N. V. of this Word, *viz* in Fence-walls. Alſo ſee *Brick-work*.

Note, In ſome Places 'tis the Cuſtom to meaſure by the Rod of 18 Foot long, in others by the

the Rod of 16 Foot: In the former Case, you must divide the Area in Feet by 324, in the latter by 356.

3. *Of their Price.*] The Price of Building of Walls is various in different Places, according to the various Prices of Materials. Mr. *Leybourn* says, (and with him agrees Mr. *Hatton*,) that the usual Price in *London*, for building a Brick and half Wall, (the Workman finding all Materials) is five Pounds, or five Pound ten Shillings *per* Rod square. And for the Workmanship only 30 *s per* Rod square, which is about 1 *s. per* Yard square

Mr. *Wing* says, That the usual Price in *Rutland* (the Workman finding all Materials) is for a Brick and half Wall 3 *s per* Yard square, [which is but about 4 *l.* 10 *s. per* Rod] for a two Brick-wall 4 *s* for a 2 ½ Brick-wall 5 *s. per* Yard Square. And for the Workmanship only (of a Brick and half Wall) 8 *d per* Yard square, which is but about 20 *s. per* Rod, Statute Measure. So that you see Mr. *Wing*'s Prices are much cheaper than those about *London*; the Reason of which, I conceive proceeds from the Cheapness of Commodities in his Countrey.

About us in *Sussex*, a Rod of a Brick and half Wall, Workmanship and Materials, will cost at least eight Pounds. For the Workmanship only, the usual Price (about us) is 24, or 25 *s. per* Rod square in a Brick and half Wall.

It should seem, that in or about *London*, Workmen do sometimes find only Mortar and Workmanship in building of Walls; for (says Mr. *Leybourn*) if the Bricks are laid in at the Builder's Charge, then 2 *l.* 10 *s. per* Rod is the usual Price. But (says he) to erect new Structures, by taking down old Walls, it may be worth 3 Pounds, or 3 Pounds 10 Shillings *per* Rod; because, in taking down the Walls, and clearing the Bricks, there is much Time spent, and also more Mortar us'd in laying them again, than in new Work.

V *Fence* —] Walls built round Courts, Gardens, Orchards, *&c* are commonly call'd *Fence-walls*. Of these, some are made of Stone, some of Flints, or Boulders, and some of Brick; Of the two former, I shall speak in the two following Numbers, *viz.* Num 6, and 7. Of the latter I shall speak here, and therein I shall say something

1. *Of their making*] These are commonly made (of Statute Bricks) a Brick and half thick

But in some parts of *Sussex* they are commonly made of a sort of great Bricks, which are 12 Inches long, 6 Inches broad, and 3 Inches thick. I have very often discours'd with the old Man who first introduced, not only those sort of *Great Bricks*, but also their necessary Concomitants, *Filaster* and *Copeing Bricks*, and the Method of making Fence-walls of 'em, V. P Bricks, Num. 3. §. 4. 9. and 13.

These Walls are but the breadth of a Brick, (or 6 Inches)

Inches) in thickneſs, only at the Pilaſters, where they are the length of a Brick, (or 12 Inches thick.) They uſually ſet a Pilaſter at every 10 Foot. I know a. Wall of theſe ſort of Bricks, (of about 9 Foot high) that has been built near 30 Years, and ſtands very well.

2 *Of meaſuring them.*] Fence walls built of Statute Bricks, are commonly meaſur'd, as is taught above, Num. 4. §. 2. But I ſhall here add. That ſome Workmen that I know, meaſure them by the Rod in length, and one Foot in heighth, which they account a Rod of Meaſure And in taking their Dimenſions they do it with a Line, going over the Pilaſters ; this for the length ; ſo likewiſe for the heighth, they meaſure it (alſo) by a Line, going over all the Mouldings, (after the manner of Joyners meaſuring their Work) even to the top, or middle of the Copeing.

I ſhall further add, That ſome Workmen (in Fence-walls of Statute-bricks) will, (if they can perſwade their Maſter to it) meaſure all that is above 1 ½ Brick thick, (*viz* The projecting of the Pilaſters, or Buttreſſes, and all below the Water-table) by the ſolid Foot, which afterwards they reduce to Rods. But this way is a conſiderable Advantage to the Workman, and a loſs to the Maſter Builder ; for it makes ⅐ part of Meaſure more than the Truth ; becauſe a Brick and half Wall is 14 Inches thick.

Fence-walls built of great Bricks, are generally meaſur'd by the Rod in length, and a Foot in heighth, (which they account a Rod of Meaſure,) the Dimenſions being taken by a Line, as was ſaid above.

3. *Of their Price.*] For the Price of Brick-walls, See (above,) Num. 4. §. 3. But ſome Workmen in *Suſſex* reckon for building of Fence-walls, (the Workmanſhip only) of Statute Bricks (a Brick and half thick) 1 *s.* 6 *d per* Rod, at a Rod long, and a Foot high, taking their Dimenſions by the Line, as was ſhewn how in the preceeding Section of this Number Sometimes they build theſe kind of Walls by the Square of 100 Foot, at 8 *s. per* Square, which is but about 1 *d. per* (Superficial) Foot

For building of Fence-walls with great Bricks, the common Price (for the Workmanſhip only) is 1 *s. per* Rod, at one Rod long, and one Foot high, the Dimenſions taken by the Line, as above.

4. *Of Coping them.*] Fence-walls built of Statute Bricks, are ſometimes coped with Stone, ſometimes with Brick : If the former, the copeing is left out in the Meaſure, and rated by it ſelf; for the Price of which, ſee *Copeing*, Num 2. If the latter, it is meaſur'd into the reſt of the Work. And this kind of Copeing is done thus; on one ſide the Wall is carry'd upright to the top, and on the other ſide there is two Courſes of Bricks ſtanding on end in an Oblique Reclining, or Slant Poſition, and a ſtretch-

a ſtretching Courſe on the top finiſhes the Wall

But Fence-walls built of great Bricks, are coped with *copeing Bricks*, of which ſee *Bricks*, Num. III. §. 3. And this Copeing is alſo meaſur'd and rated with the reſt of the Wall.

VI. *Stone.*—] Stone-walls ſerve not only for Walls of Houſes, *&c.* but alſo for Fence-walls round Gardens, *&c* Of theſe I ſhall ſay ſomething.

1. *Of Meaſuring them*] Theſe are in ſome Places meaſur'd by the Rod of 18 Foot Square: But in moſt Places (I think, they are meaſur'd by the Foot ſuperficial. Concerning meaſuring of Walls, there are theſe three Things to be further taken notice of, *viz.* That if the length of the Walls at the *ends* (of Garden, or Houſe,) be taken on the out-ſide (of Garden, or Houſe,) then the length of the Walls on the *ſides* (of the Garden, or Houſe) ought to be taken on the inſide. (2.) That when the Walls of a Houſe are meaſur'd, the Doors and Windows are likewiſe to be meaſur'd, and deducted from the whole. (3.) That in Meaſuring Fence-walls, they commonly meaſure the heighth by a Line (preſs'd into all the Mouldings) from the top of the Copeing, to the bottom of the Foundation.

2. *Of their Price.* Mr. *Wing* tells us, That Fence-walls, and Walls of ordinary Buildings, are each (only the Workmanſhip) from 16 *s.* to 3 *l.* 10 *s per* Rod of 18 Foot Square, which (ſays he) depends upon the Goodneſs of the Work. He alſo tells us, That the ſetting of Fronts in great Buildings, *viz.* Aſhlar, Architrave, Windows, and Doors, with the Ground-table, Faſcia's, and other Members, is worth from 3 *l* 10 *s.* to 5 *l. per* Rod, which (ſays he) depends upon the heighth, and well performing of the Building. The Truth is, I don't well underſtand what he means by all this Tattle; for he never tells us any thing of the thickneſs of the Walls; and beſides 3 *l.* 10 *s. per* Rod, is but little above 2 ½ *d. per* Foot; and 5 *l. per* Rod, is but little above 3 ½ *d per* Foot; either of which is certainly too little for ſuch Ornamental Work, as ſetting of Fronts in great Buildings. And then, for his Fence-walls, or Walls in ordinary Buildings; I can't ſee how the goodneſs, or badneſs of ſuch plain Work can vary the Price from 16 *s.* to 3 *l.* 10 *s. per* Rod; but ſurely it muſt be very ordinary Work that is worth but 16 *s. per* Rod, which is but little above a Half-penny a Foot

Mr. *Hatton* talks much after the ſame manner; for, ſays he, one Foot of plain Work, (as Walls, *&c.*) is worth about 8 *d* working and ſetting. He mentions nothing of the thickneſs neither.

But I ſhall leave theſe Authors in the dark, as they have left *us*, and proceed to tell *you*, what ſome experienced Workmen in *Suſſex* tell me; namely, That for building a 12 Inch Wall,

Wall, they have 2 *d per* Foot; for an 18 Inch Wall 3 *d* and for a Wall of two Foot thick, they have 4 *d per* Foot. These Prices are to be underſtood of Walls that have two fair Sides; for if they have but one fair Side, (the other ſtanding againſt a Bank,) they have a leſs Price, for in this Caſe, I have known ſome Workmen build a Wall two Foot thick, for 2 *d.* ½ *per* Foot.

VII. *Flint, or Boulder* — Walls of Flints, or Boulders, are much us'd in ſome parts of *Suſſex* and *Kent*, where I have ſeen, not only Fence walls, round Courts, Gardens, *&c.* but alſo Walls of Stables, and other Out-houſes built of them, which ſhew'd very handſome

To build Walls and greater Works of Flint, whereof we want not Example in our Iſland, and particularly in the Province of *Kent* (ſays Sir *Henry Wotton)* is (as I conceive ſays he) a thing utterly unknown to the *Ancients*, who obſerving in that *Material* a kind of *Metalical* Nature, or at leaſt a *Fuſibility*, ſeem to have reſolved it into nobler Uſe; an Art now utterly loſt, or perchance kept up by a few *Chymicks*.

Some Workmen tell me, That for building of Flint, or Boulder Walls, they uſe to have 12 *s. per* Hundred, (for ſo they phraſe it,) by which they mean 100 ſuperficial Feet; but I forgot to ask them at what thickneſs, or whether they have but one thickneſs for all their Walls. They alſo tell me, That a right and left handded Man fit well together for this ſort of Work; for they have a Hod of Mortar pour'd down upon the Work, which they part betwixt them, each ſpreading it towards himſelf; and ſo they lay in their *Flints*. They alſo tell me, That their *Mortar* (for this Work) muſt be very ſtiff, and that 'tis beſt to have a good length of Work before 'em; for they work but one Courſe in heighth at a time; for if they ſhould do more, it would be apt to ſwell out at the Sides, and run down. They alſo ſay, That in miſty Weather 'tis very difficult to make the VVork ſtand.

VIII. *Boarded*—] Sometimes VValls are boarded, particularly the VValls of ſome Barns, Stables, and other Out-houſes But of this kind of VVork, ſee VVeatherboarding.

Walling.

The making of VValls (of what kind ſoever) is call'd *Walling*; therefore for the Price, *&c.* of *Walling*, ſee *Walls.*

Walnut-tree-painting.

See *Painting.* Num. 4.

Waſh houſe.

A Room to waſh in.

Water-table.

In Stone, or Brick-walls, is a ſort of Ledge left in the VVall, ſome 18 or 20 Inches, (more

(more or leſs) above the Ground, at which Place the thickneſs of the VVall is abated, (or taken in) on each ſide the thickneſs of a Brick, (in *Brick-walls*), namely, two Inches and a quarter; thereby leaving that Ledge, or Jutty, that is call'd a Water table. Theſe Water-tables are ſometimes left plain, and ſometimes they are wrought with Mouldings; if the latter, (beſides the plain Meaſure of the Wall), they are rated at ſo much *per* Foot, running Meaſure.

Water-courſes.

Theſe are commonly rated by the Foot running Meaſure, *viz.* If the VVorkman find Materials at about 10 *d per* Foot, if he find no Materials, at about 8 *d.*

Weather-boarding.

1. *What*] A Term of *Architecture*, ſignifying the nailing up of Boards againſt a *Wall*, (ſee *VValls*, Num. 8) Sometimes 'tis us'd to ſignifie the Boards themſelves, when nail'd up. This VVork is commonly done with Feather-edg'd-boards, (ſee *Feather-edg'd.*) In plain Work they nail the thick Edge of one Board, an Inch, or an Inch and half over the thin Edge of another: But if the Work is to be a little extraordinary, they ſet an O.G on the thick Edge of every Board.

2. *Price*] The Price of plain Weather-boarding, (*viz* fitting and nailing up the Boards) is from 8 *d* to 12 *d.* the Square, according to the length and breadth of the Boards, and Conveniency of the Place. But if the lower, (*viz.* the thicker) Edge of the Boards be wrought with an O.G, it may be worth 18 *d. per* Square. This for the Workmanſhip only. But

If the Workman find the Materials, (*viz.* Boards and Nails,) it may be worth 12, or 13 *s. per* Square, or about three Half pence *per* Foot.

Weather-tyling.

1. *VVhat.*] Is the Tiling, (or Covering with Tiles) the upright ſides of Houſes.

2. *Price*] In ſome Places *Weather tiling* is done at the ſame Price as other plain Tiling. See *Tyling*, Num. 3. But in other Places they have more, in conſideration of Scaffolding; for ſome Workmen tell me they have 4 *s. per* Square for Workmanſhip only.

Well-hole.

The Hole left in a Floor for the Stairs to come up through.

White-Painting

See *Painting* Num. 7.

Wind-beam.

The ſame as Collar-beam.

Windows.

Windows.

1. *What*] Every one knows that Windows are those parts of a Building that are made to let in the Light.

2. *Situation of*—] Concerning the Situation of Windows, oberve, first, that they be as few in number and as moderate in Dimensions, as may possibly consist with other due Respects: For in a word, *all Openings are Weakenings.* Secondly, Let 'em be placed at convenient Distance from the Angles, or Corners of the Building; because that part ought not to be open and infeebled, whose Office it is to support and fasten all the rest of the Building Thirdly, Besure take great care that all the Windows be equal one with another in their Rank and Order; so that those on the right Hand may answer to those on the left and that those above may be right over those below; for this Situation of Windows, will not only be handsome and uniform, but also the *void* being upon the *void*, and the *full* upon the *full* 'twill be a great strengthning to the whole Fabrick.

3. *Dimensions of*—] In making of Windows you must be careful, not to give them *more*, or *less* Light than is needful, that is, make them no *bigger*, nor less than is convenient; wherefore you ought to have regard to the bigness of the Rooms that are to receive the Light; it being evident, that a *great Room* has need of *greater Light*, and consequently of a *greater Window*, than a *little Room*, *& è contra.*

The Apertures of Windows in middle siz'd Houses, may be 4 ½ or 5 Foot between the Jambs, and in greater Buildings they may be 6 ½, or 7 Foot, and their heighth may be double the length at the least But in high Rooms, or larger Buildings, their heighth may be a third, a fourth, or half their breadth more than double their length.

These are the Proportions for Windows of the first Story, and according to these must all the rest of the Windows in the upper Stories be for their breadth; but for their heighth they must diminish: For the second Story may be one third part lower than the first, and the third Story one fourth part lower than the second.

4. *Price of making.*] Window-frames are usually agreed for by the Light (says Mr. *Leybourn*) so that if a VVindow have four Lights, and it be double rabitted, (as the Workmen call it) it may be worth 12 *s.* that is 3 *s.* a Light for Materials and VVorkmanship. But if the Builder find Timber and Sawing, then 1 *s.* a Light is fair.

Transom-windows, (says Mr. *Wing*) are worth making (for great Buildings) 1 *s.* 9 *d. per* Light, or 7 *s per* Window. Some Workmen tell me, they make 'em for 12 *d.* 14 *d.* 16 *d.* or or 18 *d. per* Light, according to their bignefs.

Luthern Windows, fays Mr. *Wing*, the making and fetting up, are valu'd from 9 to 14 *s. per* Window, according to their bignefs. Some Workmen tell me, that (if they faw the Timber) they commonly have 20 *s. per* Window.

Shop Windows, (fays Mr. *Leybourn*) will be afforded at the fame Rate as plain or batton'd Doors. See *Doors*.

5. *Price of Painting*] The Painting of Widow-frames, (fays Mr. *Leybourn*) is not ufually meafur'd, but valu'd at 3 *d* 4 *d*. or 6 *d. per* Light, according to their bignefs, and Cafements at three Half-pence or 2 *d. per piece*, and Iron-barrs at 1 *d* or more, if very large See *Painting*.

Withs.

Thefe are us'd by Thatchers to bind their Thatching-rods to the Rafters. They are commonly fold at 6 *d* the Hundred, and a Hundred of 'em will do about three Square of Thatching; for fome Workmen tell me, That they ufe about 33,. or 34 Withs, and as many Thatching-rods, (which are of the fame Price with the Withs) in a Square; for they bind down their Straw at every Foot, or thereabouts, *viz*. at every other Lath; (for they Lath but 2 Laths in a Foot,) and each Courfe of Thatching (bound down with one length of Rods.) is about three Foot in breadth.

Xyftos.

AMong the Ancient *Greeks* was a *Portico* of uncommon Length, either covered or open, wherein the *Athletæ* ufed to exercife themfelves in running Races and Wreftling. The word is derived from *Xyein*, to polifh, it being their cuftom to anoint their Bodies with Oil before the Encounter, to prevent their Antagonifts from faftning Hold of them. The *Romans* too, had their *Xyftus*; which was a long *Ifle* or *Portico*, fometimes roofed over, and at other times open, and ranged on each fide with Rows of Trees, forming an agreeable Place for the People to walk in.

Zoccolo.

THE fame as *Plinth* It is an *Italian* word, and fignifies a fort of wooden Shoes or Sandals: From the *Latin*, *Soccus*, the Buskin wore by the ancient Actors. In *Architecture* it is a fquare Body, lefs in Heighth than Breadth, and placed under the *Mouldings* of the *Bafes* of *Pedeftals*, &c. not under the *Bafes* as *Perrault*'s Tranflator has it. It is call'd in *Englifh*, *Socle* or *Zocle*

Zophoros.

Zophoros.

The ſame as *Freeze* A large flat Member which ſeparates the *Architrave* from the *Cornice*. It is deriv'd from the *Greek*, *Zoophoros*, (*i. e.* Animal-bearing) it being uſual for Animals to be repreſented upon it. The word *Freeze* comes from the *Greek*. *Phrygion*, *i. e.* an Embroiderer, the *Freezes* being frequently adorn'd with Figures in *Bas-Relief*, ſomewhat in imitation of Embroidery.

FINIS.